# Under Ice

William M. Leary

# Under Ice

Waldo Lyon
and the
Development
of the
Arctic
Submarine

Texas A&M
University Press
College Station

Copyright © 1999 by William M. Leary
Manufactured in the United States of America
All rights reserved
First edition
The paper used in this book meets the minimum requirements of the American National Standard for Permanence of Paper for Printed Library Materials, Z39.48-1984.
Binding materials have been chosen for durability.
∞

Library of Congress Cataloging-in-Publication Data

Leary, William M. (William Matthew), 1934–
    Under ice : Waldo Lyon and the development of the Arctic submarine / William M. Leary. — 1st ed.
      p.   cm. — (Texas A & M University military history series ; 62).
    Includes bibliographical references and index.
    ISBN 0-89096-845-4
    1. Nuclear submarines—United States—Design and construction—History.  2. United States. Navy—Submarine forces.  3. Arctic region. 4. Lyon, Waldo—Career in naval architecture. I. Title.  II. Series.
v857.5.L43   1998
359.9'3834'0973—dc21                 98-19137
                                        CIP

For My Son
PETER DECKLIN LEARY
University of Virginia, Class of 2000

*May the wind be always at his back*

## Contents

List of Illustrations, *viii*

Foreword, by John H. Nicholson, *xi*

Preface, *xv*

Prologue, *xvii*

Chapter 1. The Challenge, *3*

Chapter 2. A New Ocean, *29*

Chapter 3. Icebreakers and Submarines, *55*

Chapter 4. A Whole New World, *85*

Chapter 5. Operation Sunshine, *109*

Chapter 6. *Skate*, *133*

Chapter 7. *Sargo*, *156*

Chapter 8. Closing the Circle, *185*

Chapter 9. Tactics and Weapons, *206*

Chapter 10. The Arctic Submarine, *226*

Epilogue, *249*

Notes, *259*

Bibliography, *287*

Index, *293*

## Illustrations

*Nautilus*, xxv

Bathymetric surveys conducted by NEL, *50*

Areas studied by NEL expeditions, *56*

USS *Burton Island*, *76*

*Nautilus* under massive ice floes, *123*

Barrow Sea Valley, *129*

Submarine maneuver for surfacing in a polynya, *140*

Winter cruise by *Sargo*, *161*

*Sargo's* shallow-water under-ice transit, *165*

Cruise routes, *186–87*

Areas surveyed to date, *238–39*

PHOTO SECTION ONE    *following page 108*

O-12 undergoing modifications

Sir Hubert Wilkins on deck of *Nautilus*

Waldo and Virginia Lyon

Triplane target

*Northwind* tows *Sennet*

Eugene C. LaFond, Walter H. Munk, and Graham W. Marks

Waldo Lyon and sonarman from *Boarfish*

Ice caught on deck of *Carp*

Rex Rowray

*Cedarwood*

LCVP from *George Clymer*

Robert McWethy

*Burton Island*

*Northwind* takes *Redfish* under tow

Lyon examines Nansen water sampling equipment

Conference on *Northwind*

Lyon on bridge of *Northwind*

Crew from *Labrador* erect antenna

PHOTO SECTION TWO   *following page 205*

*Nautilus* welcomed in New York

Lyon and Hyman G. Rickover

*Skate* surfaces through frozen lead

Lyon and Art Roshon

Under-ice sonar equipment on *Sargo*

*Sargo* surfaces through ice

Crew inspects damage to *Sargo*

Commander John H. Nicholson

*Seadragon* dwarfed by iceberg

Lyon, Commodore O. C. S. Robertson, and Art Malloy

*Skate* and *Seadragon* moored at North Pole

Scuba team from *Skate*

President John F. Kennedy congratulates Virginia Lyon

Arctic Submarine Laboratory

Sketch of preparation for ice break-through tests

Staff of Arctic Submarine Laboratory

USS *Whale*

Recipients of the Lowell Thomas Award

# Foreword

Control of the seas is a primary mission of the U.S. Navy. Most seas can be controlled by combinations of ships, submarines, and aircraft, but the ice-covered Arctic can only be controlled by nuclear submarines. During World War II the Arctic was an active theater of operations for both German and Soviet submarines (U-boats sank a total of twenty merchant ships *inside* the Gulf of St. Lawrence). During the Cold War the Soviets operated both attack and strategic submarines in the Arctic, and the U.S. Maritime Strategy called for sending submarines into ice-covered waters to attack Soviet submarines in the event of war. Despite the end of the Cold War, Russia still operates its submarines within and under the ice. In addition, over forty other countries possess submarines that can be modified and equipped for under-ice deployment and missile delivery. It is therefore vital for the United States to have arctic expertise and arctic antisubmarine warfare (ASW) capability against potential enemies.

Professor Leary has written *the* definitive history of U.S. submarine operations in the Arctic and the significant role Dr. Waldo Lyon played in developing the capability of submarines to operate and to fight in the Arctic. *Under Ice* describes Lyon's almost single-handed efforts commencing in 1948 to convince the navy of the importance of the Arctic. He drafted a comprehensive plan for arctic oceanography that stressed the need for exploratory submarine cruises, then battled for years to obtain support for his programs despite the lack of enthusiasm by some operational commanders. Despite years of disappointments, he managed to obtain valuable arctic experience on both diesel submarines and icebreakers so that when the opportunity arose in 1957 for *Nautilus*, the first nuclear submarine, to make an arctic cruise, Lyon was ready. He immediately proposed that he accompany *Nautilus* to operate the topside echo sounder and other equipment that would be provided by his laboratory. His request was approved, and his performance aboard proved critical when the ship's navigational equipment failed deep into the ice pack and he was able to make valuable recommendations to permit a safe exit.

The following year he was aboard *Nautilus* for its historic transpolar trip, and once again his advice on the best route and his interpretation of echo sounder readings were invaluable. The success of the *Nautilus* cruise and a *Skate* cruise conducted immediately thereafter resulted in a dramatic increase in support for arctic operations and research.

Leary describes in detail the early conventional submarine cruises, the *Nautilus* patrols, and the subsequent nuclear submarine cruises by *Skate*, *Seadragon*, and *Sargo*. He utilizes the classified patrol reports, books by the commanding officers, and, for the first time, significant inputs from Lyon's Scientific Journal and other papers that he maintained throughout his lengthy service with the U.S. Navy. Lyon's observations not only provide valuable scientific insights but also candid comments on the crews and commanding officers of the various ships and submarines associated with the navy's arctic program. In addition, Leary has gone to great pains to verify important events on the various cruises by interviewing and corresponding with many participants in the early submarine arctic voyages.

Also, due to recent declassification of portions of the submarine patrol reports covering ASW, weapons tests, and tactics, Leary is able to include information in his work that makes it even more valuable from a warfare perspective.

In late 1962 Lyon and others drafted a seven-year plan for under-ice operations. But despite Soviet press reports that a Russian nuclear submarine had reached the North Pole, interest in the Arctic was fading in the navy, partially due to the tragic loss of *Thresher* and the subsequent need for modifications to all submarines. Because submarines were not available to make arctic cruises, Lyon concentrated on sea ice break-through tests at his laboratory.

Meanwhile, a new class of submarines, the 637/*Sturgeon*, was being designed and built. Between 1967 and 1981, a number of 637-class submarines conducted ASW patrols to improve weapons and sonar capabilities in the under-ice environment. These patrols are covered in the book but in lesser detail than the earlier patrols due to continuing security constraints.

In 1993 the Undersea Warfare Systems Division of the American Preparedness Association presented Lyon with its prestigious David Bushnell Award. "Rarely," the citation read, "has a single individual so dominated an area of technology as has Dr. Waldo Lyon the field of naval arctic operations." Without question, his persistent determination has resulted in significant improvements in the capability of U.S. submarines to conduct warfare in the under-ice environment. Thanks to Leary, Lyon's accomplishments, as well as those of other arctic submarine pioneers, are well documented in this book.

Despite Lyon's many accomplishments and awards, he remains convinced that the problems of locating, approaching and successfully attacking a submarine in the ice have not been completely solved. With the last of the 637-class submarines due to be retired by the end of the 1990s, the navy will lose most of its under-ice capability. Not only did the navy never approve his requests to build an experimental arctic submarine, but it has recently directed that his Arctic Submarine Laboratory be closed. He firmly believes that with an unprecedented worldwide proliferation of submarines, there will surely come a time when the United States will need submarines able to operate effectively in ice-covered seas, including shallow waters.

The navy would do well to heed Lyon's warning and not lose its hard-earned arctic capability and expertise.

John H. Nicholson
Vice Admiral, USN (Ret.)

# Preface

I first encountered the story of Dr. Waldo Lyon and the under-ice submarine while working on *Project* COLDFEET: *Secret Mission to a Soviet Ice Station*. Richard Boyle, one of the readers of the manuscript, pointed out that my sketchy treatment of the development of the arctic submarine needed attention. In reviewing the literature on the topic for the short section of the book, I recognized that there had been little written on what seemed a remarkable phase of submarine operations. With an introduction from Mr. Boyle, I later visited Lyon in San Diego to see if sufficient documentation existed to support a book-length study of the under-ice submarine. I was delighted to learn that there was abundant historical material at the Arctic Submarine Laboratory and in Lyon's personal papers. The Lyon Papers include the archives of the Arctic Submarine Laboratory on Point Loma, California, consisting of correspondence files, technical reports, station logs, patrol reports, and other documents relating to the development of the under-ice submarine and other work done by the laboratory. This collection constitutes an invaluable archival record of the U.S. Navy's arctic submarine program. Lyon also kept a Scientific Journal when away from the laboratory. More a diary than a scientific notebook, the journal provides unique insight into the thoughts and actions of the central participant in the evolution of the arctic submarine. With Lyon's full cooperation, I set out to write a detailed account of a remarkable man and his quest for an arctic submarine.

During the course of my research, a number of individuals graciously provided information, read drafts of various sections of the manuscript, and offered encouragement. I am indebted to William R. Anderson, Dean L. Axene, M. G. Bayne, David S. Boyd, Jon L. Boyes, James F. Calvert, William A. Cameron, Edward O. Dietrich, George S. Field, I. J. Galantin, A. L. Kelln, Eugene C. LaFond, Richard B. Laning, Robert D. McWethy, David A. Phoenix, Jackson B. Richard, Jonathan C. Schere, George P. Steele, James T. Strong, Frank L. Wadsworth, William M. Wolff, Jr., Steven A. White, and William Yates. I am especially grateful to three knowledgeable former submariners who reviewed and commented on the entire manuscript: Richard Boyle, John Nicholson, and Alfred McLaren. They saved me from a number of errors. I should add that I am alone responsible for any remaining errors.

One of the pleasures of working in recent history is the opportunity to meet the individuals who participated in the events that are being documented. I have had the great fortune of coming to know many of the scientists and submariners who worked with Lyon to transform an arctic dream into reality. I only hope that I have done justice to their accomplishments.

For the past twenty years my wife Margaret has provided the needed support that has made it possible for me to pursue my historical interests. As she well knows, she has my love and gratitude. John Morrow, Franklin professor of history at the University of Georgia, is a friend and colleague who has offered encouragement and sound advice over the years; he even volunteered to translate some German material for me. James Glerum deserves special recognition for the meals, fine wine, and friendship that he generously provided during a year-long stay in Washington, D.C.

Last, but certainly not least, I wish gratefully to acknowledge the financial support that my work on this book received from a Vice Admiral Edwin B. Hooper Research Grant of the Naval Historical Center.

# Prologue

As far as scholars have been able to determine, the first mention of an under-ice submarine came shortly after Cornelius Drebble's successful experiments with a leather-covered, twelve-oared submersible boat in the Thames River in the 1620s.[1] Evidently influenced by the Dutch-born inventor's work, John Wilkins, later Bishop of Chester and a founder of the Royal Society, discussed the potential advantages of a "submerged ark" in *Mathematical Magick, or The Wonders That may be performed by Mechanicall Geometry*, published in 1648. Wilkins, an indefatigable promoter of seventeenth century England's "new science," speculated that a submerged ark "may be of very great advantage against a Navy of enemies, who by this means may be undermined in the water and blown up." It also would permit travel safe from "the violence of Tempests," from "Pirates and Robbers," and from "ice and great frosts, which doe so much endanger the passages toward the Poles."[2]

The most enduring image of an under-ice submarine, however, came not from the technological speculations of a prescient English cleric but from the fertile mind of a French novelist. In 1869, more than 200 years after Bishop Wilkins imagined the advantages of a submerged ark, Jules Verne published *Twenty Thousand Leagues Under the Sea*, featuring Captain Nemo and his marvelous submarine. Nemo's well-appointed *Nautilus* was 232 feet long, with a beam of 26 feet. (The nuclear-powered *Nautilus* of the 1950s, by comparison, was over 320 feet long and had a beam of nearly 28 feet.) Powered by electricity that Nemo generated from the sodium chloride in sea water, the submarine had a top speed of fifty knots. In a voyage to the South Pole, Nemo used *Nautilus* as an icebreaker until he reached "the Great Ice Barrier" at 67°39′ South. He then dove under the ice. Nemo calculated that for every foot of ice above the surface, 3 feet lay below. As there had been no icebergs higher than 300 feet, the downward-projecting ice could not extend beyond 900 feet. This posed no challenge for the deep-diving *Nautilus:* Nemo simply selected a cruising depth of 2,600 feet. In any event, the submarine carried "vast" supplies of fresh air, and it was equipped with a "spur" to break through the ice, if necessary.

Although the ice proved thicker than anticipated, averaging 1,300 to 1,600 feet, *Nautilus* easily cleared it. Nemo found open water on the other side of the Great Ice Barrier. Locating the South Pole on an island in this Antarctic Sea, he climbed a peak and raised a black banner with the gold letter "N," claiming the land for himself. The solitary Nemo obviously owed allegiance to no nation!

Among the many individuals inspired by Verne's tale was a ten-year-old boy in New Jersey who dreamed about building a real *Nautilus.* Simon Lake, born in 1866, early displayed an inventive temperament. At age twenty-one, he took out a patent for an improved steering device for the high-wheel bicycle; it proved the first of more than 200 patents he obtained. From the beginning of his career, however, Lake remained focused on his boyhood dreams of a submarine. In 1893 he submitted plans in the U.S. government's competition for a practical submarine. Although the contract went to the more experienced John Holland, Lake decided to press ahead with his submarine. *Argonaut I* was launched in August, 1897, the same month as Holland's *Plunger.* Powered by a thirty-horsepower gasoline engine, with tubes reaching the surface for air, the submersible had wheels for traveling along the bottom of the sea. It also included a diver's compartment in the bow that permitted direct access to the ocean, a feature obviously inspired by Verne.[3]

*Argonaut I*, Lake claimed, traveled more than 2,000 miles, including an open sea passage along the New Jersey coast from Cape May to Sandy Hook. To the inventor's great delight, Verne sent a cable of congratulations. "While my book TWENTY THOUSAND LEAGUES UNDER THE SEA is entirely a work of the imagination," he wired, "my conviction is that all I said in it will come to pass." The success of *Argonaut* I provided confirmation of this view.[4]

While he was experimenting with *Argonaut I*, Lake received an inquiry from Alfred Riedel, a friend of the great Norwegian arctic explorer, Fridtjof Nansen, about the practicality of building an under-ice submarine. Lake believed that it could be done, and he prepared plans for an arctic submarine. On February 6, 1898, a full-page, illustrated article on Lake's creation appeared in the *New York Journal*. In it, the inventor discussed his concept for a vessel that would remain in contact with the underside of the ice canopy by means of slight positive buoyancy. The submarine's sledlike superstructure would permit it to slide along under the ice while a topside wheel provided motive power. An adjustable guide would allow the submarine to negotiate uneven spots in the ice cover. Boring devices would permit access to air.

Unfortunately, the article's illustrations bore little resemblance to its contents. One showed Lake's submarine proceeding along the bottom of a mile-deep ocean in the midst of icebergs that extended downward to more than 1,000 feet. As a result, Lake believed, his plan received extensive public criticism. "Many considered the idea as fantastic," he recalled, "the ignorant imagining of some poor 'hairbrained [sic] inventor.'" For a time, there was a possibility that Joseph Pulitzer of the *New York World* might back a submarine polar expedition, but these plans soon fell through. The idea apparently was also too imaginative for the publishing magnate.[5]

Public reaction to Lake's scheme was hardly surprising, sensational illustrations aside. Only a few months before the article appeared, the public had been treated to a science fiction tale of an under-ice voyage to the North Pole by one of the most popular writers of the day, Frank R. Stockton. "The Great Stone of Sardis," which was serialized in *Harper's Monthly* from June to November, 1897, told the story of an inventor, Roland Clewe, and his submarine *Dipsey*. Set fifty years in the future (1947), the adventure featured an electric-gilled vessel that derived air from sea water and proceeded to the North Pole while trailing a telegraph wire for communications. Stockton's notion of both the submarine and the Arctic owed a great deal to Verne. In the public's view, the under-ice submarine remained in the realm of fantasy. Given the technology of the time, this was not an unrealistic position to take.[6]

Undeterred by his failure to sell *Argonaut I* and the hostile reception to his plans for an under-ice submarine, Lake organized the Lake Torpedo Boat Company in 1900 and set out to build a better submersible. The result was *Protector*, launched in 1902. Sixty-five feet long and with a beam of ten feet, the gasoline-powered submarine included such innovations as fore and aft diving planes and an early form of the periscope (which Lake termed the "omniscope"). It lacked wheels.

During the winter of 1903–1904, while conducting trials of *Protector* in Narragansett Bay, the submarine navigated under ice for a brief period, then broke through a floe eight inches thick. Encouraged by this performance, Lake fitted *Protector* with inverted sled runners over the top of the conning tower and planned to make extensive under-ice "sliding" experiments. Before these could take place, however, he sold the submarine to Russia.[7]

At the same time that Lake was contemplating an under-ice voyage to the Arctic, a German scientist was developing similar plans. In 1901 Professor Hermann Anschütz-Kämpfe drew up an elaborate scheme for a submarine trip to the North Pole. The thirty-year-old Anschütz-Kämpfe, a veteran of arctic whaling voyages, intended to use a submarine that was being built at Wilhelmshaven for a summer journey from Spitzbergen to the North Pole, surfacing in open areas (later known as polynyas) en route. The submarine—70 feet long, with a beam of 26 feet, and displacing 800 tons—would have an underwater endurance of fifteen hours at a speed of five knots. It featured an automatic depth control system, ambient light meter to measure ice thickness, and an atmospheric control system. Anschütz-Kämpfe estimated that the polar ice would have an average draft of 16 to 20 feet, with a maximum draft of 80 feet. As his proposed submarine would be able to cruise under the ice at a depth of 160 feet, he would easily clear all obstacles. He expected to find openings in the summer ice pack about every three miles. His picture of the under-ice environment on the route to the North Pole, one veteran arctic submariner pointed out in 1987, "was surprisingly accurate, even by today's standards. His plan to overcome the difficulties posed were similarly sophisticated."[8]

Anschütz-Kämpfe's project received a sympathetic reception when he presented it at a meeting of the Imperial Geographical Society in Vienna. He even had an audience with the Austrian emperor. Nothing ever came of it, however, most likely because submarine technology had not advanced to the point where an arctic submarine was possible. Nonetheless, he went on to make a lasting contribution to under-ice submarine navigation. Aware of the limitations of magnetic compasses at high latitudes, Anschütz-Kämpfe set out to develop a gyroscopic compass. He thought that it would take a year to perfect a suitable compass; it took closer to a decade of trial and error. But Anschütz-Kämpfe eventually succeeded. Before World War I, his gyroscopic compass had not only been adopted by the German navy but it also had been ordered by the British navy for twenty-one battleships and two submarines. His invention later would play an important role in under-ice voyages.[9]

While the idea of using a submarine for arctic exploration never died, nothing of significance happened until Hubert Wilkins—no relation to Bishop John Wilkins—appeared on the scene. Born in Australia in 1888, Wilkins left his father's sheep ranch as a young man to embark upon a life of adventure. After working as a motion picture cameraman with the Turkish army during fighting in the Balkans in 1912–13, he joined Vilhjalmur Stefansson's Canadian Arctic Expedition. "He taught me to work like a dog," Wilkins later wrote of Stefansson, "and then eat the dog."[10]

The two men spent many hours discussing ways to explore the Arctic. Wilkins believed that an airplane would be the best means to investigate the far north, while Stefansson argued that a submarine would be better, especially for oceanographic, biological, and geodetic studies. They agreed that greater knowledge of the polar regions was necessary to predict changes in climate. This was a subject close to Wilkins's heart, as a severe drought had ruined his father's sheep business.[11]

Wilkins left the expedition in 1916, returned to Australia, and learned to fly. As a military photographer with the Australian Flying Corps on the western front during World War I, he was twice mentioned in dispatches and won the Military Cross with Bar. After the war, he presented a plan to the Royal Geographical Society for permanent weather stations in the Arctic and Antarctic, but nothing came of the idea.

Wilkins participated in the British Imperial Antarctic Expedition of 1920–21 then returned to the far south with the Ernest Shackleton expedition of 1921–22. He gained worldwide recognition, and a knighthood, in 1928 when he flew with Carl Ben Eielson, an Alaskan airmail pilot, from Point Barrow to Spitzbergen. The 2,100-mile trip was the first transarctic flight between North America and Europe.

Wilkins had never forgotten Stefansson's enthusiasm for a submarine voyage to the Arctic. As he had flown over the Arctic Ocean, he had observed numerous open areas in the pack ice. It seemed likely that a submarine could surface in these ice lakes, or polynyas, as Anschütz-Kämpfe earlier had predicted, en route to the North Pole. In 1929, following discussions with Anschütz-Kämpfe, Wilkins raised the possibility of a submarine expedition to the Arctic with Secretary of the Navy Charles F. Adams. "He was not favorably impressed," Wilkins noted, "perhaps because I was not able to clearly state my case."[12]

Wilkins's plans matured over the next year. He envisioned a voyage from Spitzbergen to the North Pole, then a zigzag course to the Bering Strait and the Pacific Ocean, collecting scientific data along the route. He contacted Simon Lake, most likely at Stefansson's suggestion, and received a warm welcome. Comdr. Sloan Danenhower also was brought on board the project. A retired submariner, Danenhower had deep ties to the Arctic: his father had been navigator—and a survivor—of the ill-fated *Jeanette*, which had been crushed by the ice in 1881 while attempting to reach the North Pole. After Wilkins cultivated a number of influential congressmen, Secretary Adams agreed to "interpose no objection" to the transfer of an obsolete submarine to the U.S. Shipping Board for use by a polar expedition.[13]

On July 15, 1930, the U.S. Shipping Board concluded a charter agreement with Lake & Danenhower, Inc., for the use of O-12 at a cost of $1 a year for five years, with the proviso that the submarine would be used for scientific research only. Due to be scrapped under the terms of the London Naval Treaty, O-12 had been built by the Lake Torpedo Boat Company in 1918. It was a twin-screw boat, powered by two 500-BHP Busch-Sulzer diesel engines. The plant was provided with two 400-HP electric propulsion motors, which could be connected to 120 Exide storage battery cells. With each cell weighing 1,000 pounds, the battery had a total capacity of 500 amps and provided forty hours of underwater endurance at three knots.[14]

Wilkins orally agreed that inventor Lake would have final authority over the type and number of special modifications that O-12 would need for the polar voyage. At the same time, he agreed that Danenhower would select the crew and command the vessel. Wilkins's major task would be to raise funds for the expedition.[15]

While O-12 underwent extensive modifications at the Philadelphia Navy Yard, Wilkins sought money to pay for the work. In all, he managed to raise $278,000. Lincoln Ellsworth made the most substantial contribution. The wealthy aviator and polar explorer donated $70,000 to the fund and provided another $20,000 as a "loan." Hearst Enterprises paid $61,000 for exclusive rights to the story of the expedition for publication in the *New York American*, while Woods Hole Oceanographic Institution contributed $35,000. A book on the expedition brought in $3,200. Wilkins advanced $22,000 to the fund, representing lecture fees and personal savings.[16]

Wilkins also was putting together a scientific contingent for the expedition. He was fortunate to secure as leader the Norwegian scientist Harald U. Sverdrup. Professor of meteorology at the Geophysical Institute in Bergen and due to become the director of the Scripps Institution of Oceanography at LaJolla, California, Sverdrup had been chief scientist on Roald Amundsen's *Maud* expedition of 1918–21. He had conducted numerous experiments and accumulated a wealth of data as *Maud* drifted in the ice pack in a failed attempt to reach the North Pole.[17]

Sverdrup had perhaps the most insightful ideas about how a polar submarine might transit the ice pack to the North Pole. The experts agreed that there were numerous openings in the ice during July and August, he wrote to Wilkins in October, 1930; the problem was to know when the submarine was underneath one of these openings. He suggested that "a light sensitive cell" be placed under a quartz window that was being installed on the top of O-12. The cell could be connected to a recorder. The recorder would show not only when the submarine was under an opening but also could be used to measure the thickness of the ice. In addition, it might be possible to install "a special kind of sonic depth finder" that could be directed upward. This also would indicate the presence of an opening. Unfortunately, Sverdrup's advice was ignored by Lake.[18]

Lake, in fact, was determined to push ahead with modifying O-12 along the lines of the under-ice submarine that he had proposed in 1898. There would be a sled deck on top to permit it to slide underneath the ice. A pneumatically-controlled guide wheel would sense obstacles, enabling the submarine to change depth automatically. A jackknife folding periscope would be used in order to avoid damage by overhead ice. The most elaborate modification involved a telescopic conning tower with a drill that was capable of penetrating thirteen feet of ice. Scientists would be able to conduct their research with the assistance of a diving compartment and air lock in the old torpedo room.[19]

When Wilkins visited Philadelphia, he did not like what he saw. He was especially concerned about the ice drill. Wilkins later wrote that he had "objected strenuously to the design, even so far as to refuse to pay for its construction." Lake, however, had final authority, and he insisted on installing his "patented" drill. With the costs of conversion escalating to three times the original estimate, Wilkins had to keep busy with fund-raising and had no time to spend at the shipyard.[20]

After five months at Philadelphia, during which little was accomplished, Lake moved O-12 to the Mathis Shipyard in Camden, New Jersey. Modifications were to be completed by January, 1931, followed by extensive trials. The schedule fell so far behind, however, that the trials had to be canceled lest the submarine not reach the Arctic in time for a summer transpolar crossing from Atlantic to Pacific.

As modification problems mounted, the expedition came under increasing criticism. British submariners, a London newspaper reported, considered the project "foolhardy." Even if Wilkins could surface through the ice, they thought it likely that the boat would freeze in solid before its engines could recharge the batteries. Rear Adm. Richard E. Byrd, the noted polar explorer, agreed that the voyage would be extremely hazardous; however, he told the press that "If anyone can come through and accomplish the impossible, my friend Wilkins can."[21]

Finally, on March 24, 1931, work on the submarine reached the point where it could be christened. Lady Wilkins did the honors with a bucket of ice-water at the Brooklyn Navy Yard, renaming O-12 *Nautilus* as a grandson of Jules Verne looked on. The choice of the name had been Lake's, not Wilkins's. Indeed, concerned about the criticism that was being directed against his "unrealistic" plan to travel to the North Pole, Wilkins (who had not even read *Twenty Thousand Leagues Under the Sea* until November, 1930) had been reluctant to associate the expedition with the science fiction tale.[22]

Delays continued to plague the expedition, and *Nautilus* was not able to depart Provinceton, Massachusetts, for the trans-Atlantic crossing until June 4, 1931. The following week, the starboard engine broke down; the port engine soon followed. *Nautilus* sent out an SOS on June 14. Later that day, it was taken under tow by the battleship USS *Wyoming*, which happened to be in the area while on a midshipman cruise to Europe. *Nautilus* arrived in Cork on June 22, from where it was taken by tug to Devonport, England, for repairs.[23]

*Nautilus* was not able to get underway until July 28. It reached Bergen on August 5 and took on the scientific party. After numerous mechanical problems, the submarine finally made it to Spitzbergen on August 15. Two days later, it left for the Arctic Ocean. By this time,

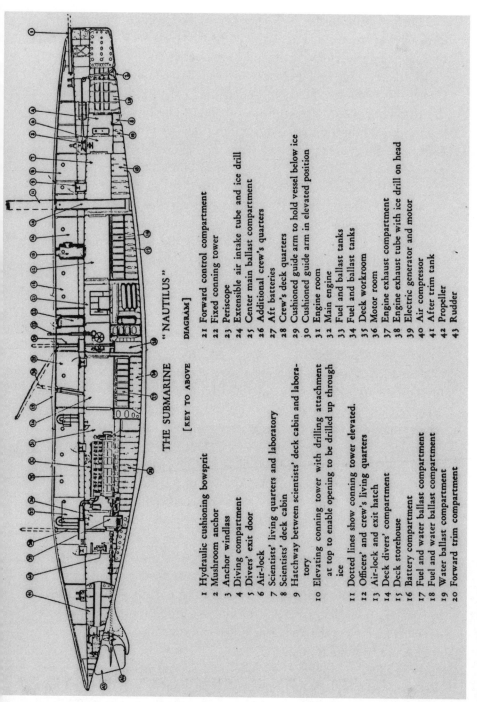

THE SUBMARINE "NAUTILUS"

[KEY TO ABOVE DIAGRAM]

1 Hydraulic cushioning bowsprit
2 Mushroom anchor
3 Anchor windlass
4 Diving compartment
5 Divers' exit door
6 Air-lock
7 Scientists' living quarters and laboratory
8 Scientists' deck cabin
9 Hatchway between scientists' deck cabin and laboratory
10 Elevating conning tower with drilling attachment at top to enable opening to be drilled up through ice
11 Dotted lines show conning tower elevated
12 Officers' and crew's living quarters
13 Air-lock and exit hatch
14 Deck divers' compartment
15 Deck storehouse
16 Battery compartment
17 Fuel and water ballast compartment
18 Fuel and water ballast compartment
19 Water ballast compartment
20 Forward trim compartment
21 Forward control compartment
22 Fixed conning tower
23 Periscope
24 Extensible air intake tube and ice drill
25 Center main ballast compartment
26 Additional crew's quarters
27 Aft batteries
28 Crew's deck quarters
29 Cushioned guide arm to hold vessel below ice
30 Cushioned guide arm in elevated position
31 Engine room
32 Main engine
33 Fuel and ballast tanks
34 Fuel and ballast tanks
35 Deck workroom
36 Motor room
37 Engine exhaust compartment
38 Engine exhaust tube with ice drill on head
39 Electric generator and motor
40 Air compressor
41 After trim tank
42 Propeller
43 Rudder

The submarine *Nautilus*. Courtesy U.S. Navy

Wilkins had realized that it was too late in the season for a transpolar crossing. He had decided to make a cruise to the ice pack, gather scientific data, and experiment with under-ice operations.

*Nautilus* encountered its first ice on August 19 at 80°20′ North. As Captain Danenhower prepared to dive under the ice, he noticed that the aft diving planes seemed to be missing! (The forward planes had been removed during the modifications; they were not necessary for Lake-style under-ice operations, and it was believed that they would be damaged by contact with the ice.) A diver confirmed that the planes were gone. Wilkins believed that a crew member had sabotaged them.[24]

Although disappointed, Wilkins pushed ahead with the scientific program. Sverdrup tried out the diving chamber on August 26. It worked well. "We spent six and a half hours in the diving compartment with no feeling of discomfort or ill effect," Wilkins reported, "in fact we enjoyed working in the comparative warmth and under pressure." Sverdrup lowered a winch-driven bottom sampler and brought up a core of bottom deposit.[25]

The scientific work continued over the next ten days. Between August 27 and September 6, Sverdrup took nine oceanographic stations in the waters northwest of Spitzbergen between 80° and 82° North. In addition to taking core samples, scientists recorded bathymetric data by echo sounder, made gravity measurements, determined the temperature and salinity of the water at various levels, and collected plankton.[26]

On August 31, in the midst of the scientific program, Danenhower came up with a technique to take *Nautilus* under the ice. He trimmed the submarine 2° down at the bow and proceeded under a three-foot-thick ice floe. "The noise of the ice scraping along the top of vessel," Wilkins later wrote, "was terrifying. It sounded as though the whole superstructure was being demolished." When an inside inspection revealed no damage, Danenhower trimmed down further and continued under the ice. An attempt then was made to use the large ice drill that had caused so much controversy during the modifications. Wilkins was not surprised when the shaft failed. Danenhower then backed the submarine out from under the ice.[27]

As mechanical problems multiplied and the weather worsened, *Nautilus* returned to Spitzbergen, arriving on September 8. Wilkins left immediately for the United States. *Nautilus* eventually made it to Bergen, where it was scuttled off the coast with the assistance of the Norwegian navy.

Although the Wilkins expedition was subjected to considerable criticism at the time, a later Naval Examining Board concluded that it had had "considerable merit." The board pointed out that the expedition had been "seriously handicapped" by the poor condition of O-12/*Nautilus*, especially its engines. Also, conversion problems for under-ice operations had caused numerous difficulties. Most of the Lake-installed equipment had failed. The ice drill had not worked, and the jackknife periscope had proved unsatisfactory due to excessive fogging. In fact, the board dismissed as unrealistic the entire Lake method of under-ice navigation. Given the numerous equipment failures, it was hardly surprising that the crew had been reluctant to take the submarine under the ice. Indeed, the board believed that "the crew were justified in their attitude and lack of enthusiasm."

While there had been little success with under-ice operations, the primary purpose of the expedition, as Wilkins frequently had emphasized, had been to carry out geophysical research in the Arctic. Wilkins wanted to contribute to the limited fund of knowledge about a remote but important part of the world. Judged on this basis, the expedition had been worthwhile. A good deal of valuable scientific data had been collected, the board emphasized, which likely would assist the navy in future operations in the area. Later events would prove the board correct on this point.[28]

Wilkins never lost interest in a polar submarine, but the concept remained too exotic for both private investors and governmental agencies. World War II, however, saw the Arctic become an active theater of operations for both German and Soviet submarines. At various times, submarine commanders ducked under the ice canopy, mainly to escape enemy escorts. There is one report that a Soviet submarine in January, 1940, traveled thirty-one miles under the ice in the Gulf of Bothnia, then surfaced through the ice. Also, in April, 1943, a German submarine (U-262) on what proved to be an abortive mission to rescue POWs made an under-ice transit of Cabot Strait into the Gulf of St. Lawrence in eastern Canada; it surfaced once through heavy ice cover and suffered extensive damage. None of these wartime submarines carried special equipment for under-ice navigation.[29]

It was not until after the war that work began on developing equipment and techniques for submarine operations under the ice. The United States Navy took the lead in this area, thanks largely to the work of a naval scientist who welcomed a challenge. In the end, Bishop John Wilkins's submerged ark would conquer the frozen north—it just took three centuries to accomplish.

# Under Ice

# The Challenge

Waldo Lyon was born in Los Angeles on May 9, 1914, his parents having moved to the west coast from Iowa City three years earlier. His father, Charles Russell Lyon, the oldest of five children, had wanted to go to college and become an engineer, but his father had insisted that he join the family-owned jewelry factory. While working in the factory, he had met and married Anna Mary Kampmeier, daughter of a Unitarian minister in Iowa City. Discontent with factory work, Lyon's father had moved to Los Angeles, where he had found employment as an auto mechanic. He later formed the Charles R. Lyon Truck & Repairing Company, which handled the upkeep and maintenance of various trucking and automobile fleets in the city.[1]

Lyon grew up in the open fields of southwestern Los Angeles, near the site of the later Coliseum. Music formed a significant part of his early life. He played the flute, while his sister, Gretchen Mary Lyon, who was seven years older and had been born in Iowa City, played the violin. His mother accompanied them on the piano. The Lyon Trio was sufficiently skilled to appear several times on local radio stations.

Lyon's passion for the flute began to wane about mid-way through Manual Arts High School, when his uncle sparked his interest in the radio fad that was sweeping the country in the late 1920s. Lyon, who learned mechanics from his father while working in the family's automotive repair shop during the summer and who liked to work with his hands, built several high fidelity speakers. Later, while in college, he formed Supex Laboratories to produce and sell his speakers to local enthusiasts.

Grandfather Lyon, who retired to Los Angeles and moved into a large house next door, came to dominate the extended family. Lyon recalls family meals, with everyone sitting at a long table, enjoying the food and conversation. Grandfather Lyon was a strict disciplinarian, and there was never any question about who ruled the household. As Lyon grew up, he would engage in friendly arguments with the older man, especially on the topic of politics. As his grandfather was a staunch Republican, Lyon naturally took the side of Franklin Roosevelt and the New Deal. Grandfather Lyon was a skilled carpenter, and his grandson spent many hours in the garage learning carpentry skills. Looking back on his youth, Lyon would remember his grandfather with affection and respect as an individual who had been "a considerable factor" in his maturation.

Lyon also developed an early interest in sports. His mother played basketball, and as a child he would accompany her to local gymnasiums. As a result of this childhood experience, he took a keen interest in the game. Although well under six feet tall, he became sufficiently skilled in the sport to play for his high school team.

It was assumed that Lyon would attend the University of California at Los Angeles, from which his sister had graduated in 1929 and then joined the faculty as a microbiologist. It cost only $18 a semester to go the university. He could live at home and take the streetcar that passed his house to school. Fortunately, his family was managing to survive the Great Depression without undue difficulty as his father and sister remained employed.

Lyon entered UCLA in February, 1932. His best subject, at least at first, was chemistry, where he stood in the top bracket of a class that included Glenn Seaborg, a later Nobel prize winner. His chemistry professor, however, told him that he really belonged in physics. "My strong point," Lyon recalled, "was building instruments and using them to measure things." This talent could be more usefully applied in physics than in chemistry.

In his junior year, UCLA and the University of California at Berkeley formed basketball teams for players in the 145-pound range. This gave the slightly built Lyon the opportunity to display the skills that he had honed in high school. He became the captain of the UCLA team and, as a graduate student, its coach. A member of a Lyon-coached team later recalled him as "friendly and soft-spoken, but quite firm when he had to be." Lyon believed that his basketball experience contributed to his later success, helping him to lead and organize laboratory task forces by being able to note and use each individual's "strong points, skills, and personality to form the needed team...."[2]

Lyon graduated from UCLA in June, 1936, as a physics major. He wanted to go to graduate school, at least in part because it was easier than finding a job during the Great Depression, but his undergraduate grades were marginal. It took the intercession of Dr. Vern O. Knudsen, head of the physics department and dean of the graduate school, to make the necessary arrangements.

Another factor in his decision for graduate school was his engagement to Virginia Bakus, who was entering her senior year at UCLA. Lyon had known Virginia since they had attended the same Sunday school as children. Virginia's father was a prominent Unitarian minister, while Lyon's mother was active in the church. The young couple began dating in 1932, when Virginia was a senior in high school. She entered UCLA the next year, majoring in physical education. They became engaged in 1936 and married in August, 1937.

Lyon's marriage came just after he finished his M.A. degree, working under Joseph W. Ellis on infrared spectroscopy. Offered an assistantship, Lyon continued into the newly instituted Ph.D. program in physics at UCLA. Virginia found part-time positions as a swimming instructor at Metropolitan High School for Adults and as a physical education instructor at playgrounds in the Latino section of Los Angeles. With a combined yearly income of $2,000, a good sum for the time, the newlyweds were able to move into an apartment in central Los Angeles.

As one of a small group of doctoral students in the new program, Lyon had an opportunity to work closely with the faculty of the physics department. One seminar on ultra short wave electromagnetics, Lyon recalled, consisted of Lyon and three professors, Ellis, E. Lee Kinsey, and A. H. Warner. Almost all the published work on the subject, which formed the scientific foundations of radar, was in German. As Lyon's mother had taught him to read and write the language, he ended up reading the papers and acting as lecturer throughout the seminar. Also, Lyon and his new wife, together with a few other students, spent several weeks each summer with Professors Kinsey and Ellis at their retreat on Lake Tahoe.

Lyon continued his work in spectroscopy, the branch of physics dealing with the analysis of light. He studied under Professor Kinsey, whom he considered "a superb physicist." Under Kinsey's direction, Lyon built a high-resolution spectrometer that enabled him to measure the absorption of water molecules in crystals. After completing all the work on his doctoral dissertation in January, 1941, he was asked by Kinsey to remain in the physics department as an instructor while waiting to receive his degree in June.[3]

In the early months of 1941, with the prospects of a teaching position at Stanford University on the horizon, Lyon had to face the question of military service. He had a low draft number and had been called up for a physical examination in January. Induction into the army in the near future seemed a certainty. At this point, however, an alternative appeared when Leo P. Delsasso, an acoustician on the faculty at UCLA and naval reserve officer, was called to active duty and assigned to the newly organized U.S. Navy Radio and Sound Laboratory at San Diego. Was Lyon interested in joining the facility, Delsasso wanted to know? Lyon was indeed interested. During his graduate studies he had taken field trips with fellow students in acoustics who were working on listening devices to detect aircraft. Although his own specialty had been optics, he felt comfortable with that branch of physics that dealt with sound. Also, laboratory work seemed infinitely preferable to the army. He accepted Delsasso's offer and started work as a monthly contract employee in May, 1941. His permanent position came through in August.[4]

Lyon joined the Radio and Sound Laboratory at a time of rapid expansion. In May, 1939, the Bureau of Engineering had recommended to the chief of naval operations (CNO) that the navy establish a radio laboratory on San Diego's Point Loma peninsula to coordinate research and development work in electronic communications. Home of a naval radio station since 1906, Point Loma seemed the ideal site for the new facility. The navy already owned most of the land on the peninsula, and the location afforded easy access to the Pacific Fleet. On June 1, 1940, with a worsening international situation following Germany's invasion of France, the secretary of the navy authorized the laboratory.[5]

Shortly before Lyon took up his position at the laboratory, Point Loma became the home of a second research organization. During the winter of 1940–41, the National Academy of Sciences had received a request from the navy to undertake a study of the problems of locating submarines by sound transmissions. The Academy reported in January, 1941, that the navy's methods of detection had made little progress since 1918. It recommended that the navy establish two laboratories, one on the east coast and one on the west coast, to focus on the research and development of sonar equipment and techniques. The navy agreed, deciding to locate the west coast facility with the Radio and Sound Laboratory at Point Loma, where it already had a sound school to train sonar operators. It would staff the facility with civilian scientists and technicians from the University of California. On April 26, 1941, the university established a Division of War Research (UCDWR) to administer the new laboratory.

As Lyon soon discovered, the work of the two laboratories overlapped, and the division between the Radio and Sound Laboratory and the UCDWR Laboratory was more an administrative convenience than a functioning reality. During the war years, the two groups of scientists would work harmoniously to develop radar, improve radio transmission and reception, and expand the navy's sonar capability.

When Lyon started work at Point Loma in the spring of 1941, Delsasso assigned him the task of organizing the new sound division. The former basketball coach immediately set out to recruit personnel for his "team." One of the first to come aboard was C. H. Milligan, a retired gunner's mate who had been recalled to active duty. Milligan, with service in submarines extending back to the navy's early gasoline-powered boats, acted as Lyon's mentor on submersibles. Another important early acquisition was Victor Battani, who had been working as the west coast representative of the Submarine Signal Company. "He knew everything there was to be known about submarine sonar at that time," Lyon recalled. Battani became the young scientist's tutor on underwater sound transmission. Finally, and most important of all, Lyon learned oceanography from Roger Revelle, a naval reservist and oceanographer at Scripps Institution, who joined the sound division after being called to active duty.[6]

Lyon was in Los Angeles on the fateful Sunday of December 7, 1941. While the prospects of war with Japan had loomed ever closer during the year, news of the attack on Pearl Harbor still came as a shock. Lyon returned to San Diego the next day and immediately installed a hydrophone at the entrance to San Diego harbor as the first defense "system" against enemy submarines. Only later was it possible to set up echo-ranging equipment. Eventually, nets and gate vessels with detection gear provided more adequate security for the navy's most important facility on the west coast.

The war years saw the Radio and Sound Laboratory, together with the University of California Division of War Research, make a number of important contributions to the navy's efforts against Japan. The San Diego facility tested the navy's first operational radar set, devised a new layout of antennas on ships that substantially improved radio reception, developed a radar beacon to assist navigation, and devised sound decoys to protect U.S. submarines under sonar attack. The laboratory's most significant achievement, at least in terms of its later importance to under-ice navigation, came with the development of QLA sonar.[7]

QLA—the initials were not an acronym—was a continuous tone, frequency modulated (CTFM) scanning sonar. Operating in the range of 36 kHz to 48 kHz and rotating at an average rate of 4 RPM, it displayed multiple targets in a radarlike fashion on a cathode ray tube, or Plan Position Indicator (PPI). Designed to detect enemy mines, it also produced an audible signal that submariners labeled "Hell's Bells." The QLA equipment, known as the Echoscope system, was first tested in February, 1943. It took over two years to perfect the navy's first sonar set that provided a plot display of multiple targets. By the summer of 1945, the laboratory had shipped forty-five sets to the Fleet. In Operation Barney, U.S. submarines used QLA to avoid the mines that guarded the entrances to Japan's Inland Sea. As a result, the navy's stranglehold on Japan's seaborne commerce became complete.[8]

While these developments were taking place, Lyon was continuing his work on harbor defenses. He learned a great deal about how submarines used currents and temperature layers to avoid detection. In his efforts to come up with countermeasures against intruders, Lyon devised acoustical equipment to measure currents at the bottom of harbors, the height of waves, and the size of wakes made by ships.

The more Lyon investigated the properties of sound in water, the more he realized that the navy lacked basic information on the physics of deep submergence. Beginning in September, 1943, Lyon began to lobby for a research facility to study undersea problems. The key piece of equipment for such a facility, he believed, would be a large, high-pressure tank. Obtaining the approval of his superiors, Lyon arranged with the Navy Yard at Mare Island to construct a tank, five feet in diameter and eleven feet long, that would be capable of working at 1,500 p.s.i., the pressure found at a depth of approximately 3,000 feet.

On September 5, 1944, as the tank neared completion, Lyon submitted a plan for a Central Sound Laboratory. There was at present, he pointed out, no facility equipped with the special instruments needed to study the physics of deep submergence. The most necessary instrument was the high-pressure tank, which would be used to test the many devices used on submarines, especially sonar projectors. It would permit researchers to observe the movement of mechanical parts of the devices and measure the strains on them produced by high hydrostatic pressure. Also required for the laboratory would be a high-penetration X-ray machine to measure the movement of internal parts of the devices while subjected to high pressure. Equipped with these kinds of scientific instruments, Lyon argued in a series of memoranda, the resulting facility would be in "a unique position among the nation's laboratories." There was no question, he emphasized, that the Fleet would benefit from the work of a Central Sound Laboratory.[9]

As plans for the new laboratory matured, Lyon continued his work on harbor defense systems. In April, 1945, he traveled to the Pacific Northwest to evaluate the defenses that protected the Straits of Juan de Fuca. Shortly after his arrival in the area, Lyon was contacted by John P. Tully, an oceanographer at the Canadian Pacific Biological Station in Nanaimo, British Columbia. Tully wanted to know if a joint U.S.-Canadian antisubmarine warfare (ASW) exercise might be conducted. As Lyon had authority to use the submarine USS *Tambor* (SS 198), and Tully had access to a Canadian ASW corvette, it seemed like a good idea.[10]

Following trials in the Straits of Juan de Fuca involving a comparison between the sonar detection of a six-foot triplane target and *Tambor*, Lyon visited Nanaimo to discuss the problems of cold water acoustics with Tully and his subordinate, William M. Cameron. Lyon had never before encountered the distortion caused to sonar by layers of colder water below the surface. The Canadians, on the other hand, were painfully aware of the way that strata of cold water could deflect sound and mask the presence of submarines. German U-boats had taken advantage of cold water layers while attacking Allied shipping at the mouth of the St. Lawrence River in eastern Canada. Intrigued by the phenomenon, Lyon promised to return at the end of June and conduct additional tests.[11]

A submarine was not available when Lyon again visited Nanaimo, so he had to make do with the triplane targets. Two ASW ships—PC-795 and HMCS *Ehkoli*—participated in the tests. Lyon certainly enjoyed the spectacular scenery on the east coast of Vancouver Island.

"Beautiful country," he noted, "precipitous cliffs, pines to water's edge, rocks and snow—bright blue water and brilliant sun." The cold water exercises, however, raised more questions than they answered. At the end of two weeks, Lyon and Tully agreed that a major experiment should be conducted later in the year.[12]

Before Lyon could return to British Columbia, the end of World War II in September, 1945, prompted discussions of organizational change at the Radio and Sound Laboratory. During a time of administrative confusion, which saw various groups and factions maneuvering for advantage in any postwar organization, Lyon had to fight to hold on to his proposed Central Sound Laboratory. On September 26, to Lyon's horror, the Joint Planning Committee, which was considering the future of the San Diego facility, proposed that his recently delivered high-pressure tank—the centerpiece of his new laboratory—be located at the Sweetwater Calibration Station, seventeen miles from Point Loma![13]

Lyon promptly responded to the challenge. The Planning Committee, he wrote to the group on October 1, did not seem to understand the objectives of the Central Sound Laboratory. The high-pressure tank was not intended as "a single instrument" for the purpose of calibrating sonar transducers. On the contrary, it was only one of the special tools of a research laboratory that would be devoted to an examination of deep-submergence physics. The Central Sound Laboratory expected to receive requests from the Pacific Fleet and other organizations for the study of "the propulsion of any device, propagation of any energy form or the study of any property of the ocean's structure." It would be a grave error to send the high-pressure tank to Sweetwater and thereby undermine the purposes of this needed naval research facility. Fortunately, the Planning Committee bowed to the logic of Lyon's arguments: his prized high-pressure tank would remain in San Diego.[14]

Having successfully survived the threat to his new laboratory, Lyon was ready to go back to the fascinating problems of cold water acoustics. In October, 1945, he set out for British Columbia with Ralph B. Doherty, a senior laboratory radio mechanic, together with their families, a twenty-foot triplane target, and miscellaneous equipment. Lyon and Doherty drove a six-and-a-half-ton GMC truck, loaded with the triplane target and other gear, while Virginia Lyon followed in Doherty's car, carrying Barbara Doherty and son Lowell, and the two young

Lyon children (Lorraine May, born in March, 1942, and Russell, born in February, 1944).[15]

Lyon spent nearly two months in British Columbia, running exercises with Tully and Cameron in the waters east of Vancouver Island. This time, thanks to persistent requests to the Bureau of Ships (BuShips), a fleet submarine—USS *Stickleback* (SS 415)—participated in the tests, along with a variety of American and Canadian ASW vessels. The main purpose of the experiment was to develop sufficient data for the construction of an analytical method to assess the effectiveness of underwater sound equipment. Operating under a variety of conditions—temperature, salinity, sea state and swell, bottom, ship speeds, and so forth—the ASW ships attempted to locate the submarine and triplane target, comparing the results of sound transmissions.[16]

Following the conclusion of the exercises, Lyon spent several days at the Biological Station in Nanaimo, working with Tully and Cameron on the data that had been accumulated and planning a joint report on the recognition of submarine targets. Tully was the "idea man" who visualized the theorem; Lyon put Tully's ideas into mathematical form and took responsibility for preparing the final report.[17]

The pleasant and productive scientific venture in British Columbia ended in early December. Lyon and Doherty packed up equipment, wives, and children, and retraced their route to the south, arriving in San Diego on December 13. It had been a great experience, Lyon later recalled, particularly for the children. "Our son," he quipped, "came away speaking like a Canadian."[18]

When Lyon reported to work at Point Loma, he found that the proposed organizational realignment had taken place. On November 23, 1945, the name of the Radio and Sound Laboratory was changed to the U.S. Navy Electronics Laboratory (NEL). The University of California Division of War Research was in the process of being terminated, with its contracts absorbed and continued by the new NEL. Also, many civilian employees of UCDWR were transferring to the civil service payroll of NEL, which now reported to BuShips.[19]

Lyon, who became head of the Surface and Subsurface Research Division at NEL, looked forward to completing the joint report on target recognition and continuing the development of his sound laboratory. His work, however, was interrupted in January, 1946, by a request from Roger Revelle, now at BuShips, for NEL to participate in Project

Crossroads, the planned testing of two atomic bombs at Bikini Atoll in the central Pacific. "This is a most unusual opportunity for the laboratory," Lyon wrote. "I am assigned the job of heading a party of seven to measure the waves created by the explosion."[20]

Lyon and his team spent February and March assembling the eighteen tons of equipment that they would need for their task. In the midst of these preparations, Tully visited San Diego to discuss the possibility of establishing permanent cooperation between NEL and the Canadian research group. The framework for the continuing relationship was sketched out in a series of meetings between March 8 and 14. Lyon was delighted with the prospect of continuing cold water sound research with the Canadian scientists, who planned to set up a new sonar laboratory on Vancouver Island. "This is the most enterprising task I have ever attempted," he wrote, "and do hope that the plan will be consummated—the opportunity of performing outstanding underwater research in that area is very great."[21]

With thoughts of colder climes on his mind, Lyon departed for the warmer waters of the central Pacific on March 20, 1946. Arriving in Pearl Harbor on March 26 on USS *Henrico* (APA 45), he learned that Crossroads had been postponed for six weeks. He saw to the unloading and storage of his equipment, then flew back to San Diego to deal with the continuing organizational problems involved in the transition to NEL.

Lyon returned to Hawaii on May 8. After loading his equipment on USS *Fulton* (AS 11), Lyon and his colleagues from NEL sailed westward on May 17, reaching Bikini Atoll seven days later. After inspecting the island's beach with Doherty and Leighton L. Morse, the three men began setting up poles in the surf to measure the waves that would be generated by the atomic blasts.

It was hard work. The poles had to be embedded in the sand, then wires had to be strung between them. "Sun very hot," Lyon reported. "Coral sand brilliantly white, water warm and clear, fighting sunburns and blistered feet." Lyon and his group next installed fathometers and other recording instruments on the target ships. Shortly before the first scheduled explosion, they revisited the ships, wound and set the clocks, and placed batteries in the recorders.[22]

Lyon was on USS *Rockbridge* (APA 228), eighteen miles from Bikini, on Able Day, July 1, 1946. Shortly before 9:00 A.M., a B-29 dropped a twenty-three-kiloton atomic bomb (that bore a picture of actress Rita

Hayworth) over the lagoon. The weapon exploded 518 feet above the surface of the water. Lyon peered through dark glasses as a giant red globe appeared on the horizon, followed by a tremendous mushroom-shaped cloud that rose two miles into the air. "It was like watching the birth and death of a star," wrote *New York Times* science reporter William L. Laurence, "born and disintegrated in the instant of its birth." Lyon, who lacked Laurence's descriptive skills, could only note: "Never will forget the tremendous awfulness."[23]

Lyon visited Bikini on July 2 to check the poles on the beach. Over the next two days, he attempted to retrieve the data from the recorders on the surviving target ships. It was difficult work. He found the cruiser *Pensacola* (CA 24) "a twisted mess." The decks had been burned to charcoal, and he had to struggle through debris to reach the recorders. Between July 9 and 18, Lyon and his scientific contingent changed and recharged the batteries on the recorders on the target ships and rechecked the poles on the beach. Finally, on July 24 Lyon went by small boat to each of the target ships, started the clocks and set the recorders.

Baker Day, July 25, found Lyon eight miles away from the center of the underwater blast. He watched through field glasses as the surface of the lagoon erupted. "The explosion is undescribable," he reported. He attempted to board the target ships the next day, but they were too radioactive. The blast, it was later determined, had released the heaviest amount of radioactivity in history. On August 1, after the ships had been repeatedly washed down, he donned canvas boots and heavy gloves and was able to secure the data from the recorders that had survived the explosion. He then went back to Bikini and surveyed the surf and wave line. The underwater explosion had unleashed the greatest waves ever known to humanity.

Vice Adm. William H. Blandy, commander of the tests, subsequently wrote a letter of commendation for Lyon's work on Crossroads. The success of the wave measurement program, Blandy noted, had been due in large part to Lyon's "tireless efforts, high standards of workmanship and far-reaching scientific knowledge." Lyon's service and devotion to duty, Blandy concluded, were in keeping with "the highest standards of the scientific and engineering profession."[24]

Following a brief visit to NEL and a report to the director on his team's work on Crossroads, Lyon collected his wife and two children and boarded a train for Seattle. From there, the family took the overnight ferry to Victoria, then went by train to Nanaimo. Tully picked up the Lyons at the train station and drove them to an auto court on the south side of town. The next week they moved into a small cabin on the beach at Departure Bay.

Virginia Lyon and the children enjoyed the visit to British Columbia. "Children had much fun with oyster hunting, swimming and sunning at Departure Bay," Lyon wrote. There were walks deep into the adjoining woods, and climbs to a rock dome that overlooked the camp. In the evenings, the family roasted frankfurters and corn on the beach. Each day Lyon walked to the Pacific Biological Station, where, together with Tully and Cameron, he "pounded away" on the report on target recognition.[25]

The two months on Vancouver Island passed quickly. Although the joint report was taking shape, much remained to be done. "Spent weeks trying to tie down Jack Tully's material," Lyon noted in November. In the midst of this work, NEL received a routine letter informing the laboratory that the navy would be conducting a major expedition to Antarctica. Was NEL interested in participating?

Lyon seized the opportunity to extend his work with cold water acoustics. Why not, he wrote to NEL Director Rawson Bennett on October 16, 1946, ask for the use of a submarine for the expedition? Captain Bennett agreed. Lyon then drafted a letter that was sent to the chief of naval operations over the director's signature. Somewhat to the surprise of both men, the navy approved the request. "It seems to me," Lyon later reflected, "it is always of great credit to our Navy that ... CNO saw fit to add a submarine to our request to the Antarctic task force for no other reason than our curiosity to see what would happen."[26]

Largely through the efforts of Rear Adm. Richard E. Byrd, Operation Highjump had been organized in the fall of 1946 to explore the largely unknown continent at the bottom of the world. In addition to learning more about the nature of Antarctica, Highjump was intended to train naval personnel and test ships, planes, and equipment under frigid conditions.[27]

On December 3, 1946, Lyon gathered some books and clothes and left San Diego for Port Hueneme, forty-five miles north of Los Angeles, where he boarded USS *Merrick* (AKA 97). Two days later, at 10:45 A.M., the transport weighed anchor and proceeded southward in company with USS *Yancey* (AKA 93). "I felt very lonesome," Lyon wrote, "dropped with baggage on a large ship on which I didn't know a soul." He quickly settled into his spacious cabin on the starboard side of the deck—"really super quarters"—and continued to work on the joint report, attempting to put the target recognition theorem into rigorous mathematical form.[28]

At 3:00 A.M. on December 11 Lyon was shaken awake and told to report to the forward crow's nest to stand the 4:00 to 8:00 A.M. watch. He knew what was coming. His instructions: scan the horizon for the chalk line of the equator. His real initiation began after a lunch of bread and water. "The usual stuff," he reported, involving electric shock, hot pepper drinks, grease in the hair, being covered with fuel oil, and being pummeled by crew members while crawling through a garbage-filled chute. "Took a good beating," he noted. "Anyway, I was now a shellback by the best Navy tradition, no holds barred."

In addition to working on the target recognition report, Lyon also tried to learn all that he could about Antarctic oceanography and the physics of sea ice. "I'd never seen a piece of ice on the sea," Lyon recalled. He had read about Canadian and German experiences with ice-filled water during World War II, but he had no "feel" for what lay ahead. He had brought along the leading textbook on oceanography, Harald V. Sverdrup, et al., *The Oceans: Their Physics, Chemistry and General Biology* (1942). The treatment of sea ice, however, was sketchy, consisting of only a few pages. He also had a copy of the U.S. Navy Hydrographic Office's *Sailing Directions for Antarctica* (1943). Again, there was little information on sea ice. The scientific investigation of the phenomenon, it was clear, had only begun.

The first iceberg was spotted on December 27. *Merrick* and *Yancey*, as ordered, proceeded to the vicinity of Scott Island and scouted the ice pack ahead. En route, a large piece of ice—a growler—passed between the two ships. It could have stove in either vessel, and no one had seen it. "A real scare," Lyon observed.

Four days later, Lyon transferred to an LCVP "and went splashing across a very cold sea" to USS *Sennet* (SS 408), the fleet submarine that had been assigned to NEL for cold water research. Lyon traded his elegant quarters on *Merrick* for a cramped space on the submarine, with a single drawer to hold all his belongings. But this failed to dampen his sense of excitement. He now had "plenty to do."

*Sennet* joined the Central Group, which had assembled off Scott Island, and proceeded into the ice pack. The coast guard icebreaker

*Northwind* (WAGB 282) led the way, followed by *Merrick*, *Yancey*, flagship *Mount Olympus* (AGC 8), and *Sennet*. The column made good progress during the first day, following leads in the ice at a speed of five knots. On January 1, 1946, however, the ice became heavier as a steady southeasterly wind closed the leads.

*Sennet* proved surprisingly ice-worthy, at least at first. Using its maneuverability, it avoided the larger pieces of ice left by *Northwind*, and smashed into the smaller ones. The boat would shutter alarmingly as it rammed its way through the ice. "The sub showed what a beating she could take," Lyon observed, "and what a beautiful icebreaker she is. . . ."

As the ice became thicker, however, *Sennet* found itself in trouble. "Of all the vessels," the historian of Highjump, Lisle A. Rose, reported, "the submarine seemed in greatest danger." Because of its low freeboard, *Sennet* soon dropped below the level of the ice pack. If the submarine became caught in a rafting situation, the ice could crawl up and over the hull, forcing *Sennet* under. The ice conditions became so threatening at one point, Lyon recalled, that Capt. J. B. Icenhower discussed abandon-ship procedures with the crew.[29]

Despite *Northwind*'s best efforts, ice battered the ships in the column, denting the sides of *Mount Olympus* and the two cargo ships, and damaging their propellers. By January 2, *Sennet* was slowing down the column. Because it was not required for the primary mission of the task force, which was to establish an airbase, Rear Adm. Richard H. Cruzan, task force commander, ordered the submarine back to Scott Island for sonar tests. *Northwind* would tow *Sennet* clear of the ice, then return and escort the other ships to the planned airbase site at Little America.[30]

The outbound trip taxed the structural strength of the submarine and the nerves of the crew. Although the towing procedure worked well enough, Lyon found the experience "harrowing." As the submarine was dragged behind *Northwind*, ice scrapped along its sides, "making a shrieking, screeching sound, something like finger nails across a blackboard, only a thousand times worse."[31]

The submarine's crew had to endure the pounding for nearly seventy-two hours. By the time *Sennet* reached open water, all the paint had been scraped off its hull, leaving bright, shiny steel. There was only minor damage, however. *Sennet*'s prow had taken a beating when the towing cable had broken, and its propellers had been bent by ice. But overall, Lyon reported, the submarine was "in good shape."

*Sennet* spent the next three weeks conducting radar and sonar experiments on the northern edge of the ice pack. In positive gradient conditions similar to those found off Vancouver Island, Lyon added to his growing store of knowledge by identifying various-sized pieces of ice on both QLA scanning sonar and echo- ranging sonar.[32]

Lyon's experiences in the Antarctic further whetted his appetite for delving into the problems of submarine operations in cold waters. The trip, he later recalled, served as "a very good tempering for one to appreciate ice and give self-confidence for any future ice operations." There remained much to be learned before ice operations could be practical, but his curiosity had been stimulated, and he had become "fully intrigued" with the problem.[33]

While returning to San Diego on board the aircraft carrier *Philippine Sea* (CVA 47), Lyon managed to complete the joint report on "Recognition of Submarine Targets." The object of the report, authors Lyon, Tully, and Cameron stated in the preface, was "to present a direct analytical method for the assessment of underwater sound equipments and doctrines; the method to utilize strictly naval equipment and procedures, and to present answers in operational terms." As it turned out, the joint report had more impact on Canadian than on American antisubmarine warfare training. The Canadian navy, Lyon later observed, used the recognition theorem procedures for many years. "It never caught on in the U.S.," he noted, "probably because I did not push it." Also, a number of high-ranking officers in the U.S. Navy resisted the idea as "not invented here." Finally, the United States had submarines available for training exercises, while the Canadians did not.[34]

The pace of Lyon's life seemed to quicken following his return from Highjump. The spring and summer of 1947 saw him active in three directions: (1) acquiring equipment for the sound laboratory in San Diego; (2) finalizing arrangements for continued work with Canadian oceanographers; and (3) seeking a submarine for experiments in the Arctic Ocean.

In May, 1947, Lyon went to Milwaukee to see the Allis Chambers Company about the purchase of a twenty-five-megavolt Betatron. This massive high-penetration X-ray device would enable Lyon to inspect sonar projectors and other equipment to a degree that had not been previously possible. The following month he visited the U.S. Navy Gun Factory in Washington to discuss the possibility of modifying a large naval rifle to serve as a high-pressure chamber, producing conditions equivalent to the ocean bottom. Personnel at the Navy Yard indicated that a twelve-inch, fifty-caliber gun would be available for modification. The gun would be cut into a short chamber, fitted with a screw plug, and drilled to accommodate a one-inch electrical cable.[35]

Upon returning to San Diego, Lyon presented a proposal to locate the growing list of laboratory equipment on a new site. The instruments, he informed his superiors, should arrive before April, 1948. "Massive and bulky" in character, they included a twelve-ton Betatron, 30,000 p.s.i. pressure chamber (a modified twenty-ton, twelve-inch rifle), a water filter plant, 700-gallon boiler, and three spectrographs. The Betatron, in particular, required special facilities to protect personnel from both radiation and noise. The "optimum choice" for a new building, Lyon suggested, was the decommissioned twelve-inch mortar pits of Battery Whistler, just south of the present NEL facilities at Building Four. Whistler had a large underground area with massive concrete barriers. No excavation would be required. The gunpits, now filled with sand, would have to be cleaned out. Other work included the erection of steel columns and overhead rails for a six-ton hoist, putting a roof over the battery pits, and constructing a concrete bulkhead across part of the north mortar pit to form a pool. "With these changes," Lyon concluded, "Battery Whistler will form an excellent laboratory for the Betatron with more than adequate protection for all personnel."[36]

While Lyon was securing equipment for the new laboratory, formal arrangements for U.S.-Canadian research cooperation were being discussed on May 7, 1947, at a conference at BuShips. Leading the Canadian delegation was Dr. George S. Field, director of scientific research and development in the Department of Defence (Naval Service). Commander Revelle, Lyon's longtime friend and supporter, represented BuShips, while E. W. Thatcher and R. D. Russell attended for NEL. The result of the conference was a letter from BuShips to the chairman of the Joint Committee on Oceanography of the Canadian Naval Service, offering to continue joint underwater research on a long-term basis.[37]

Field replied for the Canadian government on July 7. "It is our belief," he wrote, "that further joint operations of similar nature [to earlier work] would be to the considerable advantages of both the United States and Canada." Accordingly, the Joint Committee on Oceanography would authorize Canadian researchers to participate in joint operations with their American counterparts. Detailed arrangements for this cooperative effort would be worked out between the laboratories involved.[38]

Lyon, who had visited Field in Ottawa in May to discuss the cooperative arrangements, was delighted with the conclusion of a formal agreement with the Canadian government. The main focus of his attention, however, centered on a proposal to use a fleet submarine for experiments in the Arctic Ocean. Lyon knew that during the during the summer of 1946 USS *Atule* (SS 403) had attempted to penetrate beneath the arctic ice pack in Nares Strait, north of Baffin Bay, without success. The initiative had come from *Atule*'s commanding officer and had not involved any scientific element. The Atlantic Fleet boat had gone only 1,000 yards under the ice on July 28, 1946, when its periscope shears struck an ice pinnacle. The experience had dampened the Atlantic Fleet's interest in under-ice work.[39]

Also in the summer of 1946 four submarines from the Pacific Fleet had conducted "cold water maneuvers" off Northeast Cape, Siberia. Operating on the surface, the submarines had skirted the edge of the ice pack. Lawson P. Ramage, the division commander in charge of the group, concluded that the Arctic would never be an important area for submarine operations. The Russians, he believed, would lose themselves in the hostile environment.[40]

Despite the pessimism about arctic operations that could be found in most naval circles during 1946, Lyon remained optimistic about the prospects for using submarines in the far north—or at least the need to find out if it were feasible. Fortunately for the future of under-ice operations, Lyon found an ally for his work who occupied a key position. Rear Adm. Allan R. McCann was commander of submarines in the Pacific (ComSubPac). In 1931 McCann had been liaison officer at the Philadelphia Navy Yard for the conversion of O-12 to an under-ice configuration for the Wilkins arctic expedition, an experience that had left him with an enduring fascination about the possibility of operating submarines in polar regions. McCann now was prepared to support Lyon's experimental work.

Aware of McCann's interest, Lyon drafted a letter to ComSubPac, which was sent out under Director Bennett's signature on May 1, 1947. The letter pointed out that the Antarctic experiments with *Sennet* had led to several primary conclusions about submarine operations in ice-filled waters. It had become clear, for example, that further experiments with drift ice were necessary if submarines were to take advantage of ice-filled areas for escape and attack. Also, additional studies were needed of water density profiles near ice so that sound ranges could be accurately predicted. In addition, complete oceanographic studies of the arctic area would be most beneficial. Finally, the experience with *Sennet* had revealed that in ice-filled water scanning sonar was essential for navigation, and that a recording fathometer was needed topside.

In order to continue with the work that had begun with *Sennet*, NEL requested the use of a submarine and escort vessel for at least thirty days in July and August for sonar and diving operations in the Bering Sea. The submarine should be equipped with QLA scanning sonar. NEL would install a topside recording fathometer, recording thermocouple, and salinity-sound velocity meter.

Despite unfavorable press reports of the submarine operations off Antarctica, which over-emphasized the damage done to *Sennet*, NEL believed that further work in ice areas should proceed. "The strategic importance of submarine experience and an oceanographic survey of the Bering Sea," Bennett concluded, "is obvious."[41]

To Lyon's great delight, McCann enthusiastically supported NEL's request. He would provide *five* submarines, plus a submarine tender to serve as an oceanographic vessel, all under his personal command. USS *Boarfish* (SS 327), the submarine designated for the under-ice experiments, called at San Diego in June where NEL personnel replaced the QB soundhead on its starboard column with a QLA scanning sonar soundhead. Also, NEL installed a topside recording fathometer (type NK, model 808B) for upward ranging while under the ice.[42]

While *Boarfish* was being prepared for the northern voyage, Lyon and Eugene C. LaFond, NEL's leading oceanographer, were consulting with Harald Sverdrup, the chief scientist on the Wilkins expedition of 1931 and now director of the Scripps Oceanographic Institute. Lyon planned to penetrate the ice field in the Chukchi Sea, a shallow body of water some fifteen to thirty fathoms deep with a flat bottom of soft mud for great distances. Sverdrup had examined the area while serving as chief scientist on the *Maud* expedition of 1918–25, and his comprehensive report, "The Waters of the North Siberian Shelf," was considered one of the most important oceanographic surveys ever published. Lyon drew heavily on Sverdrup's arctic experience in planning the experiments with *Boarfish*. "There's a real debt to Harald Sverdrup for all of this," he later commented.[43]

Lyon flew to Alaska on July 17 to join McCann's task force. As the group proceeded northward from Adak, a party of NEL oceanographers, headed by LaFond, was taking a series of scientific measurements on the tender USS *Nereus* (AS 17). In all, LaFond and his colleagues occupied forty-five oceanographic stations between Adak and the ice pack, taking forty-one Nansen stations, thirty-seven plankton hauls, twenty-seven bottom cores, sixty-six bottom samples, and over 500 bathythermograms.[44]

*Boarfish*, under the command of Lt. Comdr. John H. Turner, reached the edge of the ice pack in the Chukchi Sea in late June. Lyon, McCann, Turner, and a group of the submarine's officers boarded a launch from *Nereus* to survey the floating blocks of ice. Measuring the height of the ice floes above the water, Lyon was able to estimate their maximum draft. He explained to McCann that the submarine would have ample room to pass underneath the ice. As it was doubtful that a submarine commander would have placed his boat at possible hazard on the word of an experimental scientist, McCann's participation in the decision to continue with the under-ice trials was essential. As Lyon pointed out, McCann "took responsibility for going under this first time."[45]

Shortly before the first under-ice dive on August 1, McCann and three of his staff officers came over from *Nereus* to participate in the historic trip. By now, *Boarfish* was located at 72° 5' North, 168° 42' West. Captain Turner recorded the depth of the water at thirty-one fathoms, with the usual flat bottom. He commenced a stationary dive at 2:39 P.M. Everything went well except for the QLA topside soundhead. During the dive, the key slipped out of the training shaft keyway, leaving the soundhead in a fixed position. Turner had to rely on the bottomside soundhead. This left a large cone of silence overhead, although the topside fathometer would provide information on the ice directly above the submarine.[46]

Turner proceeded cautiously under the ice. In the conning tower, NEL's Arthur H. Roshon, one of the inventors of the QLA sonar, watched the screen and interpreted the data for Turner. Lyon was stationed in the forward torpedo room, which was the only place that had room for the equipment, watching the indicator of the upward-looking echo sounder as it registered the thickness of the ice above *Boarfish*. The ice, Lyon reported over the intercom to Turner in the conning tower, averaged eight to ten feet thick, with some ice features as deep as eighteen feet. At 3:21 P.M., Lyon watched as *Boarfish* passed under ice that was thirty-two feet thick. "The commanding officer did a lot of running back and forth to see what was happening in the forward torpedo room," Lyon recalled, "and the same thing was happening to Admiral McCann."[47]

After more than an hour under the ice, Turner decided to surface. He brought the submarine up to 110 feet and hovered while looking through the periscope. All he could see above him was deep emerald green water. At 80 feet, he reported: "Can still see nothing, water is a light jade green overhead, can see no ice." Coming up to 60 feet, the periscope broke the surface. Turner saw ice all around, with the closest piece 10 feet away. At 30 feet he blew main ballast and brought *Boarfish* to the surface. "One pan of ice skidded over the bow," he noted, "one over the stern and one off the port side of the cigarette deck." The only casualty was an antenna stanchion on the port wing, which was bent over about six inches from the top.

Turner believed that he had been lucky to avoid even more severe damage. *Boarfish*'s periscope had come up in the midst of several large pieces of ice. The QLA, he pointed out, failed to record pieces of ice that measured 3' x 10' x 10'. As ice weighed approximately fifty-six pounds per cubic foot, these missed pieces weighed about seven-and-a-half tons. Surfacing with existing sonar equipment, he concluded, "is not recommended if one desires to keep his periscope." Obviously, better sonar would have to be developed.[48]

*Boarfish* made two additional forays under the ice, the longest extending for twelve miles. The overhead ice, Lyon reported, contained numerous downward-projecting pieces, some as deep as fifty feet. The water below the ice was stratified into distinct thermal layers, the result of a mixing of warm water from the Bering Sea, cold arctic basin water (on the bottom), and melt water (on the surface). The QLA sonar provided a sharply defined portrayal of large ice targets, both on the PPI screen and audibly. Relatively shallow targets were detected at ranges of 1,000 yards or more, with contact lost at 200 yards as the ice passed above the sound beam. Thanks to McCann's willingness to accept responsibility for the first trials, Lyon emphasized, *Boarfish* had demonstrated that extended under-ice navigation was entirely practical.

At McCann's request, Lyon prepared a memorandum on the results of the Antarctic and Chukchi Sea experiments for presentation at a conference of the commanders of the Pacific and Atlantic fleets. The primary goal of both experiments, Lyon wrote on August 14, 1947, was "the adoption of a submarine for operation in polar seas." It was based on the premise that distinct tactical advantage could be gained by a naval force that was able to operate on the fringes of the polar ice pack. The ice, he pointed out, afforded "unequalled protective cover" against present methods of detection and attack on a submarine at periscope depth or deeper. A periscope or snorkel was almost impossible to detect by radar or visually in brash ice. Sonar detection, possible only by another submarine, would be extremely difficult due to multiple echoes from the ice.

Experiments to date, Lyon continued, had demonstrated that ice was sufficiently reflective to permit under-ice navigation by sonar. Current equipment, however, needed to be modified. The under-ice navigators must be able to control the elevation of the QLA sound beam so that they could scan the ice ahead at any overhead angle in order to judge the type of ice and to locate breaks in the ice that were large enough for a submarine to surface and use its snorkel. Also, the QLA sonar should be modified to project a beam at a narrow vertical angle. The normal width of the sound beam could be used for general scanning, while the narrow beam could be employed to measure the depth of ice targets that appeared to be large or deep.

"I believe," Lyon concluded, "that the problem of submarine adaption to polar seas has been formulated and that the submarine shows promise of unusual tactical advantage." Experimental operations, he urged, should be continued if the full tactical advantage of submarines in the arctic area was to be realized. The next step would be to see if submarines could locate ice lakes (also known by the Russian term of polynyas) from beneath by use of QLA sonar, then surface into the lakes by means of stationary ascents.[49]

Working on the development of an under-ice submarine excited Lyon. "I intend to stay right in the middle of the polar undersea warfare program," he wrote to Jack Tully in October, 1947, "for it appears to be most productive, interesting and is virgin territory." Nonetheless, the prospect of leaving government service and returning to academic life remained tempting. UCLA's physics department had given Lyon until

1948 to make a decision on whether or not to rejoin the university. In order to test the academic waters, Lyon spent part of each week during the spring semester of 1948 in Los Angeles, teaching a sophomore physics class. At the end of the semester, UCLA offered him an assistant professorship, with promotion to associate professor within a year.[50]

Lyon wrestled with the decision for several months. He and his wife realized that they had come to a major crossroads in their lives. Lyon had a strong sense of loyalty and obligation to UCLA's physics department, but he was not sure if he wanted to devote his life to the classroom. While he had enjoyed teaching the physics course, he had no desire to repeat it on an annual basis. His forte, Lyon recognized, lay in designing and building instruments to study problems, not in teaching the theory of physics. His superiors at NEL had promised to free him of supervisory responsibility, allowing him to head a small research unit that would focus on development of the under-ice submarine. Also, Tully kept reminding him about the dangers posed to the United States and Canada from the lack of undersea knowledge. In the end, the lure—the challenge—of developing a special arctic laboratory and solving the problem of the under-ice submarine proved compelling. Lyon and his wife returned to San Diego and purchased a large house near NEL to remodel. "We never looked back again," he later wrote.[51]

In March, 1948, Lyon responded to a request from Admiral McCann for information on NEL's program of arctic work. NEL, Lyon wrote, had three questions under study: (1) development of a narrow-beam QLA sonar for use in identifying ice lakes; (2) study of sound propagation in water of upward refraction and extremely positive thermal layers; and (3) oceanographic studies of arctic areas for sonar and submarine charts, including bottom topography, thermal-salinity structure, horizontal currents, and the properties of ice.

The lack of information on sound propagation in positive-layered water, Lyon emphasized, "is nearly alarming." Although the Fleet had had nearly eight years of experience with negative-layered waters, it had practically none with positive layers. NEL planned to attack the problem through comprehensive sound transmission studies in the Chukchi and Bering Seas in 1949 and through field laboratory studies in the Vancouver Island area in cooperation with Canadian oceanographers.

NEL also intended to study the acoustical and physical properties of ice both in the field and in laboratory studies at the new underwater sound laboratory. "We are converting our gear in our deep submergence laboratory," Lyon explained, "to provide water at any pressure, temperature, and air content." Lyon expected to study the growth of sea ice and measure its properties. "This is a long term point of view," he noted, "which we hope will pay off some day in submarine applications."

Lyon hoped that ComSubPac could supply a submarine for the summer of 1948 to investigate ice from underneath. "Our operational objective," Lyon stressed, "is to provide a submarine with the information to permit its movement to any distance and across the Arctic Ocean."[52]

McCann, who had developed an extensive personal correspondence with Lyon since the *Boarfish* experiments, shared the scientist's enthusiasm for the under-ice program. "I was very much interested in your plans and fully concur with all you want to do," McCann replied on March 31. "I assure you of my hearty cooperation."[53]

With McCann's endorsement, the chief of naval operations approved the use of submarines for arctic experiments in 1948. "Data acquired," CNO noted, "would be valuable for planning purposes." In May USS *Carp* (SS 338) arrived at Mare Island for installation of sonar equipment. Lyon was on hand to supervise the work and discuss the upcoming voyage with *Carp*'s commanding officer, Comdr. James M. Palmer.[54]

On August 9, 1948, Rear Adm. Oswald S. Colclough, who had relieved McCann as ComSubPac, issued operational orders to Commander Palmer. *Carp*, Colclough wrote, was assigned to experimental work for NEL under Lyon's direction. "Be guided by his request regarding under ice excursions and tests as far as practical," Colclough ordered, "but take every precaution to safeguard your ship and its materials." Colclough enjoined Palmer to exercise particular care "to guard against getting so far under the ice field that you might damage your topside appurtenances or screws by a forced surfacing before you are clear." Finally, the admiral instructed Palmer to load sixteen torpedoes. "Without any specific information of increased tension," he warned, "the international situation is such that you should be prepared for any eventuality."[55]

With Lyon on board, *Carp* reached the ice pack in the vicinity of 72° North, 167° West, on September 3, 1948. Using QLA scanning sonar, Palmer located and plotted the location of several polynyas, then selected a course and speed to reach a clear area for surfacing. On his ascent, an upward fathometer detected overhead ice. On one ascent, a large piece of ice was brought up on *Carp*'s afterdeck. This pointed to the need for a second fathometer, which would be mounted aft of the conning tower.

During the course of fourteen vertical dives and ascents between September 3 and 11, Palmer perfected the technique that was necessary to use the ice lakes. The diving procedure varied from standard technique in that the negative tank was only partially flooded (12,000 pounds) to prevent an excessive down angle. Palmer controlled the submergence rate by blowing the negative tank to its normal trim level of 6,000 pounds. By starting to blow the safety tank slowly at fifty feet, he achieved sufficient longitudinal stability to prevent seesawing. On ascents, enough water was blown from the safety tank to obtain positive buoyancy of about 4,000 pounds. He controlled the bubble by using the planes to increase or reduce resistance to ascent on either bow or stern. This procedure worked well, although venting the safety tank blanked out *Carp*'s sonar gear.[56]

Following *Carp*'s successful cruise, an elated Lyon returned to San Diego and reported the results at a director's meeting on September 24. He requested that a fleet-class submarine be modified for additional under-ice experiments. Captain Bennett told him to draft the necessary letter. The under-ice experiments, Bennett informed Lyon, were classified secret, while the final objective should be considered top secret.[57]

Two weeks later, Lyon presented his request for a modified submarine at a conference at NEL with Admiral Colclough and his staff. Lyon began by noting the primary conclusions that had been derived from submarine cold water experiments since 1946. Work with *Sennet* had led to techniques for handling submarines in heavy ice, demonstrated that they could withstand heavy punishment, and showed that submarines could navigate by sonar in the vicinity of icebergs. The *Boarfish* cruise had proved that extended under-ice navigation was possible, primarily through the use of scanning sonar. Finally, *Carp* had shown that vertical dives and ascents could be made in polynyas. "The practicality of a submarine that can cross or cruise to any part of the Arctic Ocean (summer months for present)," Lyon emphasized, "is within our immediate reach. This we had not realized and we cannot now ignore pressing the possibility to a conclusion."

The problem, he continued, was now divided into two phases. First, oceanographic and sonar studies of the Arctic had to be done as available data on the area was "almost zero." NEL planned to spend the summer of 1949 in the Bering and Chukchi Seas, studying oceanography and sound propagation in an effort to secure "at least the first rough answers."

The other phase of the problem involved the modification of a fleet-type submarine for further study of under-ice operations. These modifications, Lyon stressed, were relatively simple. They would include running a topside keel to permit the submarine to rest underneath the ice; cleaning up the superstructure so that the topside sound gear could be retracted below the topside keel and the periscope dropped to the control room; installing two six-inch ice drills so that a snorkel could be placed above the ice cover for engine supply and exhaust; and making internal modifications to expedite navigation by sonar, including QLA sonars in the conning tower and five topside fathometers that would be distributed along the deck for complete overhead coverage.[58]

Although Colclough was non-committal, Lyon was sufficiently encouraged to draft a formal request to ComSubPac, which was sent on December 3 under Bennett's signature. "The all important objective," the request stressed, "is to measure a new ocean—determine its oceanography and undersea physics. The objective can be met by present sonar equipment and an experimental fleet submarine modified for the Arctic Ocean."[59]

As Lyon soon discovered, Colclough did not share McCann's enthusiasm for arctic work. Although he had been prepared to honor his predecessor's commitment to NEL for 1948, Colclough did not intend to continue with the experiments. He rejected NEL's request for use of a fleet submarine in 1949. Indeed, as it turned out, NEL would not see another fleet submarine for under-ice experiments until Colclough's relief as ComSubPac in 1952. If Lyon were to measure the "new ocean," he would have to do so with his own resources.

Lyon had accomplished a great deal by the end of 1948 in his pursuit of an operational under-ice submarine. On November 18 NEL published his Research Report 88, "The Polar Submarine and Navigation of the Arctic Ocean." In what was destined to become a prophetic document, Lyon reviewed the progress that had been made during the experiments with *Sennet*, *Boarfish*, and *Carp*, and he reached an optimistic conclusion. "The reality of a polar submarine that can navigate the *entire* Arctic Ocean," Lyon wrote in the widely distributed document, "is not only admissible, but may be an immediate practicality."

Although much work remained to be done, Lyon had laid the foundations for the future. With the support of NEL Director Bennett, he had secured the equipment for laboratory studies of underwater sound propagation, the physics of sea ice, and other important aspects of cold water research. In October, 1948, Lyon moved to Battery Whistler and began to set up the new laboratory—later named the NEL Submarine Research Facility.

With the approval of his superiors, Lyon also had done the groundwork for a cooperative program of cold water research with Canadian oceanographers. In the years ahead, this cooperative effort would pay rich dividends for both governments as American and Canadian scientists conducted numerous expeditions that revealed the nature of the far north.

The obstacles to be overcome before a polar submarine could become a reality, however, remained formidable. Although Lyon's work had received the enthusiastic backing of a few "believers," like Admiral McCann, the naval establishment in general took a more skeptical view of his objective. The official position could be found in the *Navy Arctic Operations Handbook*, prepared by the Arctic and Cold Weather Coordinating Committee of the office of the chief of naval operations and published in 1949. While noting the success of the under-ice experiments with *Boarfish* and *Carp*, the *Handbook* reached a different conclusion than had Lyon. "Development of the trans-Arctic submarine," it proclaimed, "remains in the realm of fantasy."

Lyon took the statement in the *Handbook* as a challenge. Convinced that a polar submarine was entirely feasible, he became more determined than ever to prove to the navy that what many considered a fantasy was in fact a reality.

# A New Ocean

Early in the new year of 1949, Lyon sent a copy of NEL Report 88 to Admiral McCann, his friend and supporter, who had been reassigned to the navy's General Board. NEL, Lyon wrote in a covering letter, had asked for a submarine to be modified for under-ice experiments. He foresaw a time when the under-ice submarine would be able to operate to all parts of the Arctic Ocean. But before this could happen, more information was needed on the nature of the far north. It would be "a serious mistake," he warned, "not to press for a study of the entire Arctic Ocean."[1]

Captain Bennett, commanding officer and director of NEL, agreed with Lyon's desire to learn more about the Arctic. He not only approved the scientist's request to conduct a major field trip in the summer of 1949 to study sound transmission in cold water, but also told Lyon he could use the submarine USS *Baya* (SS 318), which had been assigned to NEL in 1947 to support underwater research, and the patrol craft EPCE (R) 857. Lyon's scientific program called for two weeks of experiments in the deep waters of the Aleutian chain, followed by sonar investigations of the intensely layered waters of the Bering and Chukchi Seas. In addition, the expedition would conduct oceanographic work.[2]

Hoping to continue the close relationship with Canadian oceanographers, Lyon invited his scientific colleagues in British Columbia to participate in the expedition. In February, 1949, Fred Sanders, director of the newly organized Pacific Naval Laboratory at Esquimalt, British Columbia, visited San Diego to discuss NEL's summer program. A sonar scientist, Sanders gave his full support to a joint expedition. The following month, Jack Tully and Bill Cameron of the Pacific Oceanographic Group in Nanaimo came to San Diego to discuss the details of Canadian participation. They informed Lyon that the oceanographic vessel, HMCS *Cedarwood*, would be available for the summer. The three scientists agreed that *Cedarwood* would work independently and conduct an oceanographic program to study the general circulation of the deep Bering Sea, the currents and water mass mixing through the Bering Strait, and the currents and distribution of sea water in the Chukchi Sea.[3]

Lyon also planned to work with the Canadians to investigate the tidal current flowing between the Bering and Chukchi Seas. He intended to establish a station at Cape Prince of Wales, the narrowest point on the Bering Strait, on the site of an abandoned Civil Aeronautics Administration low-frequency radio range that had been used during World War II to guide aircraft being ferried to the Soviet Union. The Weather Bureau still had an observation station on the cape, adjacent to the native village of Wales. The Canadians agreed to conduct tests in Esquimalt of an electromagnetic method that had been developed by the British admiralty. If the tests proved satisfactory, a cable would be laid from the village of Wales out into the Bering Strait. Electrodes would be attached to the cable to measure the fluctuation in the electromagnetic field, while thermometers would measure temperature. This data would indicate the strength, direction, and temperature of the waters passing between the Bering and Chukchi Seas.[4]

*Baya* and EPCE (R) 857 departed San Diego on July 5, 1949, and proceeded to Adak, Alaska, via Esquimalt and Kodiak. *Cedarwood* was waiting at Adak when the NEL group arrived. With Tully as chief scientist and Eugene LaFond of NEL as senior U.S. scientist, *Cedarwood* departed for its task of evaluating the water structure and currents immediately south, in, and north of the Bering Strait to the edge of the ice pack.[5]

As *Cedarwood* set about its work, Lyon was making preparations to establish the field station at Cape Prince of Wales. On July 18 he flew to Seattle to supervise the loading of equipment from a Canadian tug to USS *George Clymer* (APA 27). He then proceeded by air to Adak, where *Baya*, EPCE (R) 857, and *Cedarwood*, which had returned from its oceanographic mission, were at anchor. After a conference aboard *Cedarwood* about the details of the next phase of the summer's operation, scheduled to last from July 23 to August 15, Lyon returned to Seattle.[6]

On July 26 *Clymer* weighed anchor and headed northward. Six days later, the transport passed through Unimak Pass in heavy fog and entered the Bering Sea. Early the next day, August 2, *Clymer* and *Cedarwood* rendezvoused fifteen miles west of Cape Prince of Wales.

Lyon, Morse, and the two-man crew of the LCVP that would carry the necessary equipment and cable to shore, were lowered over the side at 6:30 A.M. The flat-bottomed landing craft received its final cargo by net from *Clymer*, then proceeded with *Cedarwood* toward Cape Prince of Wales. The wind was blowing from the south at fifteen knots, while heavy fog cut visibility to 2,000 yards. The fifteen-mile voyage passed without incident. "Took a little spray," Lyon recorded in his Scientific Journal, "but not too bad." *Cedarwood*, which had begun to roll heavily, anchored 3,000 yards off the beach. The LCVP headed toward shore.[7]

The landing craft had not gotten far before its coxswain found that the direct route to shore was blocked by a series of sandbars. The LCVP then swung south to attempt a landing in the deeper water in the lee of the cape. The boat, however, struck bottom 200 yards from shore. Thanks to the skill of Coxswain Harold Bender, the LCVP managed to free itself in the trough of a wave that passed under the boat and turn seaward without broaching.

The sea grew rougher as the LCVP made its way back to *Cedarwood*. Lyon was concerned that the southerly wind might increase to gale force and they would be unable to ride out the night in the LCVP. Recognizing that there was little shelter off the cape, Lyon had wanted to use a DUKW—a two-and-a-half-ton amphibious truck—for the mission, as the wheeled vehicle could be brought ashore in the event of dirty weather. Director Bennett, however, had turned him down. The backup plan for the LCVP, in the event of bad weather, was to seek shelter in a lagoon that charts showed to be fifteen miles north of Wales.

Coming alongside of *Cedarwood* so that Coxswain Bender could obtain a chart of the region, the LCVP tore a chunk out of the Canadian vessel as sea conditions continued to deteriorate. After managing to board *Cedarwood* and secure the chart, Bender pointed the LCVP to the north and set off in search of the lagoon. It took three hours to cover the thirty miles—not fifteen miles as the chart had indicated—to the entrance to the lagoon. When they arrived, they found the entrance blocked by a twenty-foot-wide sandbar.

There was no choice, Lyon believed, except to return to *Cedarwood*, even if it meant running thirty miles in the teeth of a twenty-five-knot wind and a rough sea. The trip proved "miserable and dangerous." The LCVP crashed into each wave, coming at thirty-foot intervals, throwing cold water into the boat and over its occupants. "We fought our way back for six dragging hours," Lyon wrote in his Journal, as the LCVP gradually filled with water. When the boat's pump could not handle the water, they used a hand pump. When the hand pump broke, they bailed with a bucket fashioned from a five-gallon can. But their efforts proved futile: more water came in over the bow than they could handle.

By the time the LCVP rounded the last sand spit, with *Cedarwood* on the horizon, there was nearly two feet of water in the boat. The men were exhausted from the six-hour ordeal. With the LCVP now moving at half speed and barely making headway, Lyon felt that they had reached the limits of survival. He told the coxswain to turn toward the shore, 600 yards distant, and beach the LCVP in order "to save ourselves and perhaps salvage what we could."

"Beach looked nasty," Lyon observed as they headed toward the group of buildings that marked the abandoned Civil Aeronautics Administration station. There were three distinct surf lines, marking sandbars. The LCVP grounded on the first bar, 150 yards from the shore, but Coxswain Bender managed to fight his way across by gunning the boat's engine as waves washed over the bar. But when the LCVP reached the second bar, 100 yards from the beach, it grounded fast. Within minutes, the pounding surf killed the boat's engines, rendering it helpless.

Bender took a line from the boat and began to wade through the surf. He traveled only 25 yards, however, when he dropped into a trough over his head and had to return to the boat. By this time, the two Canadian scientists who comprised the shore party had recognized the plight of the LCVP and engaged a boat—known as an oomiak— from the local village. The native boat proved ideal for the task. Powered by an outboard motor, the sealskin craft was thirty feet long and drew one foot of water. Lyon and his thoroughly soaked companions were brought on shore and rushed to a house where they were given dry clothing and hot soup. It was now 7:30 P.M. Despite their long ordeal, they suffered no lasting ill effects. Lyon was even ready to look on the bright side. Recalling the amphibious landings of World War II, he wrote in his Journal: "Happy that no one was shooting at us when stuck on sandbar."

Although his hands were sore, Lyon began work on equipment repairs the next day. The plan had been to lay two cables west of Wales, six and three miles in length, perpendicular to the flow of the current through the Bering Strait. A pair of electrodes would be attached to the cables, one pair at one and six miles from shore, and the other pair at one and three miles. With the assistance of *Cedarwood's* whale boat, the group managed to lay a field-telephone wire cable to a distance of four-and-a-quarter miles, with the remaining cable lost due to breakage at the manufacturer's splices. This cable went into operation on August 7.[8]

The second cable, recovered from the LCVP, was successfully laid by August 8. A gale on August 10, however, put both pairs of electrodes out of action. Lyon suspected that there was a break in the shorter cable, but he was unable to locate it.[9]

Seeking a respite from the arduous task of cable-laying, Lyon would take long walks on the beach, both as exercise and to reflect on possible solutions to the many problems that arose. On one of these walks, he discovered a large walrus skull. He brought it back to the station, where he planed off the top plate. He then used India ink to pen a letter to Director Bennett about the fate of the LCVP.

"After twelve hours of fighting wind and sea," he wrote on the skull, "the LCVP was swamped and lost in the treacherous surf . . . off the beach of Wales, Bering Strait, at 1830 hours 2 August 1949. The landing was mandatory to save the lives of the personnel." The LCVP, he continued, was "demolished by a 40 knot gale and scattered among the forgotten host of debris from which this letterhead was chosen."[10]

When he returned to San Diego, Lyon presented the unique communication to Director Bennett. It was unnecessary to remind his superior that the problem could have been avoided if a DUKW had been used.

Lyon left Wales on *Cedarwood* on August 15. Arriving in Teller the same day, *Cedarwood* tied up alongside *Baya*. Lyon presided over a conference later in the day to lay out plans for the following week. *Baya* and EPCE (R) 857 were to proceed into the Chukchi Sea to conduct sound propagation studies. *Cedarwood* would return to Wales and attempt to locate and correct the problem with the cable, then survey the ice pack along the Alaskan coast before heading for a rendezvous with *Baya* on August 23.[11]

*Baya* and EPCE (R) 857 departed at 4:00 P.M. the next day. On August 17, they made sound propagation runs opposite Kotzebue Sound, then headed north in search of ice. They found it two days later at 71° 40' North, 163° 15' West. *Baya* anchored near the edge of the pack, while EPCE (R) 857 made sound propagation runs, testing low frequencies at short ranges. EPCE (R) 857 also made bathythermograph and salinity checks every thirty minutes. Temperature data revealed the presence of positive thermal layers near the ice.

Sound propagation tests continued over the next two days, with EPCE (R) 857 making high-frequency runs just south of the ice pack. The two vessels then headed for the planned rendezvous with *Cedarwood*.[12]

On August 23, as scheduled, *Baya*, EPCE (R) 857, and *Cedarwood* met off Cape Lisburne. NEL personnel transferred from *Cedarwood* to EPCE (R) 857, where Lyon held a final meeting of the expedition's senior scientists. The Canadians reported the *Cedarwood* had been unable to correct the problem with the cable at Wales; it later was discovered that it had broken 5,000 feet from the beach. After a further exchange of information, *Cedarwood* departed for additional oceanographic work in the Chukchi Sea, while *Baya* and EPCE (R) 857 sailed south through the Bering Strait.[13]

Arriving at Kodiak on August 29, Lyon called on the commander of the 17th Naval District. The scientist was appalled both by the naval officer's limited information on the Arctic and by his lack of interest in the strategic importance of the area. Lyon came away from the meeting more determined than ever to secure the kind of data on the Arctic that would prove vital for military operations in the area. "Too much weight is placed on information from natives, old time inhabitants and local characters," he noted in his Journal. "These people can give qualitative color, overall impressions, but do not have exact data for undersea or amphibious operations, e.g. detail of lagoons, study of sand shifts on bars, etc."[14]

*Baya* and EPCE (R) 857 continued southward, reaching Sitka on September 2. The two NEL vessels then left for San Diego, while Lyon remained ashore to await the arrival of *Cedarwood*. The Canadian ship appeared on September 6. Two days later, with Lyon on board, it departed for Esquimalt, cruising through the inland passage.

As *Cedarwood* sailed southward, Tully briefed Lyon on the results of the oceanographic survey that the Canadians had designated as Project Aleutians. *Cedarwood* had begun work on July 13. Two scientists on each watch provided continuous coverage of the area north of the Aleutian Islands, through the Bering Strait, to the edge of the ice pack at 72° North. Tully judged the survey a success. As he later wrote, "We are now cognizant of the outstanding oceanographic properties of these seas...."[15]

When *Cedarwood* reached Prince Rupert on the northern tip of Vancouver Island on September 10, Lyon and Tully came ashore for a conference with Field and Sanders the following day. The meeting would prove crucial to the developing relationship between U.S. and Canadian scientists. NEL's principal objective over the next few summers, Lyon stated, would be to develop the arctic submarine and make deep penetrations of the polar basin. Also, despite the recent disappointment, NEL would go ahead and establish a permanent station at Cape Prince of Wales to measure the current passing through the Bering Strait. This information was necessary, he stressed, because it could lead to accurate predictions of the amount of ice in the Chukchi and Beaufort Seas.

The Canadians, Lyon continued, should assume oceanographic responsibility for the area in the Beaufort Sea adjacent to the Mackenzie River. NEL would work in the deeper waters of the Beaufort Sea. The combined oceanographic information, together with data from the *Maud* expedition of the 1920s, Lyon stated, "should place us in a favorable position for Arctic Ocean warfare."[16]

Lyon returned to San Diego on September 13, pleased with the results of the summer's work. The expedition had conducted the first arctic studies in underwater sound with an investigation of low frequencies over long distances in deep water, and a wide range of frequencies in shallow water. Although the attempt to establish a field station on Cape Prince of Wales had proved abortive, he had learned a great deal about the conditions in the area and would try again in 1950. *Cedarwood* had conducted a highly successful oceanographic survey of both the southern and northern portions of the Bering Strait. Above all, the Prince Rupert conference of September 11 had set forth the parameters for the continuing relationship between U.S. and Canadian scientists in the search for precise information about the nature of the arctic area.

In October Lyon traveled to Washington for a conference sponsored by the Office of Naval Research (ONR) to review the past and plan the future of oceanographic research in the Arctic. The arctic conference, which would prove an annual event, brought together all the agencies interested in the Arctic, including NEL, ONR, BuShips, the U.S. Coast Guard, Woods Hole Oceanographic Institution, Scripps, and the U.S. Air Force. It was clear, Lyon noted, that NEL had made "the largest contribution" to arctic research. At the conclusion of the meeting, Lyon and Roger Revelle, now director of Scripps, were given the task of writing up the conclusions and recommendations of the gathering, which would be sent out over the signature of the chief of naval research.

Realizing the likely importance of the document, Lyon and Revelle spent many hours drafting a comprehensive plan for arctic oceanography that stressed the need for continued work on the under-ice submarine. They began by placing arctic research within existing naval policy. On September 10, 1948, they pointed out, the chief of naval operations had stated that the navy should maintain the highest state of readiness to conduct defensive and/or offensive actions in the Arctic. An accompanying long-range program for the Arctic, which was approved by the secretary of the navy, envisioned the solution within five years of the principal design and engineering problems to permit naval operations in the far north. The five-year program also called for the necessary oceanographic and hydrographic data to be accumulated in order to prepare navigational charts of the area.

Certainly, the need for information could not be denied. As Lyon and Revelle pointed out, "Almost nothing is known about the oceanography of the Arctic Ocean." There had been only 150 recorded deep soundings throughout the more than three million square miles of the ocean, and these had been concentrated in a few areas. "To attempt to map the undersea topography from these soundings," they emphasized, "is comparable to making a topographical map of the United States based on two elevations in each state." Furthermore, scientists did not know if Arctic Ocean waters moved in a clockwise or counterclockwise direction, or in a more complicated current pattern. Also, no data existed on sound propagation conditions in the central arctic basin, north of 75° North.

To obtain satisfactory answers to these and other questions, a long-term and comprehensive program of research would be necessary. Arctic oceanography could be accomplished by four means. First, the report recommended that a year-round station be established to measure currents and study beach conditions in the Bering Strait. Second, surface ships should be employed to study the area northeast of Greenland, with the recent NEL joint expedition serving as a "model" for this and other operations to the far north. Third, aircraft should be used for aerial photography and for developing techniques to land on the ice during the winter to permit oceanographers to secure data. Finally, NEL's request for a submarine to be assigned for the summer of 1950 should be given "high priority" so that the laboratory could continue its work in the Chukchi Sea.

"It is expected that exploratory submarine cruises during the summer months can be carried out in future years over a major part of the Arctic Basin," the draft report stated, "and may be the primary means of obtaining bottom topography, sediments and summer water conditions of the Arctic Ocean." It urged that "immediate consideration" be given to the modification of a fleet submarine for polar research.

Lyon and Revelle completed their draft report on November 15. Adm. Thorvalt A. Solberg, the chief of naval research, signed it without revision on December 23. That was the good news. Lyon had already received the bad news: his request for a modified fleet submarine had been rejected.[17]

In March, 1949, Admiral Colclough, ComSubPac, had recommended that Lyon's proposal to modify a fleet submarine for arctic duty be turned down. A rough estimate of the added topside weight for a "turtleback" submarine such as had been envisioned in NEL Report 88, he informed the commander-in-chief of the Pacific Fleet (CinCPacFlt), would result in the loss of one-half of the submarine's buoyancy. Also, the proposed modification to the periscope would result in "an extremely difficult submarine to handle" when submerged at periscope depth in adverse sea conditions. While recognizing the importance of obtaining information on polar waters, ComSubPac concluded that this need must be carefully weighed "to insure the justification for the expenditure of funds on such an elaborate and costly modification."[18]

In order to reach a decision on the modified submarine, the chief of naval operations had to address Colclough's query about the priority to be accorded to arctic research. "At present there is no Operational Requirement for an Under Ice Submarine," CNO wrote to the chief of naval research. Was the importance of arctic research sufficient to justify converting a submarine for this purpose? Admiral Solberg replied that the importance of arctic research "must be determined by the relative military importance of the area." Because arctic research apparently was of less strategic value at present than such projects as the optimum routing of convoys and the long-range detection of submarines, both of which required more experimental ships and personnel, it did not "seem warranted at this time to spend money on the conversion of a submarine for Arctic operations."[19]

Lyon, who had pinned his hopes on securing a modified submarine by stressing its value for arctic research, decided on another tactic. In a letter to CNO, signed by NEL Director Bennett, he pointed out that "the geopolitical consequences of the operations of an under-ice submarine have not been considered" in the decision to disapprove the modification. He argued that a modified submarine could conduct extensive cruises under the ice, including a possible polar transit. If this transit took place, the navy could allow the information "to leak out." This might cause the Soviets to expend time and effort to protect their arctic shipping and deter them from establishing an air base adjacent to the Beaufort Sea.[20]

CNO was not impressed with Lyon's geopolitical musings. While the office was "extremely interested" in all oceanographic and sonar studies in the Arctic, it informed BuShips on November 10, 1949, the under-ice submarine had less strategic importance than other projects for which adequate submarines were not available. Accordingly, "The Chief of Naval Operations does not consider the proposed submarine conversion warranted at this time."[21]

Although disappointed at the denial of his request for a specially modified submarine, Lyon still had a mandate to conduct oceanographic work in the far north. Also, there was nothing to prevent him from seeking a fleet submarine to assist with this work while continuing to develop its under-ice capability. CNO was prepared to support this. On March 3, 1950, CNO approved a summer oceanographic research expedition that would include two icebreakers and a submarine.[22]

Later that month, NEL submitted its scientific plan for the expedition to the deep-water areas of the Beaufort Sea between Point Barrow and the Mackenzie River. Reviewing the progress in developing an under-ice submarine, NEL outlined the next important steps in the process, which included the evaluation of a five-unit topside echo sounder system for vertical ascents, and possible tests of long-range propagation of explosive sounds in the arctic basin. NEL would require one week to install the echo sounding system and necessary oceanographic equipment in a fleet submarine.[23]

Lyon's ambitious plans for the summer soon ran afoul of Admiral Colclough. In view of the shortage of submarines for antisubmarine warfare training, ComSubPac wrote to CNO on April 19, "it is recommended that no submarine be scheduled for an arctic cruise in 1950." CNO again demonstrated that it was unwilling to challenge operational commanders on matters of relatively low priority. "Proposed Arctic submarine operations cancelled," CNO informed NEL on April 27, 1950.[24]

Not only did Lyon lose his fleet submarine, but a change in command at NEL meant that *Baya* also would be unavailable for arctic research. Lyon's use of *Baya* in 1948 and 1949 had caused a good deal of hard feelings among the various research groups at NEL that relied upon the submarine for their own sonar work. It had been released to him only due to the support of Captain Bennett. The sympathetic director, however, was transferred to BuShips in 1950. His replacement, Capt. Dundas P. Tucker, although generally supportive of Lyon's work, did not accord it as high a priority as had his predecessor. Also, in accordance with the wishes that Lyon had expressed when he decided to turn down the appointment with UCLA in 1948, a reorganization at NEL allowed him to give up his position as chief of a major research division to become head of the Special Studies Branch, a smaller organization that would be devoted to arctic research. While this freed Lyon from unwanted administrative responsibilities, it also gave him less bureaucratic power within NEL, and made it less likely that he could secure *Baya* for his work.

With the declining level of support within NEL, Lyon turned to a different organizational structure to support his work on the under-ice submarine. Beginning in 1950, his arctic expeditions would be organized as joint Canadian-U.S. operations and involve the use of icebreakers. These operations fell under the authority of Service Squadron One in San Diego. ServRon One came under the Pacific Fleet; it also reported to the icebreaker desk at CNO. Operational orders came down from ServRon One to NEL for compliance. And, Lyon pointed out, "my group at NEL did the complying." As a result, Lyon explained,

"We were driving the work via the Fleet, i.e., as a laboratory within the Fleet rather than within the Material Command or Research and Development structure."[25]

In early May, Lyon received more bad news when he was informed that only one icebreaker—*Burton Island* (AGB 1)—would be available for the summer's joint expedition. With no choice, he modified his scientific plan. Following a series of discussions with Roger Revelle, director of Scripps, and Clifford A. Barnes of the University of Washington's Oceanographic Laboratory, an agreement was reached on an oceanographic program that included scientists from Scripps and the University of Washington. Barnes, an experienced arctic oceanographer who had previously served on coast guard ice patrols in the Atlantic, was instrumental in securing funding from the Office of Naval Research. The most significant objective of the expedition, NEL informed ONR on May 19, would be a series of oceanographic stations in the little-explored waters off the continental shelf of the Beaufort Sea. Both ONR and CNO promptly approved the revised plan.[26]

While preparing for the summer's expedition, Lyon secured some welcome support from Washington for his under-ice submarine. On June 5, CNO issued Operational Requirement SW-01402. "There are no effective directives in existence, other than this letter," CNO informed BuShips, "pertaining to the development of equipment and techniques which would enable a submarine to operate under ice." Operational Requirement SW-01402 provided the necessary mandate to develop an under-ice submarine. At the same time, CNO noted that other research projects had a higher priority. Development of the under-ice submarine was assigned a priority of only 3A. As Lyon's supporters at CNO explained, the Operational Requirement was less an urgent mandate than something—in Lyon's words—"for us to hang our hats on."[27]

Lyon at least had reason for cautious optimism as he left San Diego for Alaska. Reaching Point Barrow on August 10, 1950, he boarded a helicopter and flew to *Burton Island*, which was anchored five miles off the beach. While the members of his scientific party brought their gear on board, Lyon conferred with the icebreaker's commanding officer, Comdr. John R. Schwartz, about the work ahead.[28]

*Burton Island* got underway the next day. As the ship worked its way due north of Point Barrow, the experienced Barnes acted as a tutor in arctic oceanography for Lyon and Gene L. Bloom, a chemical engineer from NEL. The icebreaker stopped every three miles so the scientists could take an oceanographic station. This included a hydrographic cast with a Nansen bottle that took water samples at various depths; a bathythermograph to measure temperature; bottom samples; echo soundings; a net haul; and a bird and animal count. Echo soundings revealed a valley eighty fathoms deep just north of Point Barrow, then the bottom shoaled again to thirty-four fathoms.[29]

*Burton Island* reached the edge of the ice pack at 3:00 A.M. on August 12. Lyon and the icebreaker's executive officer, Lt. Comdr. Robert D. McWethy, boarded the ship's helicopter to inspect the ice field. As the Sikorsky lifted off *Burton Island's* deck, its tail suddenly spun and barely missed colliding with the ship's rigging. "As close as I certainly wish to come to a crack up," Lyon recorded in his Scientific Journal.[30]

Once safely aloft, Lyon located several large leads through the pack in a northeasterly direction. *Burton Island* followed these leads, reaching a depth of 500 fathoms by 10:00 P.M. The ship then secured for the night (a relative term at this latitude, with its almost constant daylight).

The icebreaker continued on its northeasterly course over the next week, taking oceanographic stations every twenty miles. Reaching Banks Island on August 20, Lyon and McWethy flew by helicopter to inspect the remote landfall that had been first seen by Europeans in 1820 and named by Capt. William Edward Parry for the eighteenth century naturalist, Sir Joseph Banks. The high section of the island, Lyon reported, was "covered with lichens (red, yellow, green, rust), and sparse tundra; soil has appearance of dried mud flat with drying cracks which are filled with lichens or tundra." There were numerous rodent holes in the soil. Wildlife abounded. Lyon saw ducks, geese, arctic terns, and one snow goose—"a brilliant white large bird."[31]

As *Burton Island* started to work through an area of large ice floes, Lyon began to develop reservations about Commander Schwartz's seamanship. "Captain is either showing fatigue or symptoms of fear of ice," Lyon wrote in his Journal on August 22. "He has become reluctant to take ship up well into ice until thoroughly stopped by every possible lead. He expresses fear of becoming caught by closing ice, a position which cannot occur during summer since no horizontal pressure exists." Lyon concluded: "Sometimes, it becomes tiresome or disheartening on these summer expeditions to spend so much effort and lose results because of unjustified over-caution of commanding officers. . . ."[32]

Lyon's relationship with Schwartz was deteriorating at the same time that his friendship with Executive Officer McWethy was blossoming. A product of the U.S. Naval Academy (class of 1942, which graduated in December, 1941), McWethy had gone into submarines. He had placed *Pogy* (SS 266) into commission in the spring of 1943 and had made six war patrols on the boat. Assigned to the General Line School at Monterey after the war, McWethy had taught navigation. One phase of the class involved polar navigation, but he found that there were no texts on the subject. McWethy, who had been fascinated by the Arctic ever since his boyhood reading of the adventures of polar explorers, managed to secure orders that permitted him to accompany an air force weather reconnaissance flight over the North Pole. On August 15, 1948, the young naval officer made the 103rd crossing of the pole on the B-50 "Polar Queen." Not only did he learn a great deal about polar grid navigation, but he also began thinking about the possibility of submarine operations in the far north when he saw the extent of open water during his flight over the ice pack.

McWethy returned to California determined to seek duty on the navy's icebreakers. It took until 1949, however, before he managed to obtain the assignment. He joined *Burton Island* as executive officer in time for the year's summer cruise. He enjoyed the duty. The icebreaker's commanding officer ran a relaxed ship, and the work was an adventure. Commander Schwartz, who came on board as a replacement prior to the 1950 summer cruise, was a different kind of skipper. A rather pompous individual, Schwartz had never served before in anything smaller than a battleship. Trained during his entire naval career to *avoid* hitting things, he did his best to stay away from collisions with ice.[33]

Lyon and McWethy soon found that they shared a common enthusiasm for the use of submarines in the Arctic. The usually taciturn scientist would become a different person when discussing the subject closest to his heart. With McWethy, he had a fellow believer who was eager to share ideas on how to develop the under-ice submarine into a reality. A close friendship developed between the two men, one that would have important consequences in the years ahead.

*Burton Island* continued north and west, whenever leads in the ice permitted, as the weather grew windy and colder. The scientists took their oceanographic stations, recording a depth of 1,300 fathoms on August 23 while north of Herschel Island. Schwartz, however, grew ever more hesitant to challenge the ice pack. "Commanding officer willing to take ship only a very short distance into ice although the pack is very broken because of wind," Lyon complained on August 24. "Further example of ice fever."[34]

Relations between the senior scientist and the icebreaker's commander worsened as time passed. Finally, on August 30, as the expedition drew to a close, the two men reached an impasse. "Captain refused to enter ice," Lyon reported, "said it was dangerous and difficult." As a result, the scientists were not permitted to obtain a final station in the ice pack. "The whine of a C.O. can greatly reduce the effectiveness of such operations," Lyon noted in frustration. "There is no excuse for his refusal to enter pack."[35]

Schwartz set course for Point Barrow, arriving off the coast at 5:00 P.M. Lyon completed his reports, packed his gear, and left the icebreaker.

Despite the problems with Schwartz, Lyon was pleased with the results of the expedition. His scientific team had located the edge of the continental shelf north of Point Barrow, taken soundings in the deep arctic basin out to a depth of 2,000 fathoms, and conducted seventy-three oceanographic stations. *Burton Island* had reached 73° 47′ North, or 143 miles north of Point Barrow, and accomplished the first surface navigation of 100,000 square miles of Arctic Ocean area.[36]

Prior to leaving *Burton Island*, Lyon had worked with McWethy on a series of conclusions that had emerged from the voyage. "The only acceptable attitude," Lyon recorded in his Journal, "is that naval warfare (surface and subsurface) will occur in the Arctic Ocean." Furthermore, concepts of weapons, tactics, and logistics were likely to differ substantially from those used in ice-free areas. At the center of polar sea warfare would be the under-ice submarine.

Submarine operations in the Arctic, Lyon pointed out, would require detailed oceanographic information. Submariners would need to know the bottom topography; sonar conditions for attack and escape; ice predictions in order to chose the best areas for attack and escape; and water conditions for buoyancy control.

Further oceanographic expeditions would supply the needed information by studying bottom topography and the circulation of water masses. The gross bottom structure of the Beaufort Sea had been determined by the 1950 expedition. However, there remained a need for detailed work on canyons and shore approaches to give submariners the ability to select the most advantageous approach to the coastal shelf.

Also, additional studies were required to determine sonar conditions and to predict ice.

The eastern sector of the Beaufort Sea and the Amundsen Gulf, Lyon noted, appeared to the best areas to base submarine operations. This portion of the Arctic Ocean had the maximum annual ice-free area and provided deep-water access to the arctic basin. Further expeditions must study this area in detail, as "too much thinking is based on conditions and beliefs observed during the 1840–1900 era of exploration and whaling." Lyon also wanted to explore the western approaches to the Northwest Passage, calling for a submarine in 1951 to attempt a transit of McClure Strait.

The strategic potential of the polar region, Lyon concluded, could not be satisfactorily estimated "until the physical environs of the Arctic Ocean and weapons concepts are evaluated by direct experimentation in the area." The reality of polar sea warfare, however, "must be accepted."[37]

As usual, Lyon stopped in British Columbia en route from Alaska to San Diego in order to review with his Canadian colleagues the results of the past summer's expedition and to plan for 1951. On September 3 he raised with Sanders in Esquimalt the possibility of a joint oceanographic effort the next year with Canada's taking responsibility for coastal areas and Banks Island. Sanders recommended that Lyon visit Field in Ottawa as soon as possible because the Canadian Defence Research Board's preliminary budget would be drawn up before September 15.[38]

Lyon went on to Nanaimo on September 4 for a conference with Tully. The news was not good. The Fisheries Board, Tully informed Lyon, had placed the Pacific Oceanography Group (POG) under tight budgetary restrictions. Unless POG could obtain funds from other sources, it would have to dismiss personnel. Also, it did not seem likely that there would be a ship available for oceanographic work in 1951.[39]

The future of arctic research in the United States seemed much brighter when Lyon attended the annual arctic oceanographic conference in Washington on October 17 and 18. After a review of the 1950 expedition's results, Lyon laid out an ambitious program for 1951. The year would begin with a winter cruise by *Burton Island* to the Bering Sea. NEL then would conduct a three-ship survey expedition to the Beaufort Sea in the summer. *Burton Island* would continue the oceanographic program that had begun in 1950, working alone in the polar ice fringe in the vicinity of Banks Island. A submarine would conduct operations further north, exploring the approaches to McClure Strait. Finally, a Canadian vessel, if available, would explore the area from the shore to the 100-fathom curve along the mouth of the Mackenzie River and into Amundsen Gulf.[40]

Lyon found that the attitude in Washington toward arctic oceanography and undersea warfare "has strikingly changed during the past 10 months." Previously, the dominant sentiment had been, "Justify why you spend any effort there." Now, the attitude was, "Why haven't you done more?" The naval bureaucracy, he reported, remained "split" on the question of polar warfare, with some sections still caught up in "meaningless arguments on details of arctic strategy and tactics." Also, CNO directives had not "completely filtered through." On the other hand, he was impressed by the long-term point-of-view at CNO, which apparently had been influenced by recent studies of German submarine activity in the Arctic during World War II.[41]

Lyon had good reason for optimism as 1950 drew to a close. His supporters at CNO had assured him that a submarine would be available for 1951 "unless a war with China occurs." Also, CNO had approved the establishment of the experimental field station at Cape Prince of Wales. BuShips would supply two LVT-3s to support the station and arrange for training NEL personnel in the tracked vehicles at Camp Pendleton. Even the situation in Canada looked promising, with the Defence Research Board's agreeing to give priority to oceanographic work off the mouth of the Mackenzie River and in Amundsen Gulf.[42]

Lyon's optimism did not last long. In January, 1951, CNO advised NEL that a submarine would not be scheduled for the summer's expedition. While CNO certainly concurred in the desirability of the experimental submarine cruise that NEL had proposed, Washington wrote, the shortage of submarines reported by CinCPacFlt precluded the requested assignment. As the correspondence on NEL's request revealed, Lyon's proposal to use a submarine had been strongly endorsed by ONR and BuShips. CinCPacFlt, however, argued that the shortage of submarines since the outbreak of war in Korea in June, 1950, had severely limited ASW training. "In view of ASW readiness," CinCPacFlt informed CNO, "it is recommended that the proposed cruise be deferred." As usual, CNO had bowed to the wishes of the operational commander.[43]

"I believe we should not rest our case," Lyon informed his superiors at NEL upon learning of CNO's decision, "but send a rebuttal." The real trouble, he pointed out, "is not at Washington, but Pearl." Things certainly had changed since the days of Admiral McCann. NEL Director Tucker told Lyon to go ahead and draft a response.[44]

On February 21, 1951, Lyon's rebuttal was sent to CNO under Tucker's signature. "The Laboratory," it stated, "feels obligated to emphasize certain aspects of any reconsideration of arctic submarine cruise." As under-ice experience currently amounted to only twenty hours, additional field tests obviously were necessary. Unfortunately, arctic operations "are too often met in scientific, naval and public discussions with amazement, awe, or apathy." Views on an arctic submarine, especially, are "sharply opinionated." NEL, however, had no doubts, as the letter's final paragraph—typed in capital letters—attested:

> IT IS THE CONSIDERED OPINION OF THE SCIENTISTS OF THIS LABORATORY WHO HAVE HAD UNDER-ICE EXPERIENCE THAT A FEW SUBMARINES PROPERLY EQUIPPED WITH BOTH THE TECHNICAL INFORMATION AND THE NECESSARY EQUIPMENTS CAN EFFECTIVELY CONTROL ALL SURFACE TRAFFIC THROUGH MARGINAL ICE AREAS, WHEN SUITABLE TACTICS HAVE BEEN DEVELOPED.[45]

Lyon's arguments failed to sway CNO. It approved the scientific plan for 1951 except for the portion that involved the use of a submarine.[46]

Lyon had little time to dwell on the disappointing news as preparations began for the summer's work. Some 120 tons of equipment and supplies had to be prepared for delivery to Cape Prince of Wales; a final operations plan had to be drafted for *Burton Island*; and discussions had to be held with the various scientific and naval personnel involved in the expedition. Lyon learned that his Canadian colleagues had secured the services of CGMV *Cancolim II*, an eighty-foot coast guard vessel, for their oceanographic work. Also, he would be able to use the new Electronic Position Indicator (EPI) that had been developed by the Coast and Geodetic Survey. Combining the best features of the Shoran and Loran radio navigation systems, EPI used a master transmitter on the survey ship and two slave stations on shore to provide position fixes that were accurate within seventy-five meters at distances of up to 300 miles. This kind of precision would be invaluable in the Arctic, where navigation by sun lines could fix positions only with an accuracy of plus or minus 3 miles—at best.[47]

On July 15, 1951, Lyon departed from San Diego on United Air Lines for Seattle. He was accompanied by four associates from NEL: Gene Bloom, who would lead the experiments that were to be conducted at the Wales Station; Edward Howick, an electrical engineer; and A. Wayne Medlin and Archie C. Walker, laboratory mechanics. Upon arrival, the men were taken by navy car to Pier 91, where they boarded LST-1126. They weighed anchor at five o'clock the next afternoon and proceeded northward in company with LST-1138.[48]

The LSTs reached Cape Prince of Wales at 11:00 P.M. on July 26, and anchored 1,000 yards off the beach in eighteen feet of water. Unloading began at 8:00 A.M. the next day. In clear weather, with little swells, Lyon and Medlin drove LVTs, shuttling supplies and equipment to the beach. Unlike the situation two years earlier, the sandbars proved little challenge for the tracked vehicles. By 5:30 P.M. they had managed to off-load 150 tons. Lyon reported that he was "punchy" from driving the LVT but "very happy" that there had been no major accidents during the off-loading. The entire operation had been a happy contrast to the previous year's difficulties.[49]

Off-loading was completed the next morning, and LST-1126 departed. The NEL contingent, which had been joined by Rexford N. Rowray, who would remain as resident manager, began to set up the station. After a lunch of beans, crackers, and cherries, Bloom took on the task of building a toilet while the rest of the group sorted supplies and set up a cook stove in the abandoned CAA building.[50]

The work continued over the next three days. The weather could not have been better, with clear skies and bright sunshine. Looking to the west, Lyon could see the Diomede Islands and the coast of Siberia beyond. "Really beautiful," he recorded in his Journal, "seas very calm. Of course still air means many mosquitos which give us plenty of trouble. Everyone stiff and sore from the sudden heavy work."[51]

The major task of the expedition was to install 40,000 feet of three-eighths-inch cable. The cable was to be laid in four sections of 10,000 feet each, perpendicular to the direction of the current, and extending from the CAA station toward Fairway Rock, 30,000 feet from the beach. Silver chloride sea cells and thermal elements connected to the cable would provide data on the rate of water flow through the Bering Strait and its temperature.[52]

By August 1 cable laying was ready to begin, with the first reel mounted on an LVT; however, there was a temporary delay as the sunny days of the previous week gave way to rain, fog, and a southwesterly wind of twenty-five knots. The weather cleared the next afternoon, and work began on laying the first 10,000 feet of cable from the CAA range house to a pole in the surf.

Everything went well until the two LVTs entered the water. One vehicle, with Lyon on board, was to lay the cable, while the second vehicle would act as a splice boat to attach the end of the second 10,000 feet of line. Just as the splice boat got ready to take the seaward end of the cable, its starboard motor quit. Lyon tried to haul up the end of the cable and return it to the beach, 300 yards away, but the line was too heavy. He could only tape the end and throw it over the side, using a wooden spool as a buoy.[53]

The problem on August 2 turned out to be only the beginning. It took two weeks of hard work to complete the cable-laying chore. Not until August 17 was Lyon able to leave the station for Nome. Four days later, he reached Point Barrow and boarded *Burton Island*. NEL oceanographers Alfred J. Carsola, who had acted as chief scientist in Lyon's absence, and Gene LaFond, filled him in on the past two weeks' activity on the icebreaker.[54]

Carsola and LaFond had surveyed the deep valley northeast of Point Barrow that crossed the continental shelf into the deep water of the Beaufort Sea. They had made seven crossings of the valley, each section five miles apart and five miles long, obtaining depth and bathymetric information. Profiles of the valley showed it becoming deeper on each successive crossing to the northeast, with the northwest slope much rougher than the southeast slope. The presence of what became known as the Barrow Sea Valley had obvious implications for submarine operations in the area.

*Burton Island* arrived off Barter Island on August 23, then headed north, stopping to take oceanographic stations every ten miles. The weather was calm on August 24 as the depth increased from 420 to 1,720 fathoms, enabling the oceanographers to cover an area they had missed in 1950 because of rough seas and stormy conditions. At one point, Lyon spotted a polar bear and two cubs swimming in open water. When pressed by the ships, the group was able to make four knots![55]

On August 29 *Burton Island* began to penetrate the ice pack—but not for long. "Commander Schwartz arbitrarily decided we couldn't work any further," Lyon reported, "he still shows a lack of understanding of ice and a fear of working it. He should work the Antarctic, then perhaps he would either have a mental breakdown or learn how to work summer sea ice."[56]

*Burton Island* returned to Point Barrow, then proceeded northwest along the 100-fathom curve to outline the boundaries of the Chukchi Sea. At 11:00 A.M., September 4, the icebreaker took station in 1,120 fathoms at 76° 23' North, 169° West. This was, Lyon noted, "a most amazing penetration—certainly a record for ships in this area except for *Sedov* and *Fram* which froze in to drift across the Pole."[57]

*Burton Island* then turned east, taking stations every twenty miles. Snow and sleet storms in the afternoon and evening, Lyon reported, made the oceanographic task "very miserable."

The expedition continued to develop the bottom topography of the Chukchi Sea until September 9. Having completed sixty-one oceanographic stations, *Burton Island* headed for Point Barrow.[58]

Lyon was delighted with the data that had been obtained. He was certain that it would substantially alter the accepted view of the Chukchi Sea. "The simple picture of a large oval shaped basin was based too much on the doubtful soundings of Wilkins (3,000 fathoms, western Chukchi Sea) and 2,600 fathoms of Storkersen in the Beaufort Sea. These deep soundings are erased by this expedition." The Chukchi Sea, Lyon pointed out, appeared to be a broad range of mountains or plateaus cut in valleys and steppes, extending north from the Bering Strait to 80° North latitude—or more.[59]

Lyon flew from Point Barrow to Nome on September 11, then visited Cape Prince of Wales. "Station in good condition," he observed. The sea cell at the far end of the cable was not working, perhaps due to a kink in the line, but otherwise the instruments were operating properly. Following this inspection, he continued to British Columbia to see Sanders and Tully. Learning that *Cancolim* would be available for next summer's work, he promised that NEL would supply EPI gear to provide more accurate fixes. Lyon then returned to San Diego on September 22, having completed a productive summer of oceanographic work.[60]

The NEL expedition was the star of the show at the second annual arctic oceanographic conference in November, 1951. Cliff Barnes, the

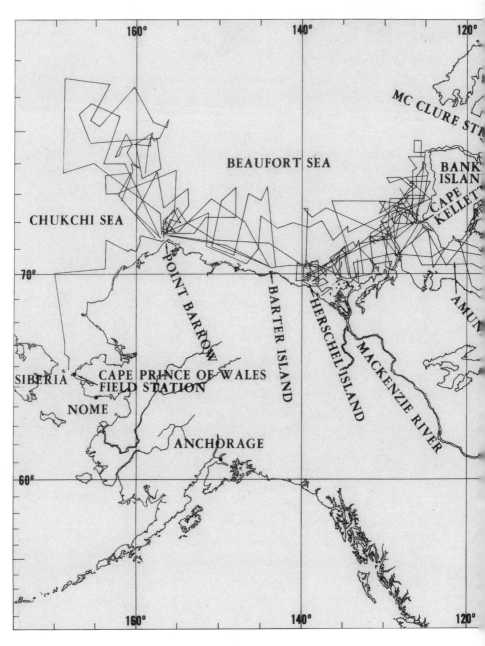

Bathymetric surveys conducted by NEL, using Electronic Position Indicator equipment, 1951–55. Courtesy U.S. Navy

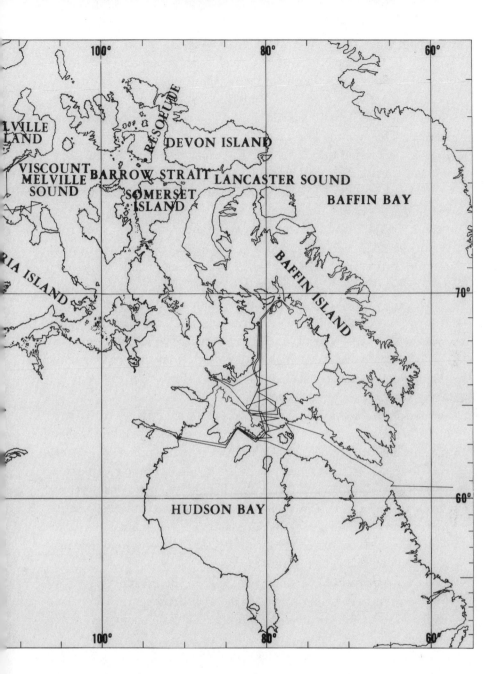

oceanographer from the University of Washington who had participated in the expeditions of 1950 and 1951, presented a report on the scientific results of both voyages to the Beaufort Sea. In 1950, he pointed out, *Burton Island* had conducted an oceanographic investigation of the continental shelf and slope of the deep arctic basin north of Alaska. "This was the first time," he emphasized, "that a modern icebreaker penetrated the ice field to the deeper water beyond the slope" and into an area "popularly considered inaccessible."

The past summer's expedition was a continuation of the 1950 work, extending the limits of the previous survey and developing additional details. Samples of water and bottom sediment had been taken from depths of 3,650 meters, more than 1,000 meters deeper than possible the previous year, thanks to a remodeled oceanographic winch with 15,000 feet of five-thirty-seconds-inch wire.

The 1951 expedition also had the benefit of EPI equipment. EPI, Barnes noted, had performed "remarkably well" and had "added immeasurably to the value of soundings by giving precise position."

"The cruise has been highly successful from an oceanographic standpoint," Barnes concluded. "The new information gained of the bathymetry, the water characteristics, and the ice conditions may well revolutionize the thinking of the Alaskan section of the Polar Basin."[61]

Lyon used the occasion of the conference to promote his ideas about the next phase of arctic research. "Primary item," he noted in his Scientific Journal, "is submarine cruise. Tremendous amount of oceanographic data now on hand. We are far behind on operational tests and development of tactics for submarine." Lyon wanted a fleet submarine for 1952 that would survey Banks Island and the approaches to McClure Strait. It also would work with *Burton Island* on sound transmission studies and conduct tactical exercises.

*Burton Island* would be employed to fill in the oceanographic gaps from previous cruises, primarily in the area between 142° West and Banks Island, and north to McClure Strait. EPI equipment, Lyon stressed, would be "a necessity."

Finally, operation of the Cape Prince of Wales field station should continue. A preliminary study of the data for 1951 indicated that the total heat transported by the water flowing north through the Bering Strait was a primary factor in determining the amount of sea ice in the Chukchi Sea. Additional data was needed to see if it were possible to predict ice coverage in the Chukchi Sea and along the northern coast of Alaska.[62]

Although the conference "strongly endorsed" Lyon's plans for 1952, he found that the pertinent desks at CNO were "very pessimistic" about the prospects of a submarine's being available. CinCPacFlt, he learned, had forwarded NEL's request for a submarine to CNO with an endorsement stating that the project would come under their "so-called research and development allotment" of thirty-two weeks a year and be assigned a priority indicator of 3A. The head of the Submarine Plan Review Section at CNO, Lyon noted, told him that the assignment to research and development with priority 3A "virtually kills the cruise."

"I came away confused as ever on the mechanics of submarine assignments," Lyon reported to his superiors at NEL. One thing, however, was clear. Although BuShips, ONR, the Research Defense Board, and the Hydrographic Office had all strongly endorsed NEL's request for a submarine, the key to the assignment was CinCPacFlt. CNO, he observed, "will go along with whatever is the real intent of Fleet Command."[63]

At the end of 1951, in a memo to the technical director at NEL, Lyon vented his frustration over the failure of the naval establishment to appreciate the importance of developing an under-ice submarine. CNO priorities, he pointed out, were based "on standing and calculated operational demands." The desks in Washington had read the same German World War II reports that he had read on submarine activities in Northern Russia. ONR, BuShips, the Research Defense Board, and other agencies had strongly supported arctic submarine development, but it had not mattered. "I feel that I have exhausted all plays through BuShips, ONR, RDB, etc.; the one remaining is directly through the Fleets, a tough selling job."

While Lyon did not intend to stop "yelling" for a submarine, he saw little hope of changing CNO priorities until "an active operational case occurs." Of course when this happened, he argued, it would be "too late." In the meantime, "we cannot stand by looking into a deep hole in our information on the sea waiting for it to occur." Although the project had a low priority, Lyon intended to turn the pool at Battery Whistler "into a miniature Arctic Sea," complete with sea ice. He planned to construct a room at the bottom of the pool that would simulate the conning tower of a submarine. In this way, he stated, "We will be able to at least approach some of the problems of ice boring and acoustical properties of sea ice." While no substitute for experiments with a submarine in actual arctic conditions, it seemed the best that could be done under existing circumstances.[64]

## Icebreakers and Submarines

Lyon had learned the hard lessons of the past when seeking support from his naval superiors for his polar submarine. Above all, he knew that the assignment of a submarine for arctic experiments depended upon the attitude of the Pacific Fleet's submarine commander. When Admiral McCann had been ComSubPac, his interest in the Arctic had led to impressive results. But his replacement, Admiral Colclough, had other priorities, and the happy cooperation between submariner and scientist had come to an end. Colclough, however, finished his tour in Hawaii late in 1951. In mid-January, 1952, Lyon met in San Diego with the new ComSubPac, Rear Adm. Charles B. Momsen, inventor of the "Momsen lung" for escaping from crippled submarines. To Lyon's delight, Momsen proved open to all sorts of experimental projects, including the under-ice submarine.[1]

On February 26, 1952, Momsen wrote to CinCPacFlt and asked permission to participate in the coming summer's Beaufort Sea expedition. Assignment of a submarine to the expedition, he pointed out, "would well justify the sacrifice of the equivalent amount of formal type training. From a broad viewpoint, operating experience in northern waters constitutes one of the finest and least available submarine type training." Projects that could be accomplished during the summer included an evaluation of NEL's topside echo-sounder system for determining the ice canopy; ice interpretation by scanning sonar; and the performance and maintenance requirements of electric torpedoes in extremely cold weather.[2]

Approval was soon forthcoming. CinCPacFlt, which had been so skeptical of arctic operations during the previous two years, now agreed that considerable training benefits could be derived from a voyage to

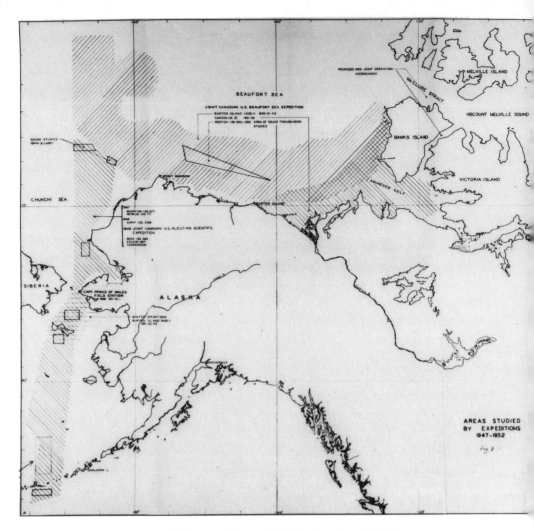

Areas studied by NEL expeditions, 1947–52. Courtesy U.S. Navy

the far north. "Participation is approved," CinCPacFlt informed Momsen on March 5, "on the basis that the period involved be charged against the submarine force type training time quota." As expected, CNO gave its consent to the recommendation of the operational commander.[3]

The early months of 1952 brought more good news for Lyon with the replacement of Commander Schwartz on *Burton Island*. He hoped that the new skipper, Comdr. Eugene H. Maher, would prove a more skilled ice pilot than his predecessor. Lyon soon had the opportunity to observe Maher in action. In early February he flew to Alaska to accompany *Burton Island* on a mission to supply the Prince of Wales field station and make a winter oceanographic study of the Bering Sea.

Picked up on the beach at Nome by helicopter on February 9, Lyon flew to *Burton Island*, which had approached the coast through the rough winter ice pack. Two days later, the icebreaker set out for the Wales station. *Burton Island* initially had no trouble carving a path through fresh ice, twelve to fifteen inches thick, as it headed north toward the Bering Strait. As it rounded King Island, however, *Burton Island* became stuck in the ice. When efforts to break free by engine power failed, Maher decided to blast his way clear.[4]

Lt. Comdr. L. A. Volse, head of the Underwater Demolition Team that had been carried on icebreakers during winter operations since 1949, placed three, fifteen-pound shaped charges along a line parallel to the ship. He located the center charge in the middle of the hummock area that seemed to be the main problem, with the other charges twelve feet on either side. The charges went off as Maher backed down, heeled, and trimmed the icebreaker. Nothing happened at first, but fifteen minutes later a crack in the ice appeared, and *Burton Island* was able to work free.[5]

Maher continued northward, picking his way through an ice field that was crisscrossed by pressure ridges. "Ship finds hard going in these ridges," Lyon noted in his Scientific Journal, "so we hunt by circuitous routes to avoid them—like a cabbage hedge or boxwood maze."[6]

*Burton Island* reached Cape Prince of Wales on February 14. Its helicopter led the icebreaker through leads in floes, some five miles in diameter, to a point close to the beach. The helicopter then shuttled instruments and supplies to the station as villagers came across the ice to visit the ship. Lyon flew ashore for discussions with Rex Rowray, the station tender, and to help unload the helicopter and store supplies. After obtaining data from recent experiments, Lyon returned to *Burton Island*.

With the supply mission completed, *Burton Island* moved across the Bering Strait toward Little Diomede Island, taking drift measurements. It was not possible to move further north, as a helicopter reconnaissance revealed that the way was blocked by large pressurized floes. "A submarine could operate," Lyon observed, "where this ship cannot."[7]

With no hope of proceeding to the north, Maher turned south toward St. Lawrence Island. He made good progress on February 15, following leads that ran perpendicular to his course. The next day, however, the ice grew thicker. *Burton Island* fought the pack all day, but the ship became beset at 10:00 P.M. The icebreaker drifted north through the night, crossing the U.S.-Soviet boundary at six-thirty the next morning. At 9:00 A.M., Commander Volse again used explosives in an effort to free the ship. This time, he detonated four shaped charges as *Burton Island* backed at full power. The icebreaker immediately broke free.[8]

It took two days, and another explosive blast, for *Burton Island* to make its way through the winter ice to Nome. Lyon left the ship well-pleased with Maher's handling of the icebreaker. The skill and determination of the skipper in the difficult winter ice conditions gave a good deal of promise for the summer's expedition.

Lyon flew from Nome to Victoria for a conference with Fred Sanders on February 23. The sonar scientist, Lyon reported, "agreed to the need for long [range] planning now for joint work and to keep [it] out of control of dog sledders." The following day, he went to Vancouver to discuss plans for the coming summer with Bill Cameron. They agreed that a survey of the Mackenzie River system was more important than one along Banks Island and should be given priority by the Canadians. Lyon then flew to Seattle for a meeting with Cliff Barnes, the University of Washington oceanographer. Plans for the summer, they agreed, depended upon the availability of EPI equipment. If EPI was not forthcoming, the expedition would then concentrate on submarine-icebreaker combat simulations and under-ice sonar transmissions.[9]

Everything seemed to be falling into place nicely for the summer's expedition to the Beaufort Sea. In mid-March Lyon learned that USS *Redfish* (SS 395) had been designated for the voyage. QLA sonar would be installed in the submarine prior to the cruise.[10]

In late April Lyon traveled to Washington, D.C., to make the final arrangements for the summer's operation. Unlike his previous trips, this time his wife and two children accompanied him. "These many years," he noted, "my family has never been able to travel with me, so for once we all went." They flew across the country on TWA, to the delight of the children. While Lyon visited the pertinent offices in the Pentagon, Virginia and the children took in the sights of the nation's capital. The family then drove to Norfolk and toured the Virginia countryside before heading for New York City and more sightseeing.

Lyon deposited his family with his wife's aunt in the Berkshires, then went back to Washington for additional discussions about the summer's work. Lyon and his family returned to San Diego in mid-May, where preparations for the 1952 expedition were in full swing.[11]

Lyon soon was immersed in the details of the operation. There were numerous conferences with ComServRon One, which had operational responsibility for conducting the expedition. Lyon had to draft the operations order and scientific plan for the summer's work. There were visits to San Francisco to supervise the installation of sonar equipment on *Redfish*. Material for the Prince of Wales field station had to be assembled and packed. Plans had to be adjusted when ONR informed NEL in late May that only two sets of EPI equipment would be available instead of the promised three. This meant that NEL would not be able to provide a set to the Canadians for their surveys.[12]

Arrangements went forward for a test of NEL's SOFAR system. This *So*und *F*iring *a*nd *R*anging system had been developed in the laboratory during World War II as a means of locating ship and aircraft survivors at sea. The survivors would drop a miniature depth charge into the water. It would sink to 3,500 feet, then explode. Hydrophones placed at the same depth would receive the sound of the explosion and fix the location by triangulation.[13]

The air force in 1952 had occupied an ice island—known as T-3—that was drifting close to the North Pole. This would afford a perfect location for a test of SOFAR under rough sea ice. SOFAR, CNO informed the air force, might provide a "simple, sure, and inexpensive means for locating aircraft down on the ice in the Arctic and polar regions." The navy would air deliver to T-3 a small box of SOFAR bombs. The personnel on the ice island would then drop each of the bombs at an exact time of day for two or three weeks, while *Redfish* listened for the sound of the explosion.[14]

The air force readily agreed to the experiment, although they preferred to deliver the box of bombs to T-3 themselves. Comdr. Jack E. Gibson at CNO told Lyon to mail the crate of bombs to Albert P. Crary, the air force's senior scientist on T-3, in care of Project Icicle, Thule Air Force Base, Greenland. Gibson, one of Lyon's supporters at CNO, then wished him well for the summer's operation. "Having an underwater vehicle should give you a great amount of satisfaction at long last," Gibson wrote. "It could be a good first step to the Navy's first, and probably the world's first under-ice submarine to be built as such. I sincerely hope so. Good ice-breaking and under-icing to you."[15]

By June Lyon had completed a revised scientific plan for the expedition. *Burton Island* would install and calibrate EPI stations at Point Barrow and on Barter Island. Sonar and undersea warfare exercises would begin about August 10, with four main objectives. The drills were designed to evaluate under-ice navigation and diving techniques; to measure sound transmission under ice; to measure ambient noise in ice fields and the sound output of an icebreaker moving through ice; and to study submarine detection and attack techniques on a ship moving through ice. Following these exercises, *Burton Island* would undertake oceanographic work in the eastern section of the Beaufort Sea to complete gaps in the data obtained during the 1949 and 1950 expeditions. *Redfish* would explore the approach to McClure Strait. Finally, the Canadian survey ship, CGMV *Cancolim II*, would continue its previous year's inshore oceanographic work along the approaches to the Mackenzie River.[16]

Lyon flew to Alaska in early August to inspect the field station at Cape Prince of Wales. Prior to his arrival, the navy had delivered 110 tons of supplies to the station as part of their annual logistical effort in the far north. The station had grown since its establishment in 1949. It now consisted of two houses, a laboratory-generator building, a large workshop, and sheds to hold a tractor and LVT. The main task for the summer would be to lay a new four-mile-long cable into the Bering Strait to replace one that had been damaged the previous winter. Lyon discussed the method of cable-laying with station personnel, then departed for Nome and Fairbanks.[17]

Lyon went on to Barrow on August 8. Three days later, *Burton Island* arrived. Lyon boarded the icebreaker, which proceeded to a rendezvous with *Redfish*. Lyon transferred to the submarine, where he rigged hydrophones in an effort to hear the first SOFAR bomb, scheduled to be detonated from T-3. The noise of the submarine, however, caused difficult listening conditions. "Absolute silent running will be necessary," Lyon concluded, "if any hope of hearing."[18]

The next day, August 12, Lyon checked out the five-unit topside fathometer system and the QLA sonar. When *Redfish* made its first stationary dive, a problem developed with two of the topside fathometers, but the remaining units still provided adequate information on the overhead ice cover. The QLA "looked great," allowing *Redfish* to surface within thirty yards of an ice floe that had been plotted on the scanning sonar. "At no time," Lyon wrote in his Journal, was there "any uncertainty of location of ice or whether clear overhead."[19]

With all under-ice navigational equipment operating satisfactorily, *Redfish* again made contact with *Burton Island*, which had finished the installation of an EPI station at Point Barrow. Lyon was disappointed to learn that the SOFAR bombs had not yet reached T-3, although they were expected to arrive within a week. The long-distance sound experiments could resume at that time.[20]

Between August 15 and 22, *Redfish* and *Burton Island* conducted a series of simulated attacks. Captain Maher came away from the experience with a high opinion of the arctic submarine. "The results of the experimental work," he later reported to CNO, "force the conclusion that the under-ice submarine is practical and that its combat potential is great. It is apparently immune from detection or attack by surface vessels operating in the ice, and is capable of detecting such vessels at great distances and attacking them at will." Lyon could not have put it any better![21]

The war-fighting exercises were followed by a series of sound transmission studies. The submarine held its position in the ice cover and listened on hydrophones that were placed under the water at 50 and 160 feet, while the icebreaker moved 100 miles away and used a Fessenden Model KE-1 non-directional oscillator, suspended 80 feet beneath the surface, to transmit a 1030 cycle sound.

By the time these experiments were completed on August 23, *Redfish* was lying in a small lake within the ice pack, with floes close on all sides. It was time to see if the techniques that had been developed since 1947 for a submarine to dive and proceed under the ice to open water would work. Both Lyon and Comdr. John P. Bienia, skipper of *Redfish*, were confident that they would.

As *Redfish* began its vertical descent, the after torpedo room reported flooding. This was news, Lyon wrote in his Journal, "which shakes anybody when you're in a submarine." The normal procedure under such circumstances would be to surface immediately. But Captain Bienia realized that the submarine was already under the ice canopy and to come to the surface likely would result in damage to the topside of the boat. He decided to wait and see if the flooding became more serious. It was soon apparent that the gaskets on the afterroom hatch were not holding; the rubber had lost its resiliency in the cold temperatures and was not properly sealing. When sufficient pressure built up in the boat, the leak stopped as metal-to-metal sealing became effective.

It took a long ten minutes, Lyon reported, to realize that "we weren't going to have a complete casualty." Bienia's decision had been the correct one, although the men in the after torpedo room, standing in several inches of water, had cause to wonder. "Considerable effort must be made to learn what error was made in gasket material supplied *Redfish* by San Francisco Navy Yard," Lyon concluded.[22]

Following the initial excitement, *Redfish* proceeded under the ice for the next twenty-two miles without difficulty. Lyon and other NEL personnel operated the special sonar equipment, interpreted the information, and informed Captain Bienia of their best estimates. After the sonar specialists identified a large open area in the ice cover, *Redfish* surfaced to recharge its batteries and replace the two gaskets in the after hatch.

In all, *Redfish* spent two days making its way out of the ice pack. Lyon and his NEL associates not only provided the information that enabled the submarine to locate ice-free areas for battery recharges, but they also used their oceanographic knowledge of the Beaufort Sea, especially temperature patterns, to identify the most direct course toward the open sea. "The diving and ascent procedures in ice lakes are completely solved," Lyon reported. "The REDFISH at no time made errors in judgment of the ice canopy overhead or in surfacing in small lakes not much larger than the submarine."[23]

After clearing the ice pack, *Redfish* set course for Cape Kellett, on the southwestern tip of Banks Island, to begin the oceanographic phase of the summer's expedition. Over a period of ten days, the submarine used EPI fixes to plot the western coastline of Banks Island while taking several oceanographic stations.

On September 5 *Redfish* was lying in an open lake amidst the ice pack, some 100 miles west of Banks Island. Lyon rigged a Brush hydrophone to a depth of 160 feet to listen for the scheduled SOFAR bomb explosion from T-3, located 900 miles away at 88° North, 96° West. "Bomb signal received very strong," he wrote in his Journal; "full swing of meter; good recording of signal."[24]

Lyon was delighted with the success of the SOFAR experiment. This one test, he believed, strongly suggested that SOFAR could be used not only to locate downed aircrews in the Arctic, but it also opened up the possibility of establishing a SOFAR network in the polar region that could be employed for navigational purposes.

As so often happened, the good news about the SOFAR test was followed by some bad news. The next day, September 6, Lyon received word that *Burton Island* had damaged its starboard propeller while trying to work free of ice off the northwestern tip of Banks Island. As Commander Maher later reported, the icebreaker had taken an oceanographic station in a small lake and was attempting to break free of the surrounding pack when the starboard screw hit a piece of free floating ice, causing the blade to sheer off. Characteristically, Maher took full responsibility for the incident. "This casualty," he wrote in his report of the expedition, "is attributed to improper ship handling."[25]

The damage reduced *Burton Island*'s maneuverability and lessened its ice-breaking capability. Lyon had hoped to spend the last two weeks of the expedition doing sounding work north of Point Barrow and in the eastern portion of the Chukchi Sea, thus completing the oceanographic and hydrographic survey of the area. With *Burton Island* out of action, however, he had no choice except to secure from the summer's operations without accomplishing the planned survey work.[26]

As *Redfish* headed for a rendezvous with *Burton Island* off Cape Kellett, Captain Bienia decided to test the impact of a torpedo against the ice. He fired at a small floe, 700' x 200' in area, and 10 feet thick. The Mark 18 torpedo exploded under the ice and tore it to pieces. Lyon concluded that a torpedo or mine could be used to open a large hole in an ice floe and enable a submarine to ascend. "This should be kept in mind," he noted, "as a safety feature for emergency."[27]

*Redfish* and *Burton Island* came together on September 8. The icebreaker dismantled the EPI station at Cape Kellett, then proceeded in company with *Redfish* to remove the station on Barter Island. The two ships next headed for Point Barrow, stopping in deep water en route "for one final cast" to check a new process of water identification in non-glass containers. It took six days to reach Point Barrow, with *Redfish*'s often acting as an icebreaker and winning high marks from Maher. The submarine, he reported, demonstrated "excellent seamanship in the ice." The fact that *Redfish*'s propellers were undamaged, he pointed out, was evidence of the skill exercised by her conning officers. In fact, Bienia had flooded main ballast tanks six and seven to keep the submarine's propellers deep in the water and clear of ice; however, *Redfish* did suffer ice damage to its four upper torpedo tube shutters.[28]

Lyon transferred to *Burton Island* shortly before the two ships reached Point Barrow. He then released *Redfish* to return to San Diego, while the icebreaker continued to Cape Prince of Wales. Upon arrival, Lyon flew by helicopter to the station to inspect the summer's work. "Station looks good," he observed. All debris had been cleaned up, an addition to the laboratory-generator building had been completed, and the new cable had been laid successfully.[29]

While *Burton Island* proceeded to Seattle, Lyon flew to Anchorage for a conference with the Weather Bureau regional office to work out the details of the transfer of the Prince of Wales station to the Navy Department. The Weather Bureau earlier had advised him that they planned to shut down the observation post on Cape Prince of Wales and wanted the navy to assume responsibility for the facility.[30]

Lyon finally returned to San Diego in late September, well pleased with the results of the summer's work. The expedition, he wrote in his summary report, had nearly completed the oceanographic and bathymetric description of the Beaufort Sea. Only the deep Beaufort Sea trench remained to be studied, a task he had been unable to accomplish due to the damage to *Burton Island*. An icebreaker would be needed in the future, he pointed out, to finish the bathymetry of the northern Chukchi Sea in the area from 155° to 161° West, by 72.2° to 74.6° North. Also, a survey should be initiated of the Western Canadian Archipelago, including Amundsen Gulf, Prince of Wales Strait, and McClure Strait.

In addition, the expedition had conducted important experiments on sonar conditions in the Arctic. It had made the first study of sound transmission under ice in water deeper than 1,000 fathoms. *Redfish* had received and recorded sound transmissions of 1030 cycles from *Burton Island* at ranges of 1 to 100 miles. Also, *Redfish* had detected a four-pound underwater explosion from a distance of 900 miles.

Above all, the summer's expedition further confirmed the viability of the under-ice submarine. Indeed, Lyon concluded, the results of experimental work since 1946 "force the conclusion that the under-ice submarine is practical and that its combat potential is great." The arctic submarine, he believed, likely would be "the supreme weapon" for combat in areas of sea ice. It was virtually immune to detection and attack by surface ships and aircraft. It could strike anywhere from the periphery of a protective sea ice cover, including Newfoundland, Spitzbergen, and the Kuriles.

CNO, Lyon pointed out, faced a decision: either direct the material bureaus to press the development of the under-ice submarine, or write-off as "operationally untenable and strategically non-essential to the Navy" all areas having ice. "It may be later than we think," he warned, "in view of the probable ten-year lapse between the issue of an initiating directive and delivery of a combat under-ice submarine to the fleet." At a minimum, CNO should order one boat to be built to evaluate its combat potential and to devise ASW countermeasures.[31]

Lyon followed up his official report on the expedition with a personal letter to Admiral Momsen at ComSubPac. "I believe it is urgent," he wrote, "to point out certain aspects of under-ice submarines which have arisen from experience with the REDFISH and icebreakers." Submarines, near and within ice fields, "can be very formidable weapons." The recent experiments had demonstrated that a submarine could make a sonar approach in ice and practically hover under a surface ship "with little danger of detection."

Although *Redfish* provided the first sonar and tactical data, Lyon pointed out the fact that this work represented almost "trivial progress" over *Carp* in 1948 in terms of developing an under-ice submarine. There was a pressing need, he advised Momsen, to modify one submarine for under-ice experiments. In 1948, he noted, he had asked for a flush-decked submarine with ice-boring snorkels. This idea had proved "too revolutionary for acceptance in one step." While a flush deck, ice-boring boat remained "the ultimate experimental goal," Lyon was prepared to work with a submarine modified with a heavy superstructure as a more acceptable interim stage.

Lyon suggested to Momsen that a fleet submarine be modified to withstand punishment from working through heavy brash ice on the surface. Also, the submarine should be able to "shoulder" its way through brash ice when coming to the surface. This involved modifying the deck and bridge shears, with topside sonars and radars retractable into the hardened deck. "We can go no further," he concluded, "until a hull is changed."[32]

Momsen proved receptive to Lyon's ideas. Indeed, ComSubPac wrote on October 14, he considered Lyon's letter "very timely." Momsen noted that he had been stressing the need for the new and revolutionary design of submarines for the past few months. He would send a copy of the letter to the Special Board that had been appointed by BuShips to study the future design of all types of ships.[33]

Shortly after Lyon wrote to Momsen, he traveled to Ottawa for a meeting with Sanders, Cameron, Field, and other Canadian scientists at the Defence Research Board. The Canadians reported that *Cancolim* had had a successful summer, working between Herschel Island at the mouth of the Mackenzie River and Cape Kellett. They considered the survey of the Mackenzie River sector complete and proposed to use a small boat in 1953 to survey portions of Amundsen Gulf, ice conditions permitting. Lyon offered the possibility of using *Burton Island* and the EPI network for the task, with Canadian oceanographers on board. Cameron thought that this was an excellent idea, and he promptly volunteered for the assignment. Lyon and the Canadians agreed that McClure Strait and the path to the east deserved the highest priority, as it offered the most probable deep channel for submarines. Prince of Wales Strait would be an alternative, but it was likely to be a shallow and tortuous passageway.[34]

Lyon also met with the Canadian director of intelligence to discuss possible Soviet submarine activity in the Arctic. This was a matter of constant concern for Lyon. He knew that the Soviet Union was far more active in the Arctic than all other nations, although Communist secrecy shrouded the full extent of their progress in both scientific and military areas. The director of intelligence informed Lyon that there were no clues at present, either positive or negative, concerning Soviet development of an under-ice submarine.[35]

When Lyon returned to San Diego, he learned that CNO had approved NEL's participation in a winter icebreaker operation in the Bering Sea with *Burton Island* and the coast guard icebreaker *Northwind*. He then prepared a plan for the summer's work. Sent to CNO for approval under Captain Tucker's signature, it followed the lines that he had recently discussed with his Canadian colleagues. "The next step in the oceanographic study of the Arctic Ocean by icebreaker," NEL's request stated, "is the survey of the western Canadian Archipelago, namely McClure Strait, Prince of Wales Strait, and Melville Sound." McClure Strait would be the most difficult area to reach due to the likelihood of encountering heavy ice, but it was the most strategic passage for submarines with its great depth and width.[36]

In early December, 1952, Lyon crossed the continent to attend the third annual arctic conference in Washington. NEL's work again dominated the past year's activity in the far north. Lyon presented the submarine sonar results of the summer's expedition to the Beaufort Sea;

NEL's Gene Bloom reported on sea ice studies and other activities at the Cape Prince of Wales station; Cliff Barnes of the University of Washington covered the oceanographic work of the expedition; and Bill Cameron discussed the Canadian contribution.

Lyon called attention to the expiration in 1953 of the first five-year increment of the navy's policy for Cold Weather Planning, and the consequent five-year plan for arctic oceanography that he and Roger Revelle had prepared in 1949 while acting as a committee under ONR. Consideration soon would have to be given, he pointed out, to an extension of the navy's program in the Arctic.

Lyon then visited BuShips to inquire about the fate of NEL's request for the summer 1953 oceanographic expedition. He found that nothing had been done. At his urging, a favorable endorsement was prepared and sent to ONR, where a second positive endorsement awaited. Also, the oral commitment of *Burton Island* and *Northwind* to the expedition had to be confirmed, and a formal invitation had to be extended to the Canadian government to participate in the summer's work. "It has been necessary to personally guide the correspondence in Washington in every expedition the past four years," Lyon noted with some exasperation.

Lyon next went to CNO, where he discussed arctic submarine warfare with the desks for Cold Weather Readiness (Op-342), Cold Weather Operations (Op-332F), and Submarine Research and Development (Op-311 and Op-316). Further meetings involved the submarine design desk at BuShips, and the Undersea Warfare Branch and sonar groups at ONR.

"Far greater interest was shown than in past years," Lyon reported upon his return to San Diego, "primarily because of the REDFISH patrol report which was under study by most desks." It was essential, he believed, to request a submarine be modified for arctic work "as soon as possible." Support must be obtained at the highest levels of CNO in order to have any hope of earmarking one hull of the new submarine construction program for arctic modifications. Direct pressure should be maintained on the appropriate desks at CNO, ComSubPac, ComSubLant, and the Special Board for Ship Design at BuShips.[37]

Lyon had made NEL's case for a modified submarine in Report 353, "An Approach to Submarine Warfare in Sea Ice," which had been completed in December and awaited a covering letter from Captain Tucker prior to forwarding to CNO. The report discussed at length the history of NEL's arctic expeditions and pointed to the strategic importance of marginal sea ice areas. The submarine, it argued, could be developed into a "formidable weapon" for operation in or near sea ice; however, the capabilities of the under-ice submarine could only be fully determined by acquiring new data from a specially designed boat.

The report offered three recommendations. First, a fleet submarine should be modified for under-ice experiments. (A section of the report provided detailed suggestions on the nature of the proposed modifications.) Second, the oceanographic survey of the Arctic Ocean should continue, including work on the study of sea ice dynamics. Finally, support should be given to the continued development and construction of laboratory facilities at NEL to simulate the under-ice environment.[38]

Lyon hoped that Report 353 would be the primary document in persuading the navy's hierarchy to modify a submarine for arctic work. "It is essential to follow through with the USNEL report," he wrote to NEL's director of research on December 30, 1952. The following week, however, he learned that Captain Tucker was having trouble signing off on the report. Portions of the report, Tucker believed, extended into tactical and strategic areas that did not lie within the assigned mission of NEL. It took over a week to work out the problems. In the end, Lyon was able to preserve most of the report intact, although he found Tucker's covering letter to be less than a ringing endorsement of his work.[39]

In Tucker's forwarding letter, a paragraph that had been drafted by Lyon and left unchanged by the director stated that six years of arctic experiments had demonstrated the "fundamental practicality" of submarine operations in sea ice. Accordingly, NEL did not intend to request the participation of conventional submarines in future arctic expeditions as there was nothing more to be gained from their presence.[40]

In sending a copy of the report to Admiral Momsen in Hawaii, Lyon hastened to assure his supporter that the statement had been inserted only "to emphasize the need for a hull change." ComSubPac should understand that NEL would be delighted to have a conventional submarine along on its arctic expeditions. There were a number of experiments on sound transmission and target approach that could be done with a conventional boat. Also, training and indoctrination of submarine crews in arctic waters was "a big need." It should be pointed out, however, that experimental under-ice work with a conventional boat might result—"through bad luck"—in serious damage to some topside gear. And this, in turn, could undermine support for arctic submarine development.[41]

While awaiting Momsen's response, Lyon headed north for the winter's icebreaker expedition. He arrived in Nome on February 10, 1953, and checked into the Polaris Hotel. The weather was clear, with a temperature of minus-20°. Lyon spent six days in the nearly deserted hotel, awaiting the arrival of *Burton Island*. The ice cover, however, was much thicker than during the previous winter, and Commander Maher was unable to break his way through to Nome.[42]

Learning that *Burton Island* would be forced to cancel its stop at Nome, Lyon decided to hire a bush pilot and fly to the Cape Prince of Wales station. He departed on the morning of February 17 in fog and low clouds. The pilot tried to stay underneath the overcast while he followed the coastline at altitudes of 400 to 100 feet. The weather continued to deteriorate as they flew north and eventually forced a landing at Tin City, some ten miles short of Wales.[43]

Lyon stayed overnight in a nearby radar station and awaited an improvement in the weather. The next morning, however, brought a full-blown blizzard, with winds of seventy to eighty-five knots and temperatures reaching minus-19°. "Physically impossible to do anything," Lyon noted in his Journal. The blizzard lasted two days. When the weather began to clear, Lyon decided to use the opportunity to return to Nome. It seemed doubtful that *Burton Island* would be able to reach Wales, and Lyon did not wish to be stuck there in bad weather. "Regret this move," he wrote, but there was no other choice. "This expedition has become one of great frustration as far as I am concerned."[44]

Shortly after Lyon returned to San Diego, he learned that Momsen had strongly recommended to CinCPacFlt that a submarine be included in the 1953 summer expedition. Projects to be accomplished included under-ice diving, navigation, evaluation of sonar, and assistance in the oceanographic study of the western portion of the Northwest Passage.[45]

Lyon responded to the news by drafting a letter to CinCPacFlt, sent out under Captain Tucker's signature, which emphasized that hull modification remained NEL's primary goal. Further under-ice cruises by conventional submarines should produce results valuable to the modification program. "However," the letter warned, "it should be clearly recognized that a conventional boat will very possibly sustain damage to superstructure, torpedo tube shutters and some topside equipments similar to that sustained by REDFISH."[46]

Lyon followed the official letter with a personal note to Momsen. If a submarine were to be provided for 1953, he wrote, "we want a standard boat with QLA scanner and wish to go right on from where we left off last summer." As it happened, Momsen had planned to assign USS *Volador* (SS 490) for the cruise—"in order to give some experience to a Guppy." The snorkel-equipped boat, however, did not have QLA sonar. In light of Lyon's wishes, Momsen agreed to change the submarine assignment and send *Redfish*.[47]

On May 18 and 19, 1953, Lyon attended the planning conference for the joint U.S.-Canadian Beaufort Sea expedition that had been convened at Long Beach Naval Station by ComServRon One. He learned that Project Lincoln, the air force plan to install a series of Distant Early Warning radar stations in the far north, had become a top priority. A survey of the ship routes for naval support of the proposed DEW Line would take precedence over all other work during the summer.[48]

The recent conference at Long Beach, Lyon wrote to Momsen on May 22, indicated that the DEW Line project "is getting serious." Surveys of beach approaches and the proposed shipping routes in the Canadian Archipelago were taking overriding priority, with oceanography and sonar work relegated to a position of secondary importance. "This is quite different from my original conception of this summer's expedition," Lyon noted.

It was ironic, Lyon pointed out, that arctic undersea warfare was "still not widely accepted as a real problem" at the same time that DEW Line sites were being envisioned that would require yearly logistical support by ships. In any event, he continued, "We hope to make the maximum gain from this summer's work because it may be our last opportunity for some time." This may have been a pessimistic assessment, Lyon admitted, but he recognized that Momsen's tour as ComSubPac shortly would come to an end, and "we have had submarines for Arctic work only when under your and Admiral McCann's sponsorship."[49]

In mid-July, 1953, word came down from CNO that "possible future establishment of additional DEW stations across arctic North America creates necessity for certain site surveys." While the air force was involved in preliminary planning for the DEW Line, it would not conduct extensive on-site surveys until the feasibility of surface resupply of the proposed area had been determined. As the sites were in the vicinity of Prince of Wales Strait, valuable hydrographic and oceanographic information might "conveniently" be obtained during the upcoming joint expedition. Wherever practical, on a not-to-interfere basis, the expedition should conduct a reconnaissance of Amundsen Gulf, Prince of Wales Strait, and Viscount Melville Sound.[50]

CNO's instructions, in fact, softened the demanding priorities of the Long Beach conference. Lyon already had planned to conduct a survey of the areas in which the air force was now so interested. At least in this early stage, the DEW Line project would not significantly impede his work.

As it turned out, the 1953 expedition proved a great disappointment. Ice conditions were the worst that Lyon had ever experienced during a summer expedition, with ice cover staying right against the beach from Point Barrow to Amundsen Gulf. While *Burton Island* was able to conduct the first-ever oceanographic survey of portions of Prince of Wales Strait, *Northwind* and *Redfish* spent long and fruitless days "struggling around the ice mess on the north coast of Alaska."[51]

On August 20 the ice became so thick that *Northwind* had to take *Redfish* in close tow. "Rough, noisy tow through ice," Lyon reported, "reminder of tows thru Antarctic ice in *Sennet*." Even worse trouble lay ahead. Five days later, while heading toward Cape Kellett through belts of old, pressurized, extremely hard ice, the shutter on *Redfish*'s number three forward torpedo tube sprang a leak.[52]

Number three tube, the middle tube on the starboard side, contained a Mark 18 electric torpedo. The question soon arose, Lyon reported, about what might happen to a torpedo under prolonged flooding. "Is there any danger of the fish arming itself by flow inside tube and how can we get fish out of the tube?" An inspection by Underwater Demolition Team swimmers revealed that the shutter had been mashed in by ice. Captain Bienia decided that the torpedo had to be removed from the tube, if at all possible.

A lengthy discussion ensued about how best to trim *Redfish* in order to expose the bow tube. As the agreed-upon plan unfolded, *Redfish* tied up to *Northwind*'s stern, using the icebreaker's tow cable. Bienia then flooded the aft ballast tanks and pumped oil from forward to aft tanks. As *Northwind* took a heavy strain on its towing winch, *Redfish*'s bow rose sufficiently clear of the water to allow a UDT swimmer to seal the cracks in the outer door of the number three tube. The inner door could then be opened and the torpedo pulled into the forward torpedo room. "Everything appears OK," Lyon reported, although it was "impossible to ascertain how many turns the arming mechanism may have turned."[53]

Two days later, Lyon decided to abandon further work and head for Point Barrow. "This year's ice conditions are many degrees worse than any previous year," he concluded. En route, *Redfish* again became stuck in the ice and had to be towed by *Northwind*. At one point in the operation, *Northwind* swung around *Redfish*, and the submarine's bow punched a hole in the icebreaker above the water line. All-in-all, Lyon noted, the summer had been "a real education in what the Arctic Ocean can do."[54]

Lyon soon put the disappointments of the 1953 expedition behind him and turned to plans for 1954. On the plus side, he looked forward to a major contribution from the Canadians. Construction of a new icebreaker, HMCS *Labrador*, was expected to be completed in time to participate in the coming summer's expedition. And he had developed a cordial relationship with the *Labrador*'s designated commanding officer, Capt. O. C. S. Robertson, who had been aboard *Northwind* as an observer during the recent Beaufort Sea expedition.

In October Lyon went to Ottawa for discussions at the Defence Research Board about *Labrador*'s deployment in 1954. The Canadians, he found, were planning to use their new icebreaker for a resupply mission to Eureka. Lyon raised the possibility of changing the assignment to permit *Labrador* to work in Viscount Melville Sound. Robertson, who attended the meeting, strongly supported the idea. Indeed, he had even grander plans in mind for his ship: Robertson was thinking about the possibility of making the Northwest Passage through the Canadian Archipelago.

Lyon came away from the meeting with a sense that *Labrador* opened new doors for his Canadian partners. "Note interest much increased in Arctic," he wrote in his Journal. Also, there had been a change from the previous Canadian role as a junior partner to the United States in arctic research to one of "leadership and acceptance of long term planning and work."[55]

Shortly after his return from Ottawa, Lyon drafted the scientific plan for the 1954 expedition. After passing for comment through the chain of command at NEL, it went out to CNO, via BuShips and ONR, under the signature of NEL's new director, Capt. H. E. Bernstein. The plan had two major objectives: a study of sonar transmission and sound detection ranges; and an oceanographic survey of McClure Strait and Viscount Melville Sound. NEL requested that a submarine be assigned to work with *Burton Island* on the sound transmission study, while *Northwind* conducted the survey of the western portions of the Northwest Passage.[56]

The first endorsement of the plan was prepared by Butler King Couper at BuShips. Lyon had met Couper, an oceanographer by training, during Operation Crossroads in 1946. "He was my contact and supporter of Arctic work for decades at BuShips," Lyon later recalled, "a key man." As expected, Couper gave the 1954 program a ringing vote of confidence.[57]

The second endorsement came from ONR. Again, the response was positive. ONR viewed a continuation of NEL's work in the Arctic as "extremely important militarily and scientifically," the endorsement stated. This time the recommendation had been written by William V. Kielhorn—"who was the key man in support of Arctic work in ONR, 1950–55," Lyon later wrote, "and reason for our ONR support in Beaufort Sea Expeditions."[58]

While things were going well in Washington, however, they were deteriorating at Pearl Harbor. As Lyon had predicted in May, the change of command at ComSubPac spelled trouble for his arctic programs. Along with submission of the 1954 scientific plan, NEL sent a request to CNO for participation of a submarine in the expedition. The letter, drafted by Lyon and signed by Bernstein, pointed to recent intelligence reports that indicated Soviet submarine activity in Baffin Bay during the past summer. It was even possible, the letter warned, that the Soviets had made a transit of the Canadian Archipelago at a time when ice conditions were extremely difficult in the Beaufort Sea. "It is assumed," Director Bernstein concluded, "that Russia leads in Arctic submarine development."

The United States, Bernstein pointed out, had only meager data on sound transmission in sea ice areas. Additional measurements were urgently needed. NEL required a submarine both for sonar work and to assist in the oceanographic study of the western portions of the Northwest Passage in 1954. And remember, Bernstein emphasized, the past summer's experience had shown that it was not possible to send a fleet submarine into sea ice areas *"unless the bow torpedo shutters are protected."*[59]

Even the specter of Soviet submarines in the Arctic failed to move the new ComSubPac, Rear Adm. George L. Russell. "Recommend submarine not participate," Pearl Harbor advised Washington, "unless CNO determines that submarine projects are of such overriding importance to warrant time and expense modifying submarine for Arctic operations." Furthermore, should CNO decide to send a submarine to the far north, CinCPacFlt suggested that NEL's own submarine, *Baya*, be converted for arctic deployment rather than modifying a fleet boat.[60]

It came as no surprise to Lyon when CNO went along with the recommendation of the operational commander. While recognizing the desirability of including a submarine in the 1954 expedition, CNO informed NEL, "the advantages to be gained are out-weighed by the adverse effects upon the combat readiness of the Naval Operation Force." Permission was therefore denied.[61]

CNO did approve the oceanographic portion of Lyon's scientific plan. As the time for the expedition approached, however, Lyon encountered problems of both funding and personnel. In May, 1954, the submarine desk at CNO advised Lyon that the coming summer's expedition had nearly been canceled due to lack of funds. Furthermore, continued financial exigencies likely would make this expedition the last one for the foreseeable future. Also in May Lyon learned that the coast guard planned to replace the experienced Capt. Richard E. Morrell as commander of *Northwind* with Capt. William L. Maloney. Maloney, who had served in New York City for the past seven years, had had no previous icebreaker duty.

Lyon turned to his old friend and supporter, Capt. Rawson Bennett, the former director of NEL who was now with BuShips, for help. "It seems I write to you whenever I have a problem," Lyon began. Because of funding problems, he explained, the year's expedition was likely to mark the end of the joint survey operations in the Arctic. It was imperative, therefore, to obtain the maximum amount of data. The expedition planned to survey McClure Strait and its western entrances—"unquestionably the most difficult and least known area for ship operations in the entire Arctic." Lyon pointed out that the skipper of *Northwind* would be the senior officer in the task force and be in operational command of the expedition. Although Captain Morrell had had long experience in the Arctic and was "tops," the coast guard intended to replace him.

Under the circumstances, Lyon suggested to Bennett, might it be possible to pay a "friendly, informal, social visit" to coast guard headquarters and explain the situation? Perhaps the coast guard could be persuaded to delay Morrell's reassignment until the completion of the expedition? "Either over-cautiousness or 'rushing-in' bravado in an area such as McClure Strait," Lyon emphasized, "can readily result in no data at great expense or very severe damage." This was no time for amateurs.[62]

The coast guard was not about to allow the navy to interfere in its personnel decisions. Lyon soon learned that Morrell's reassignment would not be delayed. Captain Maloney would take command of *Northwind* on July 1. "What a situation!" Lyon exclaimed when he heard the news. "I hope we get the breaks in weather and ice—we shall need them."[63]

The expedition, despite Lyon's fears, got off to a good start. On August 1, 1954, *Northwind, Burton Island,* and five small boats took stations at distances from one-and-one-half and twenty miles in the Bering Strait off Cape Prince of Wales to conduct a current and temperature survey. A swell running from the northwest caused the ships to roll badly, Lyon reported, and everyone in the small boats became sick. Nonetheless, "Much data were taken and a most difficult set of measurements were made."[64]

*Northwind*, with Lyon on board, and *Burton Island* then proceeded along the Alaskan coast, past Point Barrow, to Barter Island. The weather, Lyon noted, was "clear and beautiful." The ice was "not too bad," with six-tenths coverage most of the way. Reaching Barter Island on August 6, the task unit continued toward Herschel Island, remaining close to the coastline.[65]

On August 8 Lyon called a conference on *Northwind* to discuss the general plan of operations for both icebreakers. The ships would now separate. *Burton Island* would take an oceanographic section across Amundsen Gulf from Cape Bathurst to Nelson Head, on the southern tip of Banks Island, then head north through Prince of Wales Strait. *Northwind*, for its part, also would take an oceanographic section across Amundsen Gulf, from Cape Bathurst to Cape Kellett on the southwest-

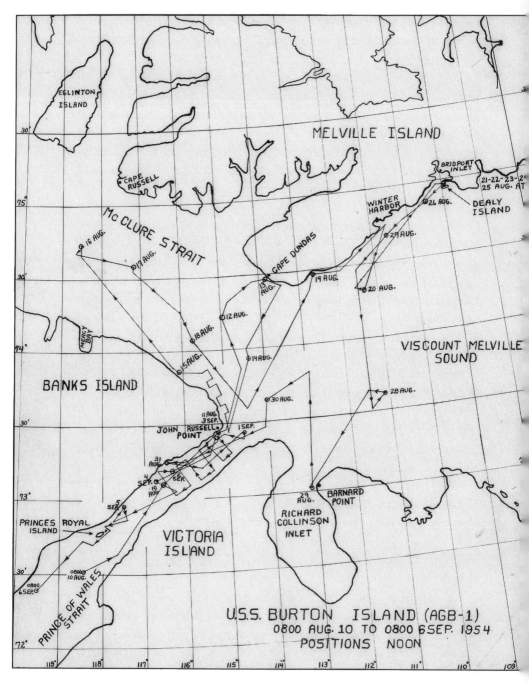

USS *Burton Island* (AGB-1) 8:00 A.M. August 10 to 8:00 A.M. September 6, 1954, positions noon. Courtesy U.S. Navy

ern corner of Banks Island. *Northwind* next would proceed up the west coast of Banks Island to Cape Prince Alfred. If ice conditions permitted, *Northwind* would then enter McClure Strait.[66]

After the ships parted company, *Northwind* turned its attention to the survey task in Amundsen Gulf. The next day, August 9, Lyon came ashore at Cape Kellett in an LCVP to unload geodetic gear for a team of Canadian geodesists who were tasked with obtaining fixes in order to compile an accurate chart of the area. Captain Maloney later came ashore by helicopter, bringing steaks and beer. The men built a large campfire to cook the meat. As they sat around the burning driftwood, a polar bear approached the helicopter. It took a good look, Lyon wrote, then ambled north along the beach "while everyone had his mouth hanging open."[67]

The following day, Lyon and NEL electronics technician Walter E. Schatzberg hiked overland to Sachs Harbor, looking for a site for a weather station. It took four hours to reach their destination. *Northwind* was waiting when they arrived.

The weather was clear, sunny, and warm on August 11, as *Northwind* worked its way northward along the west coast of Banks Island toward Cape Prince Alfred. Ice conditions became worse as they neared the cape, with nine-tenths coverage, mainly block ice and brash. Maloney used *Northwind*'s helicopter to scout out leads in the ice. The captain, Lyon noted in his Journal, was "not in a hurry," while his executive officer, Comdr. Alexander W. Wuerker, appeared "most worried." "Ships officers still have much to learn about working ice," Lyon concluded, "do not use sufficient power at the right moment."[68]

The weather remained sunny, with the temperature a balmy 50°, as *Northwind* lay in the ice south of Cape Prince Alfred on August 12. While the Canadian geodesists went ashore to take a fix, Lyon discussed with Maloney and Wuerker the decision to proceed into Mc-Clure Strait. Wuerker had been executive officer on *Northwind* under Morrell. Although Morrell had kept him busy with paperwork during ship operations, he now became the "ice authority" under the inexperienced Maloney. Wuerker strongly advised against entering the strait. "Exec. Off. extremely reluctant," Lyon reported, "always fearful." However, when the ship's helicopter showed leads in the ice running east around Banks Island, Maloney agreed to proceed toward Cape Wrottesley after passing Cape Prince Alfred.[69]

At 6:00 A.M. on August 13, *Northwind* began working its way toward

Cape Prince Alfred, thirteen miles away. Maloney moved cautiously through heavy ice, mainly large blocks and floes, all loosely packed together. Although the ice was not under pressure, the large pieces had to be broken up. "Entire operation was made far more difficult," Lyon noted, "by very inadequate icebreaker handling." Wuerker, in fact, had given incorrect advice to the conning officers on the use of power, telling them to cut the power at the moment of impact with the ice. This only created propeller drag and a great loss of momentum and control. As Lyon knew from his extensive experience with icebreakers, full power was needed at the point of impact.[70]

*Northwind* finally arrived off Cape Prince Alfred at 9:00 P.M. Maloney's night orders were to complete the oceanographic survey in the Cape Prince Alfred area, then lie to until the morning when a decision would be made on whether or not to enter McClure Strait.

By ten o'clock Maloney had turned in, leaving Lyon on the bridge with the ship's navigator. The sun was shining, and a full moon was visible. As the oceanographic work continued, Lyon managed to talk the navigator into continuing the survey further to the east. There did not seem to be a problem. *Northwind* was running between the ice and the beach, 800 to 3,000 yards from shore, in 100 to 200 fathoms.

At 5:30 A.M. on August 14, 1954, *Northwind* took an oceanographic station off Cape Wrottesley. One hour later, the ship passed Cape McClure. It was an impressive sight. The land, Lyon wrote, rose precipitously—"majestic, towering black from the water." "Cape McClure is definitely the portal to McClure Strait," Lyon observed, "and when in heavy grinding ice must be terrifying."[71]

With the way ahead clear, there was no reason for Maloney to turn around. He gracefully accepted Lyon's fait accompli and continued to Mercy Bay. At 12:10 P.M. *Northwind* reached this historic site where Capt. Robert McClure of HMS *Investigator* had been icebound from September, 1851, until he abandoned his ship in June, 1853.

While *Northwind* proceeded through the strait, Lyon read an account of the passage that had been written by Dr. Alexander Armstrong, surgeon on *Investigator*. He later copied the following paragraph into his Scientific Journal:

> "There was nothing deserving the name of bay or harbour along any part of this coast, nor any protection or shelter for ships; and exposed as it is to all the fury and violence of westerly and northwest-

erly winds, it stands without a parallel, for the dangers of its navigation, in any part of the world. The appalling evidence we were afforded of the effects of pressure, caused by stormy winds acting on a trackless icy sea, as such was we had not witnessed in any other part of our eventful voyage, and baffles all attempts at describing—mounds being piled together to the height of upwards of 100 feet. Our passage along this part of the coast was a truly terrible one—one which should never be again attempted; and with a vivid remembrance of the perils and dangers which hourly assailed us, I feel convinced it will never be made again."[72]

In sharp contrast to the hardships faced by Armstrong and the crew of *Investigator*, *Northwind* made the passage "in the bright sun of a most balmy summer day, as though on a river outing and picnic."[73]

Lyon decided to locate an EPI station at Mercy Bay, despite the fears of Maloney and Wuerker that ice might lay to the east and trap *Northwind* in the bay. The determined scientist went ashore at 2:00 P.M. and chose a sandspit at the western entrance to the bay as the site for the station. He planned to have both icebreakers survey McClure Strait, with *Northwind* working the western and central portions, and *Burton Island* the eastern and central sectors. Both ships would then retire to Prince of Wales Strait and continue the survey work.

By ten o'clock that evening the EPI had been brought ashore. "And so the end of a day I'll not forget," Lyon recorded in his Journal, "a real climax to these past 8 year's work [in] this area." And Lyon did not forget the day. Forty-two years and many accomplishments later, he wrote: "The passage of McClure Strait was the most wonderful experience in my explorations and career."[74]

On August 15, with the EPI station in operation, a small boat from *Northwind* made a survey of Mercy Bay. At 6:00 P.M., the icebreaker headed east through the strait. Shortly before midnight, it rendezvoused with *Burton Island*, which had come up through Prince of Wales Strait and into McClure Strait. Lyon and Cameron, the Canadian oceanographer who was serving as chief scientist on *Burton Island*, agreed that the bottom contour of McClure Strait was so simple that only one good section would be necessary. *Burton Island* would do the survey, running a section across the strait near Mercy Bay.[75]

The icebreakers then steamed westward, parting at Cape Vesey Hamilton. While *Burton Island* conducted the survey of central

McClure Strait, *Northwind* headed for Mercy Bay to evacuate the EPI station upon completion of the survey. *Northwind* entered the bay, which was now filled with loose pack ice, on August 17. Lyon flew to Captain McClure's beacon and cache, where the English explorer had left some coal, an anchor, barrel staves, and other odds and ends. Lyon retrieved an iron hook from the cache, which he planned to present to the Canadian archives. The EPI station, meanwhile, was being dismantled. The work was completed at 10:00 P.M., as the ice in the bay became much tighter.[76]

*Northwind* spent the next week working its way through heavy ice along the northern coast of Banks Island and into Viscount Melville Sound. On August 24 the icebreaker reached Elvira Island, just west of Stefansson Island on the eastern end of Viscount Melville Sound. Although Maloney did not attempt to enter McClintock Channel due to ice, Lyon was content to take a series of oceanographic station from Elvira Island to Prince of Wales Strait.

While *Northwind* explored Viscount Melville Sound, *Labrador* was completing its transit of the eastern portion of the Northwest Passage. The Canadian icebreaker had departed Halifax on July 14. After transiting Baffin Bay, it had resupplied the Royal Canadian Mounted Police outpost on the east coast of Ellesmere Island, then proceeded through Lancaster Sound and Barrow Strait into Viscount Melville Sound. On August 25, as planned, it rendezvoused with *Burton Island* off Dealy Island, just south of Melville Island in Viscount Melville Sound. This marked "the first time in history," the chronicler of *Labrador*'s voyage pointed out, "that two naval vessels had met across the Arctic—*Labrador* from the Atlantic and *Burton Island* from the Pacific."[77]

As *Labrador* and *Burton Island* sailed in company across Viscount Melville Sound, *Northwind* continued its oceanographic work. Maloney's handling of the icebreaker did not improve. On the night of August 26–27, while anchored off Barnard Point, Maloney decided that he could not hold his position due to the ice. The ship drifted all night, nullifying Lyon's efforts to measure the current. "I have been in the Arctic many times," the exasperated scientist reflected in his Journal. "This ship is the biggest *sissy* I've seen. What contrast to Richard Morrell last year. It is more amazing that we have accomplished as much as we have in spite of C.O. and X.O."[78]

On August 29, *Labrador* and *Burton Island* arrived at *Northwind*'s position off Barnard Point. According to plan, *Northwind* would act as the

EPI station while the other two icebreakers conducted surveys north of Richard Collison Inlet on the northwestern corner of Victoria Island and into Prince of Wales Strait.

Two days later *Northwind* again became stuck in the ice. "C.O. and X.O. went into double panic condition," Lyon reported. Fearing that the ship might be stranded in the ice for the winter, they called *Labrador* and *Burton Island* for help. "I am completely baffled as to what two other icebreakers can do to help us unless we are damaged," Lyon wrote; "it's just to have company in misery."[79]

By the time *Labrador* and *Burton Island* arrived the next day, *Northwind* had managed to free itself; Maloney advised that he no longer required help. "A sad commentary," Lyon concluded, "ample proof that some method must be found for both Navy and Coast Guard to instruct their junior officers in icebreaker operations and knowledge of arctic sea ice." These junior officers then would qualify to command icebreakers. "This is a major point of this expedition," Lyon believed.[80]

*Northwind* anchored south of Prince Royal Island, again serving as an EPI station as the other two ships surveyed the northern section of Prince of Wales Strait. The work was completed on September 5, and the ships rendezvoused for a celebration and conference. During drinks on *Labrador*, Commander Wuerker turned to Lyon and in a loud voice "objected strongly to further work and questioned the value of this and all expeditions." This display of temper, the scientist noted, had been building for weeks. After Wuerker made his angry remarks, he left the wardroom. Lyon and his associates then got down to work and made plans for the remainder of the expedition.[81]

The three ships proceeded south through Prince of Wales Strait on September 6. While the two U.S. icebreakers took sections across Amundsen Gulf, *Labrador* stopped at DeSalis Bay on the southern tip of Banks Island to conduct a magnetic station. *Northwind* and *Burton Island* then headed northward, along the west coast of Banks Island. On September 10 *Northwind* anchored off Norway Island to act as an EPI station while *Burton Island* handled the survey work.

Lyon took this occasion to vent his frustrations in a letter to Capt. Tyrvell G. Jacobs, commanding officer of ServRon One. "The success of this expedition," he wrote, "has been very considerable in spite of Northwind and not because of Northwind." While *Burton Island* was to be commended for "a job well done" and for accomplishing a great deal in the face of "incompetent task command," *Northwind* had been a constant headache. "I am sorry to report," Lyon concluded, "that our fears, expressed at the time Dick Morrell was relieved, have been proven fully justified."[82]

On September 11, to Lyon's chagrin, Captain Maloney decided not to attempt a survey of the western approaches to McClure Strait, citing deteriorating weather. Instead, *Northwind* headed toward Barter Island, where it took up position as an EPI station while *Labrador* and *Burton Island* ran oceanographic sections north of Herschel Island. Three days later, the ships met to discuss the progress of the survey work. Lyon, thoroughly disgusted with Maloney's handling of *Northwind*, took the opportunity to transfer to *Labrador* for the remainder of the expedition.[83]

During the next four days, *Labrador* worked at the edge of the ice pack, delineating the Chukchi Rise in the area between $73°30'$ and $75°$ North, and $155°$ and $165°$ West while *Northwind* and *Burton Island* anchored at a known location off the Alaskan coast and acted as EPI stations. This precision survey completed, *Labrador* turned toward Cape Lisburne. On September 21 the Canadian icebreaker arrived off Wales. Lyon bid a fond farewell to Captain Robertson, then flew by helicopter to the NEL field station.[84]

Lyon spent two days at Wales, working around the station as it prepared for winter. After a week in Alaska, he flew to British Columbia for discussions with his Canadian colleagues. On September 29 he met with Robertson on board *Labrador*, now anchored at Esquimalt, to discuss a continuation of the survey work in 1955 in the Viscount Melville Sound–Lancaster Sound area. Following brief visits to Nanaimo and Vancouver, Lyon returned to San Diego on October 1.[85]

The 1954 expedition, Lyon recognized, marked the end of "a very productive time." During five consecutive summers of icebreaker operations in the Arctic, he had gathered "a tremendous amount of information" on what had previously been a largely unknown part of the world. The expeditions had surveyed the entire area from the Chukchi Sea, across the Beaufort Sea, and into Amundsen Gulf and Viscount Melville Sound. Lyon felt familiar with the area, "both oceanographically and its bathymetry."[86]

The most recent operation had put the capstone on his survey work in the far north. For the first time in history, a ship had transited

McClure Strait and circumnavigated Banks Island. Surveys accomplished under precise EPI control included Cape Alfred to Russell Point; cuts across McClure Strait and Viscount Melville Sound; Peal Point to Cape Elvira east along the north coast of Victoria Island; the northern portion of Prince of Wales Strait; and the sector directly north of Point Barrow. In addition, a series of geodetic fixes had been taken at Cape Kellett, Cape Prince Alfred, Cape Dunday, Providence Point (Mercy Bay), Rodd Head, and Russell Point. Using photographic maps made by the Royal Canadian Air Force, cartographers employed these geodetic fixes to construct charts (within the accuracy of the star fixes) that would remain the most accurate in existence for this area until the advent of satellite navigation.[87]

The only portion of the Beaufort Sea that remained undone was the western entrance to McClure Strait. And this could have been done. In early September *Northwind* had been off Cape Prince Alfred and in a position to survey the area, which had not been open since 1951. Captain Maloney, however, had decided to withdraw toward Barter Island, claiming that the weather appeared threatening. Lyon, after returning to San Diego, had contemplated lodging a formal complaint against Maloney, but the favorable publicity given to the expedition, including a photo-spread in *Life* magazine, dictated a less direct course of action.[88]

A far greater disappointment, Lyon reflected in his Scientific Journal in October, 1954, was the limited progress that had been made in developing the under-ice submarine. Work on the polar submarine had come "little further than 1948." A "crude" five-unit topside fathometer had been tested; experiments had been run on sound propagation in arctic waters; and the many advantages of a submarine in sea ice had been demonstrated. But attempts to modify a submarine for extensive arctic experiments had come to naught.

The question, Lyon posed to himself, was: What came next? He assumed that there would be no direct support from the Pacific Fleet for an experimental under-ice submarine. Also, oceanographic survey work in the Arctic could be reduced to advice and assistance to Canada. There remained a "basic need" for the study of sea ice physics. The Soviets and the Japanese, he recognized, had a tremendous head start over the United States in this area. NEL, under the circumstances, should conduct "one intensive experiment" on the structure, mechanical properties, energy propagation, and pressure of a single sheet of ice.

Above all, as Lyon had argued in the past, work should focus on building and operating the arctic pool at NEL. Not only would the pool be used to study sea ice, but it also would provide the primary means for testing and designing equipment for the under-ice submarine. Clearly, Lyon concluded, completion of the pool represented the "most essential item" for the future.[89]

## 4

## A Whole New World

Construction of an arctic pool at Battery Whistler had begun in 1952. With no funds specifically designated for the facility, work had progressed slowly. Lyon squeezed what money he could out of the budget allocation for his section, while the director of NEL helped by allowing him to use unspent end-of-year money to purchase steel, piping, and other construction material. Most of the labor for the project was supplied by Lyon and his associates "during odd hours."[1]

Lyon hoped to attract the funds necessary to complete the pool from the Office of Naval Research. Formed in 1946 in an effort to continue the extraordinarily fruitful partnership between the academic community and the military services that had existed during the war years, ONR primarily provided research grants to civilian scientists. In 1947 ONR had established the Arctic Research Laboratory at Point Barrow "to conduct fundamental research related to Arctic phenomena." The facility served as a base for civilian scientists under contract to ONR who went north to conduct their research projects. Most of the arctic work funded by ONR had been in the area of earth and biological sciences. NEL, funded by BuShips, had taken the lead in oceanographic studies of the far north.[2]

In March, 1954, Lyon sent to ONR, over Director Bernstein's signature, a proposal to study the physics of sea ice. NEL's memorandum began by pointing out that research into the fundamental properties of sea ice had made little progress since 1949. "The lack of basic knowledge," it argued, "will be seriously felt in arctic naval warfare and is the responsibility of the U.S. Navy to correct."[3]

The memorandum went on to outline a five-year program of field and laboratory work to remedy this shortcoming in arctic knowledge.

NEL would be the logical agency to conduct such a scientific investigation, but the project needed ONR's financial support. Bernstein proposed that ONR assign the sea ice research project to NEL, contributing $80,000 to complete the arctic pool and an annual budget of $104,170 for the five-year study.

It took eight months for ONR to respond to NEL's overture. While willing to contribute $10,000 to $15,000 a year for a project to study the physics of sea ice, ONR did not believe that a large laboratory facility was warranted. Furthermore, in a slap at NEL's scientific expertise, ONR suggested "that some consideration be given to a study of the Russian and other foreign language literature on sea ice physics."[4]

A disappointed Lyon visited Washington in December, 1954, in an effort to drum up support for the sea ice physics project. Gordon Lill, his longtime supporter at ONR, told him that his original proposal had been too large and too general. Interest in the Arctic at the working level of ONR, he reminded Lyon, was in the geography branch. "Geographers," Lyon noted in his Journal, "do not think in terms of physics."[5]

Captain Bennett at BuShips urged Lyon to submit a revised proposal that would leave out all reference to construction of a laboratory facility at NEL. It was necessary, Bennett told Lyon, to force ONR's support of a modest program in order to justify a later expansion of what was obviously a necessary area of research.

Lyon took Bennett's advice and resubmitted a request for research support on January 21, 1955. NEL's original proposal, he noted, may have been "confused" by the inclusion of funding for laboratory facilities. The new proposal included only a program for basic research in sea ice physics. It was a minimum proposal, which could not be further reduced "if any progress is to be made."[6]

Lyon again stressed the need for greater knowledge of how and why sea ice forms. This information was necessary for ice prediction by the Hydrographic Office, to design ice-boring tubes for submarines, for hull design and power requirement for icebreakers, to predict sonar and radio ranges, and to determine to suitability of ice for aircraft landings.

While a large number of foreign publications were available on sea ice, Lyon pointed out that they were "purely descriptive." Furthermore, the Soviet Union had released little information since 1945. NEL, on the other hand, had studied the dynamics of sea ice at its Cape Prince of Wales field station and during icebreaker expeditions over the past several years.

NEL believed that the time had come to conduct sea ice experiments under controlled conditions. Accordingly, the laboratory at Battery Whistler was in the process of converting its pool, 75' x 30' x 14' deep, into a refrigerated pool on which a sheet of sea ice could be grown and controlled. The facility would be operated by three men, working full and part time, at a cost of $38,400 a year. In addition, there would be a charge of $38,400 in direct labor overhead, $5,000 for arctic travel, and $8,000 for material. Would ONR, Lyon asked, be willing to support this vital research to the tune of $94,800 a year for five years?

BuShips, not surprisingly, gave the request its enthusiastic endorsement. The new proposal, it pointed out, did not include the $80,000 for completing the arctic pool or the $4,000 for supplies at the Wales field station that were contained in the original request. Elimination of these items "removes a major administrative difficulty." Instead, the new plan envisioned a joint effort whereby BuShips supported the research facility while ONR funded the trained research workers in a basic scientific research program that was essential for an understanding of how sea ice was formed.[7]

ONR was not persuaded. On April 25, 1955, it informed NEL that its original offer of $10–15,000 remained the limits of its support. On the other hand, ONR no longer would tie the money to field research. "The choice between laboratory work and field work," it advised, "would rest with NEL."[8]

ONR's gesture came as small consolation to Lyon. While work on the arctic pool continued, it was clear that the facility would take several years to complete at the present level of funding.

Ironically, while Lyon's efforts to initiate a sea ice research project languished, the navy's need for basic knowledge about the subject increased. During his visit to Washington in December, 1954, Lyon had witnessed "the frantic confusion" that had followed the receipt of orders for the navy to handle the logistical task of supporting the construction of a series of Distant Early Warning (DEW) radar stations in the far north. This operation, he recognized, would be "a tremendous undertaking," with ships working in uncharted waters on the fringes of the ice pack.[9]

On May 18, 1955, NEL received a dispatch from CNO directing

that Lyon be assigned as an adviser to the naval task force that would be sailing for the Arctic in June to deliver equipment and supplies for the proposed DEW Line. The request, Lyon learned, had been initiated by his old nemesis, Captain Maloney of *Northwind*. Maloney apparently either had not heard about Lyon's discontent over his handling of the icebreaker in 1954 or had decided to overlook it. In any event, Lyon saw no good reason to spend the summer in the Arctic as an adviser. He telephoned King Couper in BuShips and objected. Couper told him that he could not refuse due to the high priority of the project.[10]

As Lyon would learn, the idea of a chain of radar stations in the far north to warn of an air attack across the North Pole had first been proposed in 1946. The scheme had languished, however, until August, 1952, when the Summer Study Group of MIT's Lincoln Laboratory urged the construction of a radar line along the seventieth parallel. Although the U.S. Air Force had funded this study, it ended up opposing the conclusions of the group out of fear that the high cost of the project would divert funds from aircraft construction. But the idea would not die. In July, 1953, a special committee appointed by the secretary of defense supported the Summer Study Group's findings and recommended that $18–25 billion be spent over a five-year period on construction of the radar network. When the Eisenhower administration learned the next month that the Soviet Union had tested a hydrogen bomb, air defense became an urgent priority. The National Security Council endorsed the DEW Line scheme in October, 1953, and President Eisenhower approved it the following February.[11]

By the end of 1954, a Locations Study Group had settled on the sites for the DEW Line, now labeled Project 572, which would extend some 3,000 miles across the northern perimeter of North America. There would be fifty-seven stations in all. Six main stations, located 500 miles apart, would be built at Point Barrow, Barter Island, Cape Parry, Cambridge Bay, Hall Beach, and Cape Dyer on Baffin Island. In addition, twenty-three auxiliary stations would be located at 100-mile intervals, with twenty-eight intermediate stations between the main and auxiliary radar sites.

Plans called for a massive construction effort to begin in the summer of 1955. The navy planned to use two task forces, one from the Pacific and one from the Atlantic, to carry more than 750,000 tons of cargo and almost 4 million barrels of fuel and oil to the building sites. More than a hundred ships would be involved in the largest peacetime operation ever conducted in the Arctic.

On June 2, 1955, Lyon boarded USS *Mount Olympus*, anchored in San Diego, for discussions with Rear Adm. George C. Towner, commander of Task Force Five. Walter I. Wittmann of the Hydrographic Office, Lyon learned, would be the chief ice forecaster for Towner on board *Mount Olympus*. Lyon had never met Wittmann, but he knew him by reputation. He looked forward to working with Hydro's leading ice expert. As it turned out, this would mark the beginning of a long and productive relationship between the two scientists.

As Lyon expected, his principal task would be to give advice on the movement of ships through the ice-filled waters off the northern coast of Alaska, especially on the route between Point Barrow and Cape Bathurst. Towner agreed, however, that in addition to his advisory duties, Lyon could take oceanographic measurements. Accordingly, the scientist would sail on *Northwind* instead of the task force flagship, *Mount Olympus*. His observations of sea temperature and salinity changes would be radioed to Wittmann on *Mount Olympus* to help with the freeze-up predictions.[12]

Lyon flew to Nome on July 2, where he boarded *Northwind* by helicopter and organized his oceanographic laboratory. The icebreaker departed for Point Barrow on July 6. It encountered heavy ice off Cape Lisburne. The edge of the ice pack, Lyon found, had remained further south than at any time during the previous nine years. It took four days for *Northwind* to work its way through belts of thick ice to Point Barrow.[13]

Ice conditions east of Point Barrow, if anything, were worse, with ice jammed up against the northern Alaska coastline. Not until August 12 did *Northwind*, accompanied by *Burton Island*, attempt to escort five cargo ships eastward from Point Barrow toward Point Franklin. It proved a daunting task. One merchant ship suffered a broken rudder; another received a ten-foot gash in its hull that flooded the engine room; and a third became stuck on a mudbank.[14]

On September 10 the navy announced that it was winning the battle to land cargo for the DEW Line. Despite fierce winds, fog, poor visibility, and heavy ice, more than 500,000 tons of equipment and supplies had been unloaded along a hundred-mile stretch between Point Barrow and the Mackenzie River. Approximately 75 percent of the operation had been completed.[15]

Eleven days later, the navy proclaimed "a great victory" in the Arctic. Vice Adm. Francis C. Denebrink, commander of the Military Sea Transportation Service, told a press conference that all supplies had been successfully unloaded. Furthermore, there had been no loss of life among the 18,647 persons aboard the 126 ships involved in the three-month operation.[16]

Lyon, in his report to Admiral Towner, suggested that specially built shallow-draft vessels be used in future supply operations along the north coast of Alaska. These vessels would be able to navigate the narrow corridor between the permanent ice pack and the shoreline. Schedules would have to be properly timed, he pointed out, and up-to-date oceanographic information would be required. Lyon recommended that scientific agencies, especially NEL, be directed to undertake an intensive oceanographic study of the area from Cape Lisburne to Hershel Island in order to provide data "in direct application of the north coast shipping problems."[17]

While everyone was supportive of Lyon's ideas, no one was prepared to fund the oceanographic study. DEW Line requirements, he learned, were expected to be greater than ever in 1956, and there would be no icebreakers available for scientific research. Also, commitments to the International Geophysical Year, scheduled to begin in 1957, continued to grow. Originally conceived as a cooperative effort by scientists throughout the world to learn more about the nature of planet Earth and a way to defuse growing international hostilities between the atomic superpowers, the IGY soon became yet another venue of competition and national pride for the United States and the Soviet Union. "The IGY has taken on aspects of an overriding priority above that of the Defense Department," Lyon noted, "with the administration determined to match the Soviet contribution in ships, men, and observation stations."[18]

On a happier note, Lyon received the first recognition for his work in the Arctic. On November 8 Admiral Mumma presented him with the Distinguished Civilian Service Award—the highest award that the secretary of the navy could bestow upon a civilian employee—at a ceremony held at BuShips. Termed "one of the nation's leading authorities on the Arctic Ocean and sea ice as a potential theater of war and on the techniques necessary to naval operations in cold weather environments," Lyon was honored for his fifteen research expeditions to the far north "during which he demonstrated courage and ability in keeping with the finest traditions of the Navy and the scientific profession."[19]

While certainly pleased with the recognition, Lyon would have preferred a more tangible commitment on the part of the navy in support of his ongoing research. But none was forthcoming.

Lyon expected 1956 to be a difficult year with his section's limited resources devoted to supporting a sea ice study at the Cape Prince of Wales field station and working on the arctic pool at Battery Whistler. Lyon's staff at San Diego, he recalled, "became iron workers, welders, etc., and did a major part of erecting the heavy steel structure of the arctic pool, sea water pump, etc." In addition, he planned to catch up on data analysis and prepare reports on past expeditions. The year, however, turned out to be far worse than he had anticipated. By the middle of 1956, Lyon found himself struggling for the very survival of the laboratory that he had worked so hard to create.[20]

In August, 1955, a planning conference between NEL and BuShips had approved a program for arctic studies that included work on sea ice physics and under-ice research projects. NEL then had submitted a budget estimate of $300,000 for what was designated as Project L6–1. The funds received from BuShips for designated projects went primarily to pay the salaries of personnel assigned to the various research tasks at NEL; money for supplies and material came out of a separate and more general allocation. Also, the director of NEL had authority to shift funds from one project to another, then justify his decision at an end-of-year conference with BuShips.[21]

Sometime before mid-year, 1956, however, NEL learned that BuShips was prepared to allocate only $140,000 for Project L6–1 in 1957. Furthermore, the director of NEL would lose his discretionary authority to shift money from one project to another. "Funds provided for this program," NEL pointed out on July 6, 1956, "constitute virtually the entire financial support for the operation of the Submarine and Arctic Research Branch of the Special Research Division. . . ." If funding were to be reduced by 50 percent, the personnel in Lyon's branch would have to be cut in half. Before proceeding with the reduction in force (RIF), NEL asked that BuShips confirm that it really wished the arctic program to be reduced by 50 percent.[22]

Lyon first heard about the funding crisis on July 24. "I am told by the C. O. here," he wrote the next day to King Couper in BuShips, "that from now on effort shall be governed by BuShips directly, i.e. by how much is placed in dollars on each problem with NEL *assuming no latitude* in moving funds from one problem to another. . . ." Lyon could hardly believe that this was in fact the case. "Is it true," he asked Couper to confirm, "that BuShips intends to precisely control problem details by money assigned? Are we, who do the work, no longer to have responsibility, choice and control to expend certain portions of funds as we believe they should be spent? Do you people really mean to now direct in ironclad rules where each dollar shall go?"[23]

The answer to Lyon's questions was "yes." Although the funding problem would not be fully resolved until NEL's program was reviewed in mid-October by high-ranking officials at BuShips, Couper responded on August 3, he believed that the cut in budget had been specially targeted in order to force compliance with the shift in priorities to air defense. Couper also enclosed a note from Capt. William I. Bull of BuShips that offered another explanation. "In this day of the 'controller,'" Bull observed with prescience, "I am afraid . . . we will have less and less flexibility with dollars."[24]

BuShips formal response to NEL's letter of July 6 came in late August. While BuShips was "fully cognizant" of the importance of the work being done by Lyon's branch, as well as its "possible far-reaching effects as it pertains to sonar operations" in the far north, the budget decision was firm. Only $140,000 would be available for arctic research in fiscal 1957.[25]

Shortly after this letter arrived, Lyon was called into a meeting with NEL's new commanding officer, Capt. John M. Phelps, and technical director, Franz N. D. Kurie. There was no choice, he was informed, except to prepare for the worst. Kurie told Lyon to evaluate the personnel in his section in the event a RIF became necessary.[26]

Lyon sent Kurie the RIF memorandum on September 10. Of the twenty-two members of his staff, sixteen worked on arctic research while six operated laboratory equipment. Under the budget allocation of $140,000, eight of the sixteen arctic research employees would have to be let go, including three physicists and a glaciologist. With the reduced staff, Lyon would be able to maintain the Cape Prince of Wales field station, which he believed should be given top priority, and continue to operate the laboratory at Battery Whistler at a minimum level. Canceled would be all sonar work for an arctic submarine, all liaison work with Canada, all work on an Alaska north coast shipping lane, and all construction on the arctic pool. In fact, there would be no money even to attend scientific meetings.[27]

Lyon asked Kurie to hold the memorandum in strict confidence. "These opinions are not known by any member of the Arctic Research Branch," he cautioned, "and if improperly guarded would critically disrupt, perhaps irreparably, the working morale of the Branch, which has fairly good internal group loyalty of long standing."

As Lyon recognized, implementation of the RIF program would cause serious—perhaps fatal—damage to the operation of the arctic research facility that he had struggled to develop over the past decade. Fortunately, his superiors at NEL also recognized the danger. Phelps and Kurie in fact went so far as to "lose" the instructions from BuShips that would have put the RIF into effect, hoping that Washington might still be persuaded to reconsider the funding decision.[28]

The delaying action soon bore fruit. In early October NEL received an urgent request from Code 1500 (Nuclear Propulsion) at BuShips to use the Betatron at Battery Whistler to x-ray heavy steel castings for pump volutes of the nuclear power plant cooling systems designed for *Skate*-class submarines and a commercial power plant at Shippingport, Pennsylvania. No technique had been developed to inspect the twenty-inch stainless steel sections for flaws before machining the pump volutes. "Apparent that considerable political pressure on everyone to get these castings done," Lyon noted. "RADM [Hyman G.] Rickover is primary pusher and responsibility."[29]

Later in October Lyon learned from Commander McWethy, now assigned to the submarine desk in CNO (Op-311), that the nuclear-powered submarine *Nautilus* might be assigned in summer, 1957, to dive under the ice north of Spitzbergen for a distance of 500 miles. This possibility had arisen because Capt. Eugene P. Wilkinson, commander of the world's first nuclear-powered submarine, was using his "driving personality" to get *Nautilus* to the Arctic. "It just seemed like a natural to try a nuclear-powered submarine under the ice," Lyon recalled.[30]

Lyon certainly was prepared to do everything he could to promote the idea. "If the opportunity should occur," he wrote in late November to Admiral Bennett, now chief of naval research, "would you plant at proper high level the idea that NAUTILUS (SSN-571) should be sent across the arctic basin next summer, or at least make a very extensive run in the basin?" In addition to the operational experience that would be gained from such a voyage, the news of it might prompt the Soviet Union to release information about their submarine activities in the Arctic, a subject about which "our intelligence is nil." "Of course," Lyon assured Bennett, "we are ready to play our part if ever the assignment is made."[31]

With continued pressure from Admiral Rickover to x-ray the pump volute castings and a stirring of interest in the arctic submarine project, Captain Phelps was able to forestall the budgetary cuts for Project L6-1 during the winter of 1956–57. The delay would soon pay rich dividends. Two events, one in March and one in April, 1957, brought a dramatic change in the fortunes of Lyon's arctic projects. The culmination would come in late summer, thanks largely to the efforts of Wilkinson and McWethy.

On March 1, 1957, Lyon received a phone call from BuShips requesting his appearance at the Pentagon on March 5. When he reached Washington, Lyon learned from McWethy that Adm. Arleigh A. Burke, chief of naval operations, had laid on his subordinates a demanding requirement to draw up an arctic program for the navy. In February, McWethy explained, Senator Henry M. Jackson (D-Wash.), influential chairman of the subcommittee on military applications of atomic energy, had quizzed Burke about the navy's efforts in the far north, especially the role of submarines. Burke had found a response prepared by Op-31, the office of the assistant chief of naval operations for undersea warfare, to be "cursory." Scheduled to appear on Capitol Hill on March 8, he had given his staff ten days to put together a briefing on all past work and future plans, followed by a comprehensive arctic submarine program. The request had ended up on McWethy's desk in Op-311.[32]

Lyon immediately got to work on the task. "For three days," he reported to Captain Phelps, "I literally lived with Op 311 day night and day." The briefing was ready on time. It stated, Lyon noted, that "we have done nothing since 1953 except for the Beaufort Sea–McClure Strait surveys and the design/construction of our arctic pool." Lyon and McWethy also put together a rough draft of the comprehensive program, which Op-311 would put in smooth form for presentation to Admiral Burke by March 11. At the heart of the Lyon-McWethy draft, designed to be issued as an OpNav Instruction to implement the U.S.

Navy's arctic and cold weather program of 1955, was a call for field tests of a specially modified arctic submarine.[33]

Lyon recognized that there was still a long way to go before an arctic submarine became a reality. The draft would have to survive "the usual round of initials" by the various CNO desks. Also, "it remains to be seen whether final intent by CNO is to initiate the program." And, in the end, even if initiated, the "real problem" would be funding.

Before leaving Washington, Lyon called on Capt. Frank G. Law in BuShips, who was working on the funding for Project L6–1. Law, Lyon reported, expressed "very real enthusiasm" for NEL's arctic work; however, his sentiments were not shared by many others at BuShips. In light of the current "panic" about arctic submarines, Law planned to endorse NEL's request for additional money to CNO "to add fuel to the fire." Law believed that emergency funds could be obtained to modify a submarine, but these funds could not be used to support NEL's long-term research programs.

The best news of all, Lyon told Phelps when he returned to San Diego, concerned plans for *Nautilus*. With the "enthusiastic help" of Captain Wilkinson, action had been initiated to send the nuclear-power submarine on an arctic patrol in the summer. Although approval for and the exact nature of the mission depended upon the current discussion about arctic submarines and the results of sea trials of *Seawolf* (SSN 575), Lyon concluded: "This is the first glimmer of action since 1951 when Chas. Momsen succeeded in assigning REDFISH to us."[34]

Progress in Washington had been good, McWethy advised Lyon on April 4. The original draft of the Lyon-McWethy proposal had started its way through the chain of command on March 13. On March 28 McWethy had defended it in a conference with Vice Adm. Harry D. Felt, vice chief of naval operations. "He kept repeating that it was a fine idea," McWethy reported, "and he didn't want us to think he was against it." After slight modifications, the proposal had gone for comment to fleet commanders, BuShips, ONR, and Hydro.

"I think the NAUTILUS trip may be the key," McWethy concluded. The plan for a summer cruise had been signed out by the deputy chief of naval operations to CinCLantFlt on April 2. As both ComSubLant and Wilkinson were in favor, "things look good." If the cruise succeeded, McWethy predicted, "everyone will want to jump on the bandwagon." On the other hand, if *Nautilus* ran into problems on the voyage, "the program will probably be set back about 5 years." But this was not likely to happen. "No reason to expect any trouble," McWethy wrote.[35]

Four days later, on April 18, BuShips forwarded to NEL, "as a matter of possible interest," a letter from Rear Adm. Elton W. Grenfell, ComSubPac. Grenfell had reported in May, 1956, that several of his submarines had experienced severe icing of their snorkels during cold weather operations in the North Pacific. BuShips had recommended that a coating of silicone grease be applied over the areas subject to icing. This solution, Grenfell now advised, had failed. Five ships had used the silicone grease during the past winter. "In every case," he noted, "severe snorkel induction icing was experienced."[36]

The ability of snorkel-equipped submarines to operate in cold weather areas, Grenfell informed BuShips, was a "mandatory requirement." The present snorkel system did not work. When ice built up in the air passage, sufficient air could not be drawn into the submarine to operate even one engine. The submarine had to submerge the snorkel mast for several hours to remove the ice. "ComSubPac is extremely concerned about this operational deficiency in our present submarines," Grenfell stressed, and "earnestly requests prompt and thorough action by the Bureau of Ships to obtain a satisfactory solution to this problem . . . at the earliest practicable date."[37]

BuShips responded at once to Grenfell's plea and appointed Lt. Comdr. Carvel Hall Blair as the project officer to find a solution to the icing problem. An experienced submariner who had served in USS *Odax* (SS 484) and USS *Grampus* (SS 523), Blair arranged for the Portsmouth Navy Yard to modify a head valve assembly by installing heated electrodes. When he learned that Portsmouth was having problems with the modification due to the scanty knowledge about how and where the ice was forming, he decided to visit NEL to obtain background information on Lyon's test facilities.[38]

Blair inspected Lyon's arctic pool on April 10–11. He came away with the conviction that the pool was unique in the United States and would be the ideal site to test the Portsmouth-designed deicer. However, as he learned from Lyon, the pool would not be completed before May, 1958, under current funding arrangements. When he returned to Washington, Blair recommended to his superiors that BuShips provide $114,000 from Research and Development and/or Design, Development, and Test funds to complete the pool in December, 1957.[39]

It did not take long for BuShips to come up with the money. "I have managed to shake loose $75,000 to work on the pool up to July 1," Blair informed NEL in late April; "more will be available as required." The money would be transferred from SSN-593 Design, Development, and Test.[40]

On May 3 CNO threw its weight behind the drive to find a solution to the snorkel icing problem. Atlantic Fleet submarines, it informed BuShips, had also experienced icing difficulties during recent operations. "The serious military deficiencies in the snorkel system must be overcome at the earliest possible date," CNO advised. Accordingly, it directed BuShips to proceed with interim steps to arrive at a satisfactory solution to the problem by late summer, and to develop a long-term testing program to reach a factory solution at the earliest possible date. BuShips was to inform CNO "of difficulties encountered and assistance required."[41]

BuShips advised CNO the following week that Portsmouth was designing snorkel modification kits for installation in the field. The kits would include heated electrodes, heating elements for the head valve, and possibly ultrasonic vibrators. The design would be completed by May 15, with delivery of twenty-five kits to begin in August. "The greatest obstacle to solution of this problem is the lack of knowledge of where and how fast ice is forming," BuShips pointed out. *"This information is urgently needed,"* it emphasized, "but probably cannot be obtained until the [NEL] Arctic pool is completed."[42]

The same day, BuShips sent NEL a copy of the May 3 letter from CNO and inquired about the earliest date that tests might begin in the arctic pool. CNO's letter, it observed, indicated the high priority that had been attached to the project. NEL replied that tests might start as early as October. "Construction of the arctic pool," it assured BuShips, "is being accelerated."[43]

As the snorkel icing problem brought new urgency to the completion of NEL's cold-weather test facilities and caused the budget crisis to fade, plans to use *Nautilus* for under-ice experiments matured. On May 1, 1957, following an enthusiastic endorsement by ComSubLant, Rear Adm. Charles W. Wilkins, CinCLantFlt, recommended to CNO that *Nautilus* be scheduled for an arctic cruise that would last from August 19 to September 19. CinCLantFlt also planned to send a diesel-electric submarine on the operation as a safety measure.[44]

Lyon was delighted with the news. He believed that the proposed cruise should involve an "extreme and roving penetration of the arctic basin, with vertical ascents in polynyas to fix position, obtain oceanographic stations and demonstrate missile launching or air search." Lyon, writing to Blair in BuShips in mid-June, emphasized that "extreme" and "roving" were the key words. Otherwise, little would be gained over the *Redfish* expedition of 1952.[45]

Lyon also passed along to Blair two charts that "should be of immediate interest to the NAUTILUS cruise." Given by the Soviets to the Canadians during a recent IGY meeting, one chart showed the oceanographic stations that had been taken by Soviet aircraft after landing on the ice during the past several years, while the other was a bathymetric picture "supposedly constructed from data obtained from the stations." It was clear from these charts that the Soviets were far ahead of the United States in exploring the arctic basin.

"It does seem to me," Lyon wrote, "that this reported USSR activity emphasizes the need for the NAUTILUS cruise." A submarine reconnaissance of the deep arctic basin would enable the United States "to catch up fast" on the Soviets, assuming that *Nautilus* would be able to fix its position with accuracy—and assuming that the Soviets were not secretly operating submarines under the ice.[46]

On July 1, 1957, CNO assigned Project FL/A188/A4–3, "Assist in the Development of Techniques and Equipment for Operating Submarines in Cold Weather and Ice Areas," in support of Operation Requirement SW-01102 ("Submarine Under Ice Operations"), the Lyon-McWethy proposal, which had finally been approved on April 23. Separate tasks, CNO advised CinCLantFlt, would be assigned on a calendar-year basis. Task One, given Priority A, would send *Nautilus* into the Arctic to investigate the capabilities of a nuclear-powered submarine in under-ice operations; study the effect of cold weather on submarine material, especially the snorkel; provide oceanographic and hydrographic data, including information on the thickness and distribution of the polar ice pack; and conduct sound propagation and communication tests. Penetration of the polar ice pack to the "vicinity" of 83° North was desired "at the discretion of the Commanding Officer."[47]

The skipper of *Nautilus* for the under-ice cruise would not be Captain Wilkinson, who had done so much to promote the idea of using the nuclear-powered submarine in the far north, but Comdr. William R. Anderson, who relieved Wilkinson in June, 1957. Fortunately for

Lyon's plans, Anderson also had been "gripped with the polar bug," as he put it. During the winter of 1956–57, the prospective commanding officer of *Nautilus* had begun to think about under-ice operations. He had read widely in the literature of arctic exploration while "keeping an ear to the ground as to what was happening in the Pentagon—productively—thanks to Bob McWethy." Anderson realized that he had to maintain a low profile due to the substantial opposition to under-ice work by *Nautilus*. "My own feeling," Anderson recalled, "was that nuclear submarine operations beneath the ice should begin where Sir Hubert Wilkins had made his ill-fated attempt, in the deep waters between Greenland and Spitzbergen. I envisioned a series of summer probes, each longer in length than the previous one, the whole operation to be stretched over a period of years."[48]

Lyon, as he had written earlier to Blair, had more ambitious plans for the summer. On June 30 he visited *Nautilus* in San Diego to discuss with Anderson the need for a deep and roving penetration of the arctic basin. He then followed this up with a detailed letter. The objective of the patrol, he stressed, should be "to gather data most pertinent to evaluating the submarine warfare potential of the Arctic Ocean." The need to do this was especially important in light of the intense activity by the Soviets in the Arctic during the past twelve years. "I visualize the maximum, and only real pay off," he emphasized, "will come from the patrol if the greatest possible area of the arctic basin is covered." Lyon further proposed that he and Rex Rowray accompany *Nautilus* to operate the topside echo sounder and other equipment that would be provided by NEL.[49]

Anderson needed little persuading. On July 5 he sent a copy of Lyon's letter to ComSubLant with his endorsement. "These objectives appear sound in principal," Anderson wrote, "and are so concurred in by CO NAUTILUS." The NEL topside array and scientific instruments would be loaded aboard *Nautilus* prior to departure from San Diego, then installed when the submarine reached New London in mid-July.[50]

Preparing *Nautilus* in a timely fashion for the summer cruise proved a challenging task. The major responsibility fell on the shoulders of Commander Blair in BuShips, who had been designated as liaison officer by CNO for Task One. Blair had immediately put together a program to procure and install the necessary equipment on *Nautilus*. Known as SCAMP, for Submarine Cold Weather and Arctic Material Program, the project had to be completed within a few months on a limited budget and with the details of the operation classified as "confidential."

Given these constraints, Blair focused on the availability of "off-the-shelf" gear. NEL supplied the upward-beamed fathometer unit, which had last been used on *Redfish* in 1952 and then placed in storage. Additional bottomside fathometer transducers, to back up the ship's main fathometer, came from standard navy stocks. Blair obtained a newly designed Mark 19 Sperry gyrocompass, with high-latitude modifications, which would provide primary directional information. Sextants, compasses, radio direction finder and other equipment were begged, borrowed, and purchased. By the time *Nautilus* reached New London on July 21, Blair was able to report that "the list [of equipment] was completely checked off and the gear ready to install."[51]

While technicians worked on *Nautilus*, Anderson decided to make an aerial survey of the area of intended operations. On August 5 a party of ten that included Anderson, *Nautilus* navigator Lt. William G. Lalor, Jr., and Comdr. Leslie D. Kelly, Jr., C. O. of *Trigger* (which had been assigned to accompany *Nautilus*), boarded a navy Super Constellation radar warning plane and flew to Thule Air Force Base in northwestern Greenland. The next day, the group flew over the ice pack north of Spitzbergen. Anderson kept his movie camera trained on the ice below. "It looked cold and rugged," he recalled. Anderson observed numerous openings in the ice. "The flight," he reported, "gave us a great deal of confidence as to our ability to operate beneath the pack and surface in openings." After surveying nearly a thousand miles of the proposed route, the aircraft headed for Iceland. Anderson and his party returned to New London the next day.[52]

By mid-August, all was ready. Before departing, Anderson sought a clarification of his orders to proceed to "the vicinity" of 83° North. In other words, Anderson wanted to know, "Could I go the North Pole?" His superiors, he was told, were confident that he "would do the right thing." Anderson concluded that the decision to go the North Pole was his to make. "I secretly hoped that we could," he recalled. However, as far as the mission of *Nautilus* was concerned, "going to the Pole was desirable but not a critical objective."[53]

Shortly before *Nautilus* set off on its under-ice adventure, Rear Adm. C. W. Wilkins suffered a heart attack. ComSubLant would not be dockside to bid Anderson and his crew bon voyage, but he did send an eloquent letter. It read:

I want to wish you and your people every success in the cruise ahead. There are those who look on the operation askance and with skepticism. I am not one of those. I believe it is a venture of great promise, in both the fields of national defense and science. I am sure the information you will collect will be of great value. The operation itself is one that appeals to the imagination and the venturesome spirit within men's souls. You will be pioneers and trail blazers. Your findings may lead the way to under-ice navigation capabilities for nuclear-powered submarines that may make it possible for them to go any place in the ice regions where the interest of national defense requires. I know you have done careful planning for this operation, that all foreseen hazards have been taken into careful account, and that every possible preparation has been made to insure a successful operation, which will add to Nautilus' laurels and the glories achieved by her officers and men.[54]

*Nautilus* got underway at eight o'clock the morning of August 19, 1957. Anderson made a trim dive to 100 feet at 5:00 P.M. and discovered a leak in the packing of number one periscope. "A hundred-million dollar home and the roof leaks," quipped a crewman. The next day, after repairing the leak, Anderson dove to 700 feet, conducted several high-angle dives and ascents, then settled into a cruising depth of 300 feet. Lyon was impressed with the performance of the nuclear-powered submarine. "Have been cruising at 15–18 knots," he wrote in his Journal on August 21, "300 feet deep without fighting high seas and consequent strain on personnel or working of ship's hull." The problems of underwater navigation and communication, however, remained to be solved.[55]

On August 28 *Nautilus* passed east of Iceland. "So another wedding anniversary comes and goes," Lyon noted; "the 20th, while again in the field like the many years before." Since 1945, he had been home for a wedding anniversary only in 1956.[56]

Two days later *Nautilus* rendezvoused with *Trigger* off Jan Mayen Island. After Anderson made contact on the underwater telephone with Les Kelly, the two submarines proceeded due north at fourteen knots, with *Nautilus* remaining beneath the surface. *Nautilus* finally surfaced at 12:30 P.M. GMT on September 1 after eleven days and more than 4,000 miles underwater. "Need more be said in contrast to a conventional boat," an impressed Lyon observed.[57]

The two submarines reached the edge of the ice pack eight hours later. The operational plan called for *Nautilus* to make a test dive under the ice for approximately 150 miles, then return to open water and rendezvous with *Trigger*. As *Nautilus* slipped beneath the ice, Lyon took his station in the forward torpedo room. Each of the five topside fathometers was connected to a recording instrument that contained a roll of paper approximately ten inches wide. Two pens traced lines on the moving rolls. One pen displayed information obtained from the ship's depth gauge on the theoretical surface water level, while the other pen used data from the upward-scanning fathometer to draw a picture of the underside of the ice. *Nautilus* did not have QLA scanning sonar, as the equipment had gone out of inventory. This was not considered a problem as the submarine would be operating in deep water and would not encounter any ice extending downward to 300 feet.[58]

As *Nautilus* made its way northward, Lyon's topside echo sounders showed a large area of brash and block ice, mostly 6 feet thick, with a maximum of 15 feet. Shortly after reaching the turn-around point, Anderson decided to surface in a polynya. As the submarine rose to the surface at a rate of 1 foot per second with 5,000 pounds of positive buoyancy, Lyon's instruments indicated an area overhead that was clear of ice. Watch officers looking upward through both periscopes reported that they could see sunlight and "fluffy clouds." Suddenly, *Nautilus*'s upward progress came to a shuddering halt. "Flood negative!" Anderson ordered. The "fluffy clouds" had been blocks of ice.[59]

When *Nautilus* reached open water at 3:00 P.M., Anderson surfaced to inspect the damage. Number two periscope had been broken beyond repair, and number one scope had been badly bent. The incident drove home two important lessons. The submarine's fragile aluminum sail and topside configuration, Lyon noted, "places undue and unnecessary sensitivity to locating every small piece of ice." If the number three echo sounder, located on the top of the sail, had been shifted to high-speed scale, it might have detected the floating pieces of ice in the polynya. An even more important lesson had been learned: *never* raise both periscopes while operating submerged under ice.[60]

The situation created a major problem for Anderson. *Nautilus* was scheduled to participate in an important NATO exercise at the end of the under-ice mission. As the submarine could not operate effectively in the war games without a periscope in working order, he would have no choice except to abandon the arctic phase of the patrol and head for England to make repairs.

Lt. Paul J. Early, engineer on *Nautilus*, held out a ray of hope. He believed that there was a slight possibility that the number one periscope could be straightened. Anderson told him to give it a try. Hydraulic jacks were brought to the bridge and pressure applied on the stainless steel barrel of the scope. As the barrel responded to the pressure, however, it split open two feet down from the periscope glass. Fortunately, *Nautilus* carried crew members with skills to repair the ship's nuclear plant at sea. One of these skills involved the welding of stainless steel materials and pipes because the nuclear plant was fabricated with special types of stainless steel. Thanks to the talented machinist mates who possessed these special welding skills, the crack in the periscope barrel was repaired.

This left one major problem to be resolved. The air from the periscope fitting had to be suctioned out and a charge of dry nitrogen gas inserted in order to avoid moisture condensation on the periscope optics. Undaunted, Early and his men ran a hose from the ship's steam-condenser vacuum pumps and suctioned the air from the periscope fitting, then shot in a charge of dry nitrogen gas. After fifteen hours of labor under extremely difficult conditions, the periscope was again serviceable. "It was the most amazing repair job at sea I had ever witnessed," Anderson recalled.[61]

Anderson and Lyon then discussed the question of whether or not to continue with the under-ice operation. Lyon pointed out that an extended under-ice voyage was necessary to assure the continuation of the arctic submarine program and to give impetus to the Navy Department's policy on the matter. "There is always the school," Lyon observed, "that will immediately use any item to discourage any further effort to solve the arctic basin operation problem." Anderson agreed completely. The voyage northward would continue.[62]

*Nautilus* completed a surface run to the edge of the ice pack at 7:00 A.M. on September 3, then dove under the ice and began its run to the north. As the submarine proceeded along the 0° meridian at a speed of fifteen knots and depth of 300 to 350 feet, Lyon and Rowray exchanged watches on the topside echo sounder recorders every four hours. Over the next twenty-eight hours, the recorders showed a number of polynyas overhead. Anderson slowed to five knots to explore one of the larger open areas, but he decided not to surface. The time allotted for arctic operations was growing short, and he was mindful of the damage that had been caused by the first surfacing.[63]

By 11:00 A.M. on September 4, *Nautilus* had reached 85° North. Although the ship's magnetic compass and Sperry Mark 23 auxiliary gyrocompass were behaving erratically, the new Sperry Mark 19 Master Compass was performing well. Anderson believed that the North Pole, only 300 miles or twenty hours steaming away, "seemed within our grasp."[64]

Just as *Nautilus* approached 86° North, however, a blown fuse interrupted the power supply to both gyrocompasses. Although power was restored within a minute, both compasses now showed erratic readings. Anderson knew that it usually took a gyrocompass four hours to establish its equilibrium. How long it would take at such a high latitude was anybody's guess.

In order to gain a reference direction, Anderson shifted the Mark 23 auxiliary compass to a free directional gyrocompass and aligned it on a true meridian as determined by the magnetic compass. The only problem, Anderson noted, was that the magnetic compass "was slowly swinging back and forth through an arc of 60 degrees."[65]

It took six hours for the Mark 19 gyrocompass to align on a meridian. By this time, *Nautilus* had crossed 87° North. Anderson initiated a gradual 180° turn and headed south. "I regretted that we had not reached the Pole," he later wrote, "but considered that to go any further would expose the ship and the crew to undue risks."

In retrospect at least, the failure of both gyrocompasses could have been avoided. As Frank L. Wadsworth, then a junior officer on *Nautilus* later pointed out, it had been a mistake to have had both compasses powered off the same electrical circuit. The failure of any one circuit could occur at any time. If *Nautilus* had lost only one gyrocompass, Wadsworth contended, "The whole nature of the expedition would have changed. It is my opinion that *Nautilus* could easily have continued to the North Pole." As it was, another valuable lesson had been learned.[66]

The immediate problem, Lyon observed as *Nautilus* headed south at eighteen knots, was that "our position is uncertain." Ever since entering the ice pack on September 3, navigation had been done by dead reckoning. The ship's course had been determined by the gyrocompasses, and the distance traveled along the course by propeller turn-count. Dead reckoning gave only an approximate position at best. The failure of the gyrocompasses greatly compounded the problem of location. The ship's navigators, Anderson reported, received "a barrage of friendly jokes about liberty in Alaska and, inevitably, Murmansk." But it was no joke.[67]

"The major problem of the Arctic Ocean is being demonstrated," Lyon wrote in his Journal on September 5, "namely to determine position." Lyon tried to ascertain the location of *Nautilus* by reference to the bathymetric chart that had been done by the Soviets. The bottom contours, however, failed to match. He then turned to the data that Harald Sverdrup had taken during the Wilkins expedition of 1931. "We were coming south," Lyon recalled, "and the only thing we could check against was whether the bathymetry and the soundings below us seemed correct, whether the temperature of the water seemed correct, and whether the ice cover looked appropriate."[68]

On the afternoon of September 5, *Nautilus* suddenly ran into shallow water; it was clear that something was wrong. "We had noted that the ice was much more compact than we expected," Anderson noted, "and that the water temperature was much colder." Lyon believed that the submarine was in the cold current that flowed along the east side of Greenland out of the Arctic Ocean. *Nautilus* likely was west of its dead-reckoning position, headed toward Greenland. Anderson agreed and ordered a course change to the southeast.[69]

There was one way to check their location, and that was to come up into a polynya and take a celestial fix. When the upward-beamed echo sounders showed a promising open area, Anderson decided to surface. As he neared the surface, however, Lyon's instruments revealed a uniform layer of fresh ice overhead. Given the fragile sail of *Nautilus* and with only one periscope barely functional, Anderson elected not to risk an attempt to break through the ice.[70]

At 3:30 A.M. on September 6, the upward-beamed echo sounders began to show an area of brash ice and frequent polynyas. The water temperature was 31°. An hour later, the temperature rose briefly to 39°, then dropped back. Lyon realized that the submarine was moving into the mixed area of the Greenland outflow and the warmer Atlantic inflow. *Nautilus* finally reached open water and surfaced at 10:00 A.M., having spent seventy-four hours and nearly a thousand miles under the ice pack. Sunlines revealed that the submarine had traveled further west than expected, placing *Nautilus* in the ice-filled Greenland cold stream.

The return track from 87° North, as Lyon had thought, had been toward Greenland before changing courses "just enough to swing east around the northeast tip of Greenland."[71]

*Nautilus* rendezvoused with *Trigger* at midnight. The next day, September 7, Anderson held station at the edge of the ice pack while the conventional submarine made an under-ice excursion. He took the opportunity to unclog the garbage ejector tube, which appeared to have something caught in its outer door. Anderson put pressure in the battery compartment so that the tube could be examined from inside. The pressure leaked into the torpedo room, causing Lyon, who had been suffering from a cold, considerable pain in his ears and nasal passages. In the end, Anderson found that the ejector gasket had been cut, likely by tin cans, and could not be repaired.[72]

Following this incident, Lyon reflected on the equipment problems during the cruise. Not only had the garbage ejector failed, but the gyrocompasses, drain pump, and periscope air jet deicer had also experienced problems. On the other hand, the "reactor/steam plant has operated 25,000 miles without loss of power due to failure of components."[73]

On September 8 *Nautilus* made a dive to the northeast across an area that had been surveyed by Sir Hubert Wilkins in 1931. Anderson came up in a polynya, the first—and only—successful vertical ascent during the cruise. *Nautilus* remained in the ice lake for two hours, taking photographs, before continuing its dive.[74]

Two days later *Nautilus* proceeded to the rendezvous point for *Trigger*. Upon arrival, the conventional submarine was nowhere in sight. Anderson ran toward the coast of Spitzbergen and took a radar fix to make sure of his location, but *Trigger* was not to be found. After searching all day, on the surface and submerged, the two submarines finally made contact at 10:00 P.M. *Trigger*, it turned out, had missed the rendezvous point by forty miles.

"Note large amount of time lost during this Arctic operation just to effect rendezvous," Lyon noted in his Journal. But this was not unexpected. In fact, it had been one of the reasons why he had objected to the inclusion of *Trigger* on the expedition. Although the diesel-electric submarine had traveled sixty-one miles under the ice, one of the longest ever excursions by a conventional submarine, "it's kind of obvious it didn't have much meaning" under the shadow of *Nautilus*. The assignment of *Trigger* as a safety measure, Lyon believed, was an example of "old-fashioned thinking, based on diesel boats alone." If *Nautilus* had gotten into trouble on her under-ice venture, *Trigger* would not have been able to help.[75]

On September 11 and 12 *Nautilus* steamed south at flank speed—twenty-three-and-a-half knots at 300 feet. The submarine surfaced at noon, September 13, and continued toward Scotland in a rough sea. *Nautilus* arrived off Rothesay at 4:00 P.M. on September 14 and tied up alongside the submarine tender USS *Fulton*. Lyon removed his equipment from *Nautilus* and put it on board *Fulton* for return to the United States. He and Rowray then made their way leisurely through Scotland and England before flying to New York on September 20.[76]

The nuclear-powered submarine, Lyon recognized, "opened the door to a whole new world." All thought of employing conventional submarines for under-ice work immediately vanished. *Nautilus* had proven that "*unlimited* movement" under the ice was now a reality. "The transarctic submarine," he wrote in his senior scientist's report of the voyage, "which five years ago was often called fantastic, is now a demonstrable fact. . . ." As a result, "the Arctic Ocean becomes an area for the submarine forces."[77]

A number of problems remained to be solved, however, before the capability of *Nautilus* would be fully exploited. The patrol had demonstrated that *Nautilus* could not surface in small polynyas without risk of damage. Clearly, the hull and fragile superstructure of the submarine would have to be modified for arctic operations. Also, a better method of determining position under the ice had to be found. In addition, a major bathymetric survey of the arctic basin was necessary so that accurate bottom charts could be compiled. *Nautilus* had taken a major step in this direction, collecting a continuous and "uniquely extensive" record of the bottom contours as well as the distribution and thickness of sea ice. Finally, the topside sonar system required modification. Lyon envisioned an array of fixed vertical sound beams with controllable variable beam widths.

"The NAUTILUS," Lyon concluded, "opened the entire Arctic Ocean to a new era of high speed submarine operations." His vision of an operational under-ice submarine seemed close at hand. Several problems needed attention, but these did not seem to pose major obstacles.

Looking to the future, he set new challenges. The objectives of the under-ice program, Lyon believed, should be to develop a submarine that would be capable of using the Arctic Ocean as a regular transit area between Atlantic and Pacific, and to extend the operating season in the far north from summer to all year.

O-12 undergoing modifications for Wilkins's expedition. Courtesy U.S. Navy

Sir Hubert Wilkins
(with pipe) on deck of
*Nautilus,* August, 1931.
Courtesy U.S. Navy

Waldo and Virginia Lyon on the day after their wedding, August 29, 1937. Courtesy Waldo Lyon

Triplane target for cold-water sonar exercises, October, 1945. Courtesy U.S. Navy

*Northwind* tows *Sennet* through heavy ice in the Ross Sea, January, 1946. Courtesy U.S. Navy

Oceanographers Eugene C. LaFond, Walter H. Munk, and Graham W. Marks on *Nereus*, July, 1947. Courtesy U.S. Navy

Lyon and sonarman from *Boarfish* check equipment during first under-ice dive, August 1, 1947. Courtesy U.S. Navy

Ice caught on deck of *Carp* during vertical ascent, September, 1948. Courtesy U.S. Navy

Rex Rowray examines Nansen bottle on deck of *Baya,* July, 1949.
Courtesy U.S. Navy

Canadian oceanographic ship *Cedarwood*, 1949. Courtesy Royal Canadian Navy

LCVP from *George Clymer* abandoned in surf off Cape Prince of Wales, August 2, 1949. Courtesy U.S. Navy

Robert McWethy on bridge of *Burton Island,* August, 1950. Courtesy Robert McWethy

*Burton Island* attempts to break through thick ice by using heeling tanks, February 16, 1953. Courtesy U.S. Navy

*Northwind* takes *Redfish* under tow, August 20, 1953. Courtesy U.S. Navy

Lyon examines Nansen water sampling equipment on *Northwind,* August, 1954.
Courtesy U.S. Navy

Conference on *Northwind,* anchored off Barnard Point, August 29, 1954.
*Left to right:* Dr. Donald Rose and Capt. O. C. S. Robertson of *Labrador;*
Dr. William Cameron; William L. Maloney; Lyon; Comdr. E. A. Trickey of *Burton Island;*
Alexander Wuerker. Courtesy U.S. Coast Guard

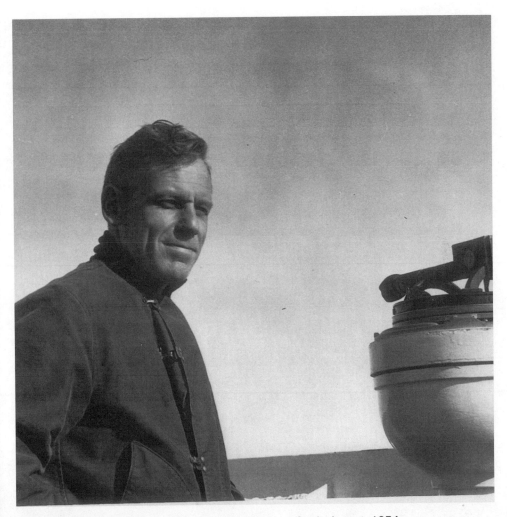
Lyon on bridge of *Northwind* prior to entry into McClure Strait, August, 1954.
Courtesy Waldo Lyon

Crew from *Labrador* erect antenna for EPI station, 1955. Courtesy Royal Canadian Navy

## Operation Sunshine

On October 29, 1957, the Navy Department issued a press release on the under-ice voyage of *Nautilus*, terming it "one of the most incredible adventures in naval history." The *Washington Post and Times Herald* buried the story on page six of its second section; the *New York Times* gave it a few paragraphs on page nine. As it happened, the big news in October had come earlier in the month with the appearance of *Sputnik*. "The Soviet launch of the world's first artificial earth satellite," historian Robert A. Divine has written, "created a crisis in confidence for the American people." Newspapers were filled with stories of how the Soviet satellite signaled a weakness of American science and threatened the nation's security. *Nautilus*, Lyon observed, became "lost in *Sputnik*'s shadow."[1]

Lyon experienced first-hand the impact of *Sputnik* when he visited Washington in early November to discuss plans for 1958. Shortly after returning from England, Lyon had proposed that *Nautilus* be sent to the arctic basin in 1958 to sample sea ice distribution and obtain bottom bathymetry along the north coast of the Queen Elizabeth Islands and Greenland. This information, he had argued, needed to be obtained "as rapidly as possible" in order to exploit the new capabilities that had been demonstrated by *Nautilus*.[2]

CNO, he learned on November 7, had grander plans for *Nautilus*, including the possibility of a round-the-world cruise via the arctic basin. "Major objective is, of course, political," Lyon noted, "but maximum payoff in scientific information is to be planned and is expected." Lyon was told that he would be alerted informally to any decision that might be taken. As he knew that directives and instructions never caught up after a decision had been made, he decided that NEL must proceed as if the operation would go forward. Lyon came away from the meeting with a sense of "the overwhelming pressure throughout Washington to produce trump cards in the world political game."[3]

Commander Anderson also was eager to return to the Arctic. "I feel that it is extremely important," he wrote to CNO on December 4, "for the Navy to determine in the clearest possible terms and at the earliest possible date, the precise manner in which our own nuclear submarines might exploit the Arctic area." There was, he emphasized, a "pressing need" for additional scientific investigation. "NAUTILUS," Anderson pointed out, "has a crew trained in under-ice work and eager for more." He proposed a cruise in summer, 1958, to survey the strategic area in which ballistic missile submarines might be staged; to examine the feasibility of a transpolar transit; to work on the problem of surfacing in ice; and to conduct experiments with very-low-frequency (VLF) radio reception. Also, Anderson concluded, "for propaganda purposes," *Nautilus* could include in the summer program a crossing of the North Pole.[4]

Two weeks later, the commander of Submarine Squadron 10 (ComSubRon 10), which had been assigned primary responsibility for accomplishing CNO's project to develop an arctic submarine, sent his comments on Anderson's proposal to the commander-in-chief of the Atlantic Fleet (CinCLantFlt). Anderson's request seemed "feasible," and his objectives were "generally sound." Present plans called for the nuclear-powered submarine *Skate* (SSN 578) and the diesel-electric submarine *Halfbeak* (SS 352) to conduct cold-weather operations in the Arctic next summer. *Nautilus* had been scheduled for overhaul from June 23 to October 28, 1958. As it now appeared that this overhaul might be canceled, ComSubRon 10 recommended that *Nautilus* be added to the coming summer's expedition. The benefits derived from two nuclear-powered submarines operating in company, he argued, would be far greater than those from two separate cruises.

ComSubRon 10 believed that the two major objectives of the summer's operations should be to solve the in-ice surfacing problem, and to conduct an extensive survey of the ice coverage and bathymetry of the arctic basin. "Penetration of the North Pole," he concluded, "is not recommended at this time since it does not support the above objectives."[5]

ComSubRon 10's endorsement of *Nautilus*'s inclusion in the summer, 1958, arctic program came as no surprise to Anderson. As he pointed out to Lyon, the response had been prepared by McWethy, an avid supporter of under-ice submarines who was now chief of staff to ComSubRon 10. While the proposal still had to make its way up the chain of command, Anderson believed that it would be approved. In the meantime, he told Lyon, "I am working hard to get a piece of prototype inertial navigation gear."[6]

McWethy shared Anderson's optimism. "Things are coming along pretty well," he informed Lyon on December 30. "We are in a position to get things on the road." McWethy enclosed a copy of a long-range program that had gone to the commander of the Atlantic Fleet's submarines (ComSubLant). It represented, he said, a combination of the ideas that he and Lyon had discussed earlier in the year, plus the recommendations contained in Lyon's senior scientist's report of the recent *Nautilus* cruise.

The Soviet Union, McWethy pointed out in his "Proposed Program for Arctic Submarine Operations," actually had a longer coastline than the United States. It extended along the northern border of the country from 170° West longitude at the Bering Strait to 30° East longitude where it met Norway. Previously, this long coastline had been protected by the arctic ice pack. *Nautilus*, however, had demonstrated the new vulnerability of the Soviet Union. When the under-ice capability of the nuclear-powered submarine was combined with the development of the submarine-launched Polaris missile that was now underway, "the former protection becomes less than useless."

There were two immediate objectives for arctic submarine development. First, nuclear-powered submarines must have freedom of movement in the arctic basin across the entire north coast of the Soviet Union. They must be able to reach and maintain their position without detection and be able to fire their missiles within a few minutes of receiving orders. Second, all submarines, nuclear and conventional, must be able to operate effectively in fringe ice areas.

The program for 1958 would begin in January with the dispatch of three conventional submarines into Cabot Strait. The purpose of this operation would be to evaluate the area for possible winter training in the years to come. In the summer two nuclear submarines would proceed in company to the east coast of Greenland, run through the Denmark Strait under the fringes of the ice pack and rendezvous with a conventional submarine. The two nuclear submarines then would conduct under-ice operations to the north of Spitzbergen. Their mission would include investigating in-ice surfacing techniques; developing and evaluating procedures for coordinated operations by nuclear submarines; testing torpedoes; providing oceanographic and hydrographic data; testing communications and sound propagation; evaluating the protection afforded by the ice pack; and investigating the problems of submarine versus submarine warfare in ice areas.

A program of "measured acceleration" would build on the experiences of 1958. The next year would include winter training in the Cabot Strait and Bering Sea areas, an early spring operation in the Greenland Sea, and a summer cruise in the Beaufort Sea, followed by a transit to the Pacific Ocean by way of McClure Strait. In 1960 operations would take place year-round to train crews, test new equipment, and extend oceanographic surveys. The year also would feature a transit from Atlantic to Pacific "across the North Pole." The program would culminate in 1961 with a regular schedule of arctic submarine training and the testing of new equipment and weapons, especially the Polaris missile.[7]

The new year found Lyon concerned with the short-term program. After receiving letters from Anderson and James F. Calvert, commander of *Skate*, he decided to circulate a memorandum that set forth his views on the proposed summer's operations. Sent to McWethy, Anderson, Calvert, and Couper and Blair in BuShips on January 13, 1958, it urged everyone "to push all angles to get both boats assigned." The gain in information, Lyon argued, "would be terrific." It would be possible to obtain data on bottom bathymetry, ice characteristics, and water structure that never could be secured with one submarine. Also, a start could be made on the challenging problem of submarine versus submarine warfare in the Arctic.

Lyon proposed to equip *Nautilus* with a five-unit topside echo-sounder array that would include a video presentation in the combat control center. There would be time to put only a three-unit array on *Skate*, but he believed that it would be sufficient. As the 1957 *Nautilus* voyage had revealed the difficulties of obtaining precise information on overhead ice at the depth and high speed of nuclear submarines, Lyon intended to use a new high-resolution topside recorder to give an expanded record of the ice profile. Also, he would install on both submarines a precise depth recorder to obtain more complete bottom data.

Lyon noted that in ComSubRon 10's letter to CinCLantFlt on December 17, 1957, "a crossing of the North Pole is not recommended."

He agreed that such a transit would contribute little to securing essential scientific data about the arctic basin. "However," he pointed out, "though perhaps I am not in the proper echelon to consider the question, it seems very pertinent for propaganda and international purposes to make the crossing. We perhaps have been remiss in not considering propaganda advantages, e.g. Sputnik."[8]

As it happened, the advantages of a polar transit were being considered at "the proper echelon." After *Nautilus* had returned to New London in late October, 1957, Anderson had gone to Washington to brief CNO on the NATO exercise and the arctic operation. Leaving the Pentagon, he had run into Capt. Peter Aurand, naval attaché to President Eisenhower. This chance meeting had led to a briefing for the White House staff. "I came away from that meeting," Anderson recalled, "convinced that the White House had more than a casual interest in the Nautilus and her polar cruise."[9]

In January, 1958, Anderson received a phone call from Washington, ordering him to report as soon as possible to Rear Adm. Lawrence R. Daspit, director of undersea warfare (Op-31). When he arrived at the Pentagon, Anderson was taken to a meeting with Daspit; Capt. Frank Walker, head of the submarine warfare branch (Op-311); and Comdr. Marmaduke G. Bayne. Anderson was told that Admiral Burke had suggested to President Eisenhower that *Nautilus* cross between the Pacific and Atlantic Oceans via the North Pole. The president's reaction had been "enthusiastic." Daspit wanted to know if Anderson thought the operation would be feasible. Anderson assured him that it could be done, but he wanted to hear from Lyon.[10]

Lyon was summoned to Washington for a meeting with Daspit, Walker, and Anderson on February 13. The White House, they informed him, wanted *Nautilus* to make a polar transit from west to east to coincide with the opening of the World's Fair in Brussels in June. Would this be feasible? There followed a discussion of depths, routes, ice coverage, and other essential operational details.

Privately, Lyon harbored reservations about the timing of the voyage. As he well knew from past experience, submarines would encounter two types of downward-projecting ice. In the deep arctic basin, pressure ridges were formed when one floe of ice pressed against another. This created ridges of ice that rose above sea level to a maximum height of approximately 30 feet. Below sea level, the ice keel could extend downward to more than a hundred feet. Because of the great depth of the arctic basin, the deep-diving *Nautilus* would easily be able to avoid these obstructions.

Another type of pressure ridge, however, was created when ice piled or rafted up on shorelines. With warmer weather, the rafted ice would be carried away from the coast with the retreating ice pack. *Nautilus* would encounter these ridges while crossing the shallow Bering Strait and Chukchi Sea. Based on his experiences during summer icebreaker operations, Lyon expected to find rafted ice with drafts of 40 to 50 feet. Ice projecting below the surface to any greater depth would be "an extreme rarity."

In the end, Lyon decided not to express his reservations at the meeting. There was no point, he believed, in raising doubts based on an extreme case. Instead, he gave the proposed voyage his unqualified endorsement.

At this point, the issue of secrecy arose. The White House wanted no publicity for the operation lest it fail and cause embarrassment comparable to the abortive launch of Vanguard the previous December. The memory of that technical and public relations disaster was all too fresh at 1600 Pennsylvania Avenue. Designed to place an earth satellite into orbit, Vanguard had been hailed as America's answer to Sputnik. Two months after the successful launch of the Soviet satellite, the world's press had assembled at Cape Canaveral to watch the navy-sponsored project place a grapefruit-size sphere into space and retrieve America's technological honor. Two seconds after the liftoff, however, the rocket tumbled to Earth in a fiery ball of liquid oxygen and kerosene—a spectacular sight for the nation's television audience. The talk of "kaputnik" had stung officials in the White House, and they wished no repeat of this type of painful episode.

Given the sensitivity of the White House, a "cover" story had to be developed for the polar effort. The group decided on a public announcement that *Nautilus* would be sent to the west coast via the Panama Canal to familiarize antisubmarine warfare units of the Pacific Fleet with the capabilities of a nuclear submarine. After calling at San Diego, San Francisco, and Seattle, *Nautilus* would return to the east coast through the Panama Canal. In reality, Anderson would head north from Seattle, pass through the Bering Strait and into the Chukchi Sea, enter the deep waters of the arctic basin, and cross the North Pole en route to the Atlantic Ocean.[11]

The plan needed the approval of Vice Adm. Thomas S. Combs, Burke's deputy for operations. At nine o'clock the next morning, Daspit, Walker, Anderson, and Lyon convened in Combs's office in the Pentagon. The proposed voyage, Lyon reported, was again "fully discussed." "My position was categorical," he noted in his Scientific Journal, "that crossing of Chukchi was possible without meeting ice greater than 50 ft.—tho I have reservations that we do not know the whole story of break up and movement of landfast pressure ridges [on the north coast of Alaska]. These have always plagued our thoughts but should be avoidable by scanning sonar as BOARFISH [in 1947]."

Based upon the study group's recommendation, Combs gave his assent. He told Daspit to inform Admiral Rickover, Admiral Mumma at BuShips, and Rear Adm. Frank B. Warder (ComSubLant) about the proposed operation. Otherwise, access to the information was to be tightly held within a small circle of naval officers and White House officials.

Daspit, Anderson, and Lyon met with Rickover later in the morning. "Rickover in his usual emotional display objected," Lyon noted, "but it was difficult to see to just what he objected because of so many parenthetical displays against OpNav." In any event, Rickover's permission was not being sought; he was being informed.[12]

Three days later, on February 17, Lyon went to BuShips for a conference with Commander Blair about equipment for *Nautilus* and *Skate* for their scheduled summer arctic cruise. Called into Admiral Mumma's office, he was told that Daspit had just telephoned with the news that the transarctic voyage of *Nautilus* had been given final approval. The code name for the cruise was to be Operation Sunshine.

Lyon now had to work out the first of what would prove to be a continuing number of cover stories to insure that *Nautilus* was prepared for the voyage without betraying the secret purpose of Operation Sunshine. The submarine, he told Blair (who was not privy to the secret), would be in New London from March 15 to April 18. It then would depart for the west coast. Because there would be no time after it returned to New London to prepare for the planned summer arctic operations, all under-ice equipment would have to be installed on *Nautilus* during the March–April period.[13]

Lyon then telephoned San Diego and asked that all work on arctic submarine equipment at NEL be stepped up to meet the new deadline. Obviously, it would be easier to explain this work, as well as his absence from NEL in June, if Captain Phelps and Dr. Kurie could be briefed on Operation Sunshine. Lyon asked Commander Bayne if this might be possible. After checking with Daspit, Bayne told Lyon that permission had been denied. "This restriction on working personnel," Lyon later noted, "made planning of the conduct of the operation extremely difficult...."[14]

Lyon visited New London in early April, 1958, to discuss the "special mission" with Anderson. Installation of equipment for the "summer" cruise to the Arctic, he learned, was going well. Anderson, after much effort, had managed to secure an inertial navigation system for the voyage. The North American Autonetic N6A autonavigator had been obtained from the U.S. Air Force, where it had been employed as a navigation system for the recently terminated Navajo cruise missile. The N6A had been taken out of mothballs and modified from three-hour use to a continuously operating submarine system. North American Autonetic engineers would be on board *Nautilus* to operate and maintain the equipment during the transarctic voyage.[15]

Anderson told Lyon that he wanted to conduct an air reconnaissance of the proposed route through the Bering Strait and into the Chukchi Sea prior to departing the west coast. Lyon suggested that he and Anderson hire a bush pilot for the survey flight, using the cover story that the reconnaissance was needed for a program to establish field stations along the Alaskan coast from Cape Prince of Wales to Point Barrow as part of NEL's continuing sea ice study. Anderson would pose as Lyon's assistant. They agreed that Lyon would prepare a set of bogus travel orders for Anderson under an assumed name. Lyon would have to travel under his own name as he was likely to meet people he knew.[16]

Lyon went on to Washington and discussed the proposed reconnaissance mission with Bayne in Op-311. Bayne told him that in order to maintain the cover story he would have to pay for the travel and hiring of a bush pilot out of his own pocket. Bayne would issue a sealed directive so that Lyon could recover the expenses at the completion of Operation Sunshine. "This operation is becoming more and more story book," Lyon noted in his Journal, "most fascinating to live through."[17]

Returning to San Diego, Lyon sought Dr. Kurie's help in obtaining an identification card of a former NEL employee. "No reason given," Lyon explained, "other than for special job to be able to enter yard without my name appearing on any list." Kurie supplied an old ID card for Charles A. Henderson. Lyon, using skills that any intelligence officer would admire, slit open the card with a razor, mounted Anderson's photograph and fingerprint, then resealed it.[18]

*Nautilus* departed New London on April 25, 1958, and passed through the Panama Canal in early May. On its first dive after the transit, a fire broke out in the engine room. Thick, acrid smoke quickly filled the compartment. While two crewmen, wearing two of the four special smoke masks carried by the submarine, attempted to find the source of the smoke, Anderson ordered *Nautilus* to the surface. With the smoke clearing, the masked crewmen managed to locate the problem. The insulation around the port high-pressure turbine had become soaked with oil after three years of operation and had ignited.

"Our greatest potential danger beneath the ice pack—fire—had risen up to slap me in the face," Anderson wrote. If the fire had broken out while the submarine was under the ice, it might not have been able to make its way to the surface. Anderson decided to devote more attention to fire hazards. He also was determined to obtain special breathing apparatus for the crew prior to heading north.[19]

*Nautilus* reached San Diego on May 12. Lyon and his associates immediately began work to complete installation of all special under-ice equipment. *Nautilus* would carry the usual five-unit topside echo-sounder array. In addition to the paper recorders, however, a five-gun oscilloscope would provide a video presentation of the information.

NEL also installed the experimental variable-frequency topside echo sounder to measure the draft of the ice canopy. Mounted on the top of the sail, this equipment had a frequency range of 30 to 200 kHz with nominal beam widths from 3° to 18°. With a pulse rate of twelve per second, the echo sounder was expected to measure the thickness of overhead ice within six inches or less. Finally, *Nautilus* would have a Sperry-Rand Automatic Depth Control System, together with a Sperry Depth Detector that would drive a recorder to give a continuous record of depth within six inches.[20]

In all, Lyon hoped that the equipment would be able to trace an accurate ice profile and provide precise bathymetry while *Nautilus* moved at twenty knots and 600 feet. What the submarine lacked, however, was forward-scanning sonar. The QLA sets that had been used on earlier under-ice experiments with conventional boats had gone out of stock. Lyon had been reluctant to press a search for scanning equipment lest he compromise the secrecy of Operation Sunshine. Ostensibly, *Nautilus* was being prepared for a summer voyage north of Greenland. If Lyon had launched a search for QLA sonar for *Nautilus*, "it would immediately have been an indication of what was up. So we just had to be careful and we just couldn't push to try to get one of those systems." Lyon would have to hope that *Nautilus* would not encounter any extremely deep-draft ice on her transit of the Bering Strait and Chukchi Sea.[21]

Before *Nautilus* left San Diego, Lyon gave Anderson the "Henderson" identification card and made plans to meet in Seattle in early June for the flight to Alaska. Lyon then made airline reservations in his own name and that of Rex Rowray, who thought that he was going to Alaska on a special mission but in fact was scheduled to make the transarctic crossing. When Anderson arrived at the airport, Lyon would simply tell the airline that "Henderson" would be replacing Rowray on the flight.

Lyon also received a call from Bayne about a special plan for Anderson. The White House, he learned, had asked that Anderson be taken off *Nautilus* by helicopter in the vicinity of Iceland, following the transarctic crossing, so that he could be flown to Washington. White House officials wanted Anderson to be standing beside Eisenhower when the president announced the success of Operation Sunshine. Anderson had protested; he wanted to remain with *Nautilus*. Although the navy hierarchy had supported Anderson, Bayne told Lyon to advise him that they had been overruled "from above." Also, Lyon was to fly to Washington and act as a courier for both the operational orders for *Nautilus* and the special instructions for Anderson. Anderson agreed to meet Lyon at the Los Angeles airport at 6:00 P.M. on May 22 to receive the documents.[22]

While *Nautilus* headed up the coast of California for a stop at Mare Island Navy Yard en route to Seattle, Lyon flew to Washington for a final meeting with Combs, Daspit, Walker, and Bayne. After reviewing all procedures for the voyage, Lyon received the operational orders for *Nautilus* and the special instructions for Anderson. "Best wishes and good luck said by all," he noted in his Journal, "and final reminder that decision remains in field as to whether transit can be made."[23]

Lyon then crossed the country on TWA, arriving in Los Angeles in time for his six o'clock meeting with Anderson. When *Nautilus*'s commander failed to show up, however, Lyon continued to San Diego. Anderson called from Los Angeles at 8:30 P.M., explained the delay, and arranged to meet Lyon at San Diego airport at 11:00 P.M. Anderson then flew down in civilian clothes, collected his papers while chatting briefly with Lyon on a nearby park bench, and returned to Los Angeles on the same plane.[24]

In order to explain his absence from NEL, Lyon had to concoct a cover story for his superiors. He told Captain Phelps that his recent trip to Washington had been to receive instructions for a "special mission of the highest classification." This mission involved a survey of the defense requirements in Alaska. He expected to be absent from San Diego for most of June. Also, Rex Rowray would be accompanying him.[25]

On June 2, as planned, Lyon and Rowray left San Diego for Seattle. Upon arrival, they took a limo to Tacoma and checked into the Olympus Hotel. It was at this point that Lyon told Rowray about their true destination. He then borrowed a heavy shirt and shoes from Rowray for Anderson's use in Alaska.[26]

The next day Lyon met Anderson at the Seattle airport. The two men—Lyon and his assistant "Henderson"—flew to Fairbanks on a Pan American Stratocruiser. On June 4 they departed at 7:00 A.M. on a Wien C-46 for Nome. After a five-hour stopover, they continued to Kotzebue, arriving at 3:00 P.M. As arranged, bush pilot Ernest Cairns was waiting with a Cessna 180 for the survey flight.

The weather was sunny and warm when they took off at 6:00 P.M. Cairns flew over Kotzebue Sound to open water south of Point Hope, then swung north around Cape Lisburne. He landed on the beach at Point Lay and refueled from a cache of gasoline drums, then continued along the coast to Point Barrow, arriving at 10:30 P.M. "Ice picture about as expected," Lyon reported. "Outstanding feature is broken, relaxed cover off Chukchi Sea, showing no heavy rafting." The reconnaissance confirmed for Lyon that the best course to the arctic basin lay through the central Chukchi Sea past Herald Shoal rather than along the northern coast of Alaska and through the Barrow Sea Valley, where heavy ice remained just off shore.[27]

Lyon and Anderson spent the night at the Wien Hotel in Barrow village. The next afternoon they were waiting in the small terminal building next to the airstrip when Max C. Brewer walked in. Director of the Arctic Research Laboratory at Point Barrow, Brewer knew Lyon well. In order to avoid an awkward discussion about his presence in Barrow, Lyon quickly ducked out the door. "Do not believe he saw me," Lyon remarked with relief.

The two men finally left, undetected, at two o'clock that afternoon. Arriving in Fairbanks, Lyon closed his account with Wien. The charter had cost $369.53. "Certainly got a lot of info out of that money," he noted, "particularly indoctrination and feel of Andy for the ice and geography of the area." They departed at 9:15 P.M. for Seattle. Landing in the early morning hours of June 6, Anderson took a taxi to *Nautilus* while Lyon headed for Tacoma.[28]

While Lyon and Rowray took a long walk through a residential area of Tacoma on June 7, Anderson dealt with a persistent leak in the ship's condenser system that threatened to delay—or even cancel—the voyage to the Arctic. When all other solutions failed, Anderson ordered a group of sailors to don civilian clothes and go to service stations throughout Seattle to purchase cans of Stop Leak, a sealant used to fix automobile radiators. They brought back 140 quart cans. Under the supervision of Lt. Comdr. Paul J. Early, engineering officer, they poured half of it into the condenser system. "What a scene!" Anderson recalled. "A hundred-million-dollar nuclear-powered submarine, the most advanced ship in history, and Early's men pouring in $1.80 cans of Stop Leak." But, as Anderson pointed out, it was no laughing matter. If the leak could not be fixed, the polar transit might have to be aborted. "Incredible as it may seem," the relieved skipper reported, "the Stop Leak treatment worked."[29]

There remained the task of smuggling Lyon and Rowray on board *Nautilus*. The two men drove a rental car to Seattle on the morning of June 8, then took a taxi to Pier 91. Lt. Shepherd M. Jenks, navigator of *Nautilus*, met them at the gate with a staff car. Taken to dockside, they stayed out of sight until signaled by the watch to board. As all hands gathered in the crew's mess for a briefing on the fictional trip to the east coast via the Panama Canal, Lyon and Rowray hastened on board, where they were locked in Jenks's tiny cabin.

*Nautilus* was supposed to get underway at 3:00 P.M. An electric motor failure on the propeller shaft lubrication pumps, however, delayed its departure until after midnight. Lyon and Rowray remained in "solitary confinement," with three pieces of bread, some fruit, and empty bottles for relief purposes. The submarine finally left the dock at 12:24 A.M. on June 9. As Lyon and Rowray were freed from confinement, Anderson used the public address system to announce the true destination of *Nautilus*. The reaction of his men, Anderson recalled, "confirmed my long-held opinion that no better crew had ever taken a ship to sea."[30]

It took five days for *Nautilus* to reach St. Lawrence Island, which guarded the approach to the Bering Strait. Anderson planned to use the western, or Siberian, side of the island to reach the Bering Strait. Not only was the water deeper on this side, but it was also believed that the shore ice receded earlier in the year than on the Alaskan side of the island.[31]

Lyon and Rowray were now on four-hour-on/four-hour-off shifts, monitoring the ice-detection equipment. At 10:00 P.M. on June 14, just after *Nautilus* rounded Northwest Cape of St. Lawrence Island, the submarine began to meet sea ice. The ice cover increased rapidly, with one piece of rafted ice extending thirty feet below the surface. "Some worry about the possible thickness of these old ridges," Lyon reported, "since no experience nor observations on origins of this ice."[32]

*Nautilus* was already in shallow waters, and even shallower stretches lay ahead. "An unnerving question arose in my mind," Anderson recalled. "What if we met a piece of Siberian shore-ice just ten feet thicker than the last one?" After conferring with Lyon, Anderson ordered a 180° turn. He had learned that "ice is an enemy to be respected." During the 1957 voyage to the Arctic, Anderson reflected, "we had run into battle like an eager young fighter ready to whip the world and literally got both eyes blackened. This year I was determined to prevent serious injury, no matter what the cost."[33]

*Nautilus* ran south until reaching open water at 7:00 A.M. Anderson and Lyon spent most of the morning discussing their next step. One option would be to abandon the attempt to penetrate the Bering Strait; *Nautilus* could proceed to Pearl Harbor for "repairs" and wait for ice conditions to improve. Lyon, however, believed that the best course would be to attempt a passage around the eastern side of St. Lawrence Island. Past experience suggested that this route should be free of ice past King Island or Cape York in the Bering Strait. But he had one important reservation. "A major factor," he noted in his Journal, "is I don't feel sufficiently familiar with this western sector of St. Lawrence to be certain of maximum thickness of fast ice that might possibly break away; should not expect it over 40 feet, but have never seen the Russian sector. . . ." After considering all the options, Anderson decided to follow Lyon's advice.[34]

*Nautilus* again turned northward at noon. Anderson came up to periscope depth of sixty feet and proceeded up the eastern side of St. Lawrence Island at ten knots. The water was shallow: he had only fifty to sixty feet of water between his keel and the sea bottom. Using his periscope almost constantly, Anderson was relieved to find the area free of ice. The only problem was posed by a number of large logs that *Nautilus* had to dodge.[35]

*Nautilus* passed King Island and entered Bering Strait at midnight on June 16, after a run of thirty-six hours. The visibility was good, and Anderson could make out Cape Prince of Wales, the Diomedes, and the Soviet coast. "Strait looked very average conditions," Lyon noted, "which I know so well." By noon on June 17, *Nautilus* was well out of range of both Soviet and U.S. radar stations. Anderson came to the surface, and in clear, sunny weather proceeded north across the Chukchi Sea. The prospects for successfully completing Operation Sunshine seemed as bright as the weather.[36]

*Nautilus* reached the edge of the ice pack just west of Point Hope at nine o'clock that evening. Anderson dove in the shallow waters and carefully made his way under the ice toward the deep waters of the arctic basin. It was a tight squeeze. The water in the Chukchi Sea was 160 feet deep. *Nautilus* measured 55 feet from keel to top of sail. Anderson proceeded at a keel depth of 120 feet. This placed *Nautilus* 40 feet from the bottom of the sea and allowed 65 feet for the top of the sail to clear downward-projecting ice.

At 10:50 P.M. *Nautilus* passed under a piece of ice that projected 47 feet below the surface. "This was entirely unexpected," Anderson noted in his patrol report, "and did not show up as such a deep piece on sonar before passing overhead." As he eased down to 140 feet, placing *Nautilus* only 20 feet from the bottom of the sea but increasing the distance between sail and surface to 85 feet, Sonarman First Class Alfred A. Charette reported two massive ridges ahead. Anderson ordered dead slow as *Nautilus* approached the first ridge. He watched as the recording pen of the topside echo sounder moved downward, then receded as *Nautilus* cleared the ice by 25 feet. "I breathed easier," Anderson reported.[37]

An even more formidable barrier lay ahead, however. "I stared in disbelief at the picture of it on the sonar," Anderson wrote in his diary of the patrol. As the recording pen swept downward, he "honestly expected" to feel the shudder and jar of steel against ice. The pen neared

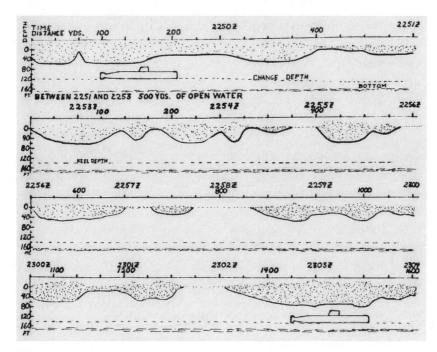

*Nautilus* under massive ice floes, Chukchi Sea, June 17, 1958. Courtesy U.S. Navy

the reference line that corresponded to the top of the sail, then sprang upward. *Nautilus* had cleared an 80-foot pressure ridge by a scant 5 feet![38]

"It took only a second's reflection for me to realize that Operation Sunshine had already totally and irrevocably failed," Anderson recalled. "Not even Nautilus could fight that kind of ice and hope to win." With shallower water and perhaps deeper ice ahead, Anderson believed that he had no choice except to exit the ice pack and seek permission to try again when ice conditions improved.[39]

As *Nautilus* headed south, Lyon recorded his disappointment in his Scientific Journal. "Cannot but admit a deep feeling of defeat and regret in view of decision to turn back," he wrote. "It cannot be denied that it is a reflection on my advice and my lack of successful estimating ice conditions...." Lyon believed that he should have foreseen the possibility of meeting a heavy piece of rafted ice and warned Anderson of the likelihood. If he had, Anderson might have been less shocked when suddenly confronted with the situation. "There is undoubtedly an open question of the decision to immediately return based on this occur-

**OPERATION SUNSHINE** 123

rence without further looking," Lyon judged, "but in view of first experience this area for this command and desire always to take assured success and certainty, is probably best decision to retire. Only hope we get a chance to try again."

If *Nautilus* failed to receive permission to attempt a later transarctic crossing, Lyon feared for the future of the entire arctic submarine program. "At the moment," he concluded, "I face the question of how to discuss this decision [to turn back] which is a complex of my advice to C.O. and my holding back of original advice that deep ice was possible but unlikely, and Andy's tendency to quick retreat when something occurs which he did not expect."[40]

Anderson, of course, had more on his mind than the success of the arctic submarine program. "Our operations order stated," he pointed out, "that the safety of the ship and men is paramount." Lacking QLA forward-scanning sonar to identify downward-projecting ice ridges in his path, Anderson's decision to abort the mission was no doubt the correct one.[41]

A sorely disappointed Anderson was relieved when a message from Admiral Burke was received on June 19. The chief of naval operations, it stated, concurred "entirely" in Anderson's "prudent action" in withdrawing from the ice pack and heading to Pearl Harbor. *Nautilus* could attempt the transarctic passage at a later time. Meanwhile, Burke advised, the security of Operation Sunshine should be maintained.[42]

Lyon could not have been happier with the news. A voyage across the Arctic Ocean during the second phase of Operation Sunshine, he believed, "is expected to be easily made." *Nautilus* would be in the Arctic during the summer months "for which we have maximum information and operating experience." Ice conditions would permit Anderson, if necessary, to follow the north coast of Alaska and enter the deep waters of the arctic basin through the Barrow Sea Valley. This passage, in fact, would duplicate the transit that *Redfish* had made through the area in 1952.[43]

As *Nautilus* headed toward Pearl Harbor, Anderson emphasized the need to maintain secrecy. "I cannot impress upon you too strongly," he wrote to all hands on June 21, "the grave responsibility which rests on each of you individually to carry out this order. Not only is it necessary for each of you to 'forget' entirely everything that has happened or been divulged to you regarding this operation, but you must each also actively participate in maintaining a plausible cover story for what we have been or will be doing. I cannot imagine a situation requiring greater discretion, common sense, alertness and loyalty." The security of the "TOP SECRET — SENSITIVE" operation, which had been shared by only a small group of high-ranking officers and civilians, had now been entrusted to the entire crew of *Nautilus*.[44]

When *Nautilus* reached Pearl Harbor on June 28, Lyon was spirited off the submarine without being seen. He was taken to Barber's Point Naval Air Station, where he boarded a DC-6 and was flown to Washington. In a conference at the Pentagon on June 30, he reviewed the first phase of Operation Sunshine with Combs, Daspit, Walker, and Warder. They told Lyon that they understood the situation and agreed with Anderson's decision to withdraw.

When the discussion turned to plans for *Skate*'s summer cruise to the Arctic, Lyon noted that "a feeling of contest has arisen between NAUTILUS and SKATE." Commander Calvert apparently had managed to persuade Admiral Warder, ComSubLant, that *Skate* should make an east-to-west crossing of the arctic basin via the North Pole. CinCLantFlt, however, opposed the plan because of the operational requirement for nuclear submarines to work with the Fleet, as well as the need to survey the Soviet side of the arctic basin. Lyon supported the CinCLantFlt view. "This seems most important and strongly endorse this eastside study first."[45]

Lyon worked all the next morning with Bayne and Anderson, who had arrived from Pearl Harbor, on a new operational plan for *Nautilus*. After numerous discussions in the afternoon with Daspit and Walker, agreement was reached on the second phase of Operation Sunshine. Lieutenant Jenks, navigator of *Nautilus*, would be sent to Alaska to make a reconnaissance flight of ice conditions. When he reported that the route to Point Barrow was clear, *Nautilus* would head north and enter the arctic basin via the Barrow Sea Valley.

There was a final meeting with Admiral Combs at the end of the long day. After he gave his assent to the plan, Lyon went out to dinner with Anderson and his wife. Everything seemed to be set for the second phase of Operation Sunshine.[46]

Calvert, however remained set on a polar crossing. One of his officers, Lt. David Boyd, had been on board *Nautilus* to learn about the operations of the inertial navigation system that also would be carried by *Skate*. Boyd, who had expected a training cruise to the west coast, got caught up in the first attempt to cross the Arctic Ocean. When he returned to *Skate* in early July, he told Calvert about the true nature of the voyage.

"I am heartsick about the results of your trip," Calvert wrote to Lyon on July 8, "but it redoubles my desire to have SKATE have a go at it." Calvert asked Lyon's help to allow him to make the transarctic crossing in the opposite direction. While Admiral Warder supported the idea, Calvert noted, Admiral Daspit was the "main doubter." Could Lyon call upon or drop Daspit a note indicating that he thought Calvert's plan would succeed? "I have a fresh [nuclear] core," Calvert pointed out, "plenty of time, the equipment, and all I need is the go-ahead. It could be a terrific thing for the Navy and for the country."[47]

Lyon found himself "in the middle between NAUTILUS and SKATE." Not only did he support *Nautilus*, but the decision to hold *Skate* to the original plan for the summer cruise had already been made at the meeting in the Pentagon on June 30. The result, Lyon recalled, was "a strain or coolness" in his relations with Calvert over the next few months.[48]

Lyon, of course, had other things on his mind in early July. He had to maintain his cover story, lest the security of Operation Sunshine be compromised. He advised his superiors at NEL that the first phase of his "special mission" to study Alaskan defense requirements had been completed. Several tasks, however, remained to be done that would require his absence from San Diego. The schedule was not yet firm, but he would be on one-day standby beginning July 12.[49]

Lyon also had to select a replacement for Rowray. It had been decided in Washington to send Rowray with *Skate* so that Calvert would have on board an experienced operator of the NEL under-ice recording equipment during his summer cruise to the Arctic. As a replacement, Lyon picked Archie Walker to accompany him on the "special mission" to review Alaskan defense requirements.

In mid-July, based on ice reports from Jenks, Anderson set the departure date for *Nautilus* from Pearl Harbor as July 21. Lyon and Walker left San Diego for Honolulu via Los Angeles on the 19th. Before they boarded the airplane, Lyon explained the true nature of the mission to Walker and offered him a chance to beg off. "He agreed completely without reservations to continue," as Lyon had expected.[50]

When Lyon and Walker reached Hawaii at 7:30 A.M. on July 20, Anderson and Jenks met them at the airport with some bad news. The

Mark 19 gyrocompass on *Nautilus* required repairs, which meant that the sailing would be delayed. Lyon and Walker checked into the out-of-the-way Alexander Young Hotel and tried to avoid areas where Lyon might be recognized. They took a walk on Waikiki Beach in the afternoon, then went to see the movie *Around the World in 80 Days*. Lyon found the film "a fitting prologue to starting our cruise across the [arctic] basin which now has been underway since June 2." The adventure to date had had "comical aspects equal to Phineas Fogg's problems on his journey."[51]

Lyon received a phone message at 1:00 A.M. on July 22 that a car would pick him and Walker up at five o'clock that morning. The two men reported on board *Nautilus* at five-thirty and were sequestered out of sight of visitors in the chiefs' quarters. Although difficulties with the gyrocompass delayed departure until the evening, Lyon experienced a "much better imprisonment than last June." Not only was he given meals, but he also received "head privileges."[52]

*Nautilus* finally got underway at 8:00 P.M. for a high-speed run to the Bering Strait. To Anderson's delight and pride, the security of Operation Sunshine had remained intact. His crew had displayed a remarkable professionalism, a tribute to both his leadership and their character. "No Commanding Officer," Anderson wrote at the end of the patrol, "has ever been blessed with such a group as that now serving in NAUTILUS."[53]

As the submarine hurried north at twenty-two-and-a-half knots, Lyon and Walker checked out the NEL under-ice equipment. The variable frequency sonar passed inspection, as did four of the five topside echo sounders. Only the number three unit, located on the sail, failed to operate properly. "Just cannot seem to keep this unit in shape," Lyon commented.[54]

On the afternoon of July 27, after one of the most sustained high-speed runs in the career of *Nautilus*, the submarine passed into the shallow waters of the Bering Sea. The next morning, *Nautilus* reached the southern corner of St. Lawrence Island. Anderson proceeded at periscope depth along the western side of the island, which was now free of ice. "Now necessary to be very careful during snorkel, radio, radar, and periscope runs," Lyon noted, "to be sure to avoid contact by land radar stations, ship or air patrols."[55]

*Nautilus* passed through Bering Strait at four o'clock the morning of

July 29 and headed north into the Chukchi Sea, following the 169° line of longitude. Everything was going according to plan. "No troubles," Lyon observed.[56]

Early the next morning, *Nautilus* ran into brash ice and fog. With visibility down to 1,500 yards, Anderson was finding it difficult to see the smaller pieces of ice through his periscope. As *Nautilus* was now beyond radar range, he decided to surface, hoping to make better time. The submarine continued on the surface until it encountered pack ice at 72° North. Anderson then turned east until he reached the 165° line of longitude. He again headed north, hoping to reach a maximum water depth of thirty-five fathoms. If so, he could proceed directly into the arctic basin and avoid the detour through the Barrow Sea Valley. Lyon was optimistic: the ice conditions were similar to those encountered by *Boarfish* in 1949.[57]

By the morning of July 31, Anderson had reached a depth of thirty-one fathoms before again being stopped by ice. "No passage here!" he noted in his patrol report. "From Northwest through North to Southeast the sea is filled with brash and block ice, and what definitely looks like the pack is directly beyond." Anderson decided to turn south and proceed toward Point Barrow. He would follow the original plan, swinging into the coast of Alaska along the 71° line of latitude, then entering the Barrow Sea Valley.[58]

When just off Point Franklin, Anderson managed to get an accurate position fix by a quick sweep of his radar. With ice ahead, he dove *Nautilus* at 4:00 A.M. on August 1. The sky was clear, with the sun beginning to rise and the moon setting. "As Alaska faded slowly into the moonset," Anderson wrote in his patrol report, "NAUTILUS set course for deep water, North Pole, and Atlantic Ocean." He proceeded through the Barrow Sea Valley that NEL oceanographers Alfred J. Carsola and Eugene LaFond had surveyed in 1951, following the northeast axis of the valley before turning north to enter the deep waters of the arctic basin. With a feeling of relief, Anderson increased depth to 600 feet and sped toward the North Pole at eighteen knots.[59]

*Nautilus* continued north on August 2. "Dr. Lyon remained glued to his sonar equipment hour after hour," Anderson observed, "watching his recording pens trace the contour of the underside of the ice." The ice pack, Lyon reported, averaged 5 to 8 feet in thickness with many pressure ridges, some with keels below 100 feet. The sea bottom remained at 2,000 fathoms until above 76° North. At this point, Ander-

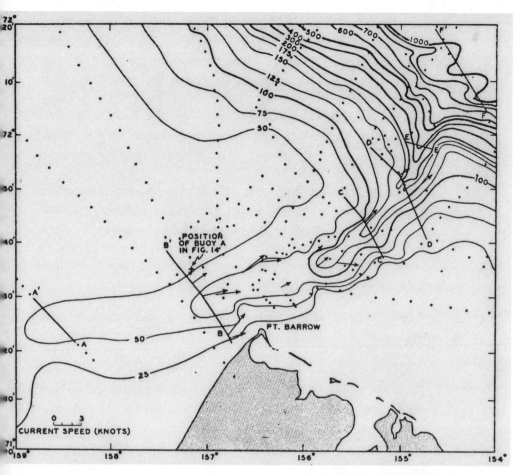

Barrow Sea Valley, bathymetry and current flow. Chart used on board *Nautilus*, 1958. Courtesy U.S. Navy

son recalled, the fathometer "suddenly spiked up to 1,500 fathoms, and then, to my concern, to less than 500." Crossing uncharted waters, Anderson did not know what to expect. "I camped alongside the fathometer for several hours, intently watching the rugged terrain as it unfolded beneath it." The bottom of the arctic basin, however, remained a safe distance below *Nautilus*. By 8:00 P.M. the submarine was 452 miles from the North Pole. "So far," Lyon noted, "ship seems to be running OK."[60]

Anderson drove *Nautilus* toward his goal on August 3, running beneath the ice pack at 600 feet and twenty knots. As *Nautilus* passed over the Lomonosov Ridge, less than a hundred miles from the Pole, Lyon noted that the depth recordings were in agreement with those on the

Soviet bathymetric chart that had been obtained from the Canadians in 1957. Finally, at 7:15 P.M. (11:15 P.M. GMT), *Nautilus* crossed the North Pole. Seconds before it reached the top of the world, Anderson spoke on the ship's public address system to the crew. "In a few moments," he said, "Nautilus will realize a goal long a dream of mankind—attainment by ship of the North Geographic Pole. With continued Godspeed, in less than two days we will record an even more significant first: the completion of a rapid transpolar voyage from the Pacific to the Atlantic Ocean."[61]

At the time of the crossing, Lyon was exhausted, having been closely monitoring the recording equipment for the past forty-eight hours. "Too tired to write," he scrawled in his Journal; "falling asleep over words."[62] Not until five o'clock the morning of August 5, when the topside echo sounders indicated open water and *Nautilus* came to the surface, did Lyon find time to express his thoughts. He wrote:

> "Thus ending the transit from Pt. Barrow, a goal demonstrated and obtained after all these years of work toward an Arctic Submarine to disprove the OpNav statement of position in 1950, i.e. the development of a transarctic submarine remained in the realm of fantasy. Surely this completion will open the door to the development and exploitation of the Arctic Ocean by U.S. Navy submarines. At this point I should feel high elation but feel satisfaction and dead-dog tired; perhaps later some emotional impact will come—now just thankful the gear held out to collect the data which at first look suggests we may get entirely new view of amount and thickness of sea ice across the Arctic Ocean."[63]

While on the surface, Anderson sent a radio message to Admiral Burke reporting the success of Operation Sunshine. The news, Lyon noted, should have come as "a tremendous relief" to Daspit, Bayne, and Mumma. No doubt, the Eisenhower administration also received the word with delight! *Nautilus* then submerged and ran south at 400 feet and twenty-three knots toward the scheduled rendezvous off Iceland. "I shall get some real sleep," Lyon wrote in his Journal.[64]

The rendezvous was accomplished on August 7. Offered a chance to go with Anderson to Washington, Lyon declined. "I hope I have made right move by refusing to accompany C.O. on this trip to Wash. D.C.," he noted. Instead, Lyon stayed with *Nautilus* as it proceeded toward Portland, England.⁶⁵

While *Nautilus* was still at sea, Anderson appeared at the White House on August 8 and received the Legion of Merit from a grateful president. In addition, Eisenhower announced, *Nautilus* had been awarded the first-ever peacetime Presidential Unit Citation. All who participated in the historic cruise would be entitled to wear the appropriate red, yellow, and blue striped ribbon with a special gold clasp in the form of an "N." Speaking to the press, Eisenhower emphasized the peaceful implications of the voyage. "This points the way," he said, "for further exploration and possible use of this route by nuclear powered cargo submarines as a new commercial seaway between the major oceans of the world."⁶⁶

While Eisenhower chose to speculate on the commercial implications of transarctic submarines, other observers drew different conclusions from the voyage. For Hanson W. Baldwin, respected military correspondent of the *New York Times*, *Nautilus*'s successful crossing of the polar region had "immense strategic importance." It meant, he argued, "that utilization of the Arctic Ocean for military purposes is now possible for the first time in history." Admiral Grenfell in Pearl Harbor not only agreed, but he also spoke for many when he termed the voyage "America's answer to Sputnik."⁶⁷

The ceremony in Washington marked the beginning of three weeks of festivities. When *Nautilus* reached England on August 12, it was met by fireboats, tugs, and all manner of small boats. The U.S. ambassador and the first lord of the admiralty were dockside to greet the crew. "Reporters all over the place," Lyon noted.⁶⁸

And England was only the prelude to an even grander welcome when *Nautilus* reached the United States. *Nautilus* surfaced off Ambrose lighthouse at the entrance to New York's harbor at 9:55 P.M. on August 24, having completed a record submarine crossing of the Atlantic in six days, eleven hours, and fifty-five minutes. Admiral Rickover, who had been conspicuously absent during ceremonies in Washington on August 8, came aboard the next morning as President Eisenhower's representative. The submarine then made a parade run through the harbor, proceeding up the Hudson River to Forty-second Street before turning back to enter the Brooklyn Navy Yard. Only a heavy rain marred the festivities.⁶⁹

Two days later, as the sun shone brightly, New York City gave the officers and crew of *Nautilus* a gala welcome. Shortly after noon, bands from the army, navy, and air force, together with the marine corps drum and bugle corps and a color guard from the coast guard, began marching down Broadway from Bowling Green to City Hill. Anderson and Rickover (in full dress whites) rode behind the bands in a cream-colored convertible, while the crew of *Nautilus* followed in twenty army jeeps. Some 250,000 spectators lined Broadway, cheering the motorcade as flurries of ticker tape and confetti rained down.

After a brief greeting at City Hall from Mayor Richard Wagner, the group was whisked off to a stag luncheon at the Waldorf-Astoria Hotel. There, Mayor Wagner presented gold medals to Anderson and Rickover, and the city's scroll of "distinguished and exceptional service" to the crew. "Everyone had a big time," Lyon reported; "New York certainly knows how to organize such affairs."[70]

In his senior scientist's report on the *Nautilus* cruise, Lyon emphasized that the transpolar crossing "culminates many years of experimentation with submarines under ice, and surely demonstrates unequivocally the on hand reality and capability of the transarctic submarine." The exploratory phase had now ended, Lyon wrote, and the door was now open to the "systematic development" of the under-ice submarine and "the consequent control of the Arctic Ocean by the U.S. Navy."

Having demonstrated the feasibility of the arctic submarine, Lyon argued, the next objective was "to obtain all season capability of the Arctic Ocean and its approaches." To reach this goal, a suitable hull configuration must be developed that would allow submarines to surface within the polar ice pack at all times of the year. Also, there was a need for a bathymetric and sonar survey of the Arctic Ocean, as well as the development of under-ice submarine weapons and tactics.

Only when submarines could operate in the Arctic on a year-round basis, Lyon emphasized, would the polar submarine become an operational reality. He was optimistic that this objective could be reached within a short time. Indeed, the recent voyage of *Skate*, during which Calvert had surfaced in a polynya near the North Pole, had demonstrated even more than *Nautilus* the growing capability of the under-ice submarine. Much remained to be done, but the way ahead was now clear.[71]

## 6 Skate

Although the publicity generated by the transarctic voyage of *Nautilus* marked a major turning point in support for the under-ice submarine program, Lyon knew that the patrol had done little to advance the technical development of the project. In a speech in San Diego on September 23, 1958, Lyon noted that *Nautilus* had been ordered to complete a polar transit as expeditiously and secretly as possible so that it would have a maximum impact on the world scene. At the same time, the "real scientific work" in the Arctic was being done by *Skate*.

Newspapers across the country picked up the story of the speech, which they reported under the headline NAUTILUS 'STUNT' SEEN. Lyon, the accounts claimed, had said that the *Nautilus* cruise had been primarily a hunt for publicity. This interpretation of Lyon's remarks did not go well with his superiors. On September 25 Admiral Mumma sent Lyon a copy of the article from the *Baltimore Sun*, together with a sharp note. "I can assure you," Mumma wrote, "that those of us who were in on the early planning of the program, as well as yourself, had ideas that it had more value than merely a stunt."[1]

Lyon was chagrined. The reporter, he replied to Mumma, had taken his remarks out of context. Lyon sent along the complete text of the speech, which he also passed to Admiral Bennett, Commander Anderson, and the navy's Office of Information. "I, too, believe that the Navy, we, had done a top job, both in operational results and in research on the ocean," he assured Mumma. "A reporter said it otherwise for his own reasons."[2]

Lyon came away from this embarrassing incident determined to give fewer speeches. While his remarks had been taken out of context, the story had contained a strong element of truth. As Lyon later acknowledged, "SKATE did make a greater scientific contribution than NAUTILUS." *Skate* not only had surveyed a far more extensive area of the arctic basin, but it also had perfected the surfacing procedure in polynyas that *Carp* had first developed in 1948. Lyon had learned, however, that the time was not right for this kind of candid appraisal.[3]

As Lyon had pointed out, the summer cruise of *Skate* had advanced the under-ice submarine program in a number of important ways. Before the White House had seized upon the notion of sending *Nautilus* on a transpolar voyage to highlight American technology in the wake of *Sputnik*, the navy's plan for under-ice operations in the summer of 1958 had envisioned a wide-ranging survey of the arctic basin by two nuclear submarines, with frequent surfacing in open-water areas within the ice pack. While *Nautilus* was making its transarctic dash and grabbing the headlines, *Skate* was following the original program.

In preparation for *Skate*'s arctic patrol, Commander Calvert had sought information on the ice pack through an aerial reconnaissance mission. On July 15, 1958, a navy Lockheed P2V patrol plane departed Thule Air Force Base, Greenland, to make the requested survey. It carried Capt. Eugene Wilkinson, the long-time arctic enthusiast who now commanded Submarine Division 102; Lt. Comdr. John H. Nicholson, *Skate*'s executive officer and navigator; Lt. Albert L. Kelln, *Skate*'s gunnery officer; and Walter Wittmann, the ice expert from the navy's Hydrographic Office.[4]

The P2V, flying at 2,000 feet, proceeded up Smith Sound, Kane Basin, Kennedy Channel, and Robeson Channel until it cleared the narrow waters that separated northwestern Greenland and Ellesmere Island. The aircraft then turned left and followed the 60° West line of longitude toward the North Pole. As Kelln, from his position in the Plexiglas nose of the P2V, took photographs and movies of the scene below to be shown to *Skate*'s crew, the other observers looked for open areas in the ice pack that measured more than 100 feet and 350 feet and were within 500 yards of the aircraft's track. In all, they counted ninety-eight leads or polynyas in 465 miles. The longest distance between open areas was twenty miles. Obviously, there would be ample areas of open water in which *Skate* could surface.

*Skate*, the lead ship in the first class of nuclear submarines, was much smaller than *Nautilus* at 267 feet 8 inches in length and displacing 2,848 tons submerged (versus 323 feet 8 inches and 4,092 tons for *Nautilus*).

It got underway from New London on July 30. As the twin-screw submarine proceeded toward Spitzbergen at 300 feet and fifteen knots, Calvert encountered problems with the four-transducer NEL topside echo-sounder array. During earlier sea trials, the number two unit, located on the forward deck, became inoperative when the pressure seal of the junction box gave way and the transducer had flooded. Pressure change, Calvert believed, had produced severe chafing of a cable within the pressure seal, causing it to fail. Repairs had been made, and the unit performed normally thereafter. Three days after *Skate* left New London, however, number one, number three, and number four transducers failed. Lacking the necessary spare parts, it proved impossible to repair the units. For ice profiling, *Skate* would have to rely on the remaining NEL echo sounder and an EDO Corporation 255-BM topside fathometer that was mounted in the sail.

Calvert soon had more than equipment problems on his mind. On August 9, as *Skate* neared Spitzbergen, he received word that *Nautilus* had crossed the North Pole seven days earlier. Although Calvert had known about the secret orders for *Nautilus*, the news came as a shock and disappointment to the crew as they believed that *Skate* would be the first submarine to reach the pole. Executive Officer Nicholson recalled that he was "just crushed" when he heard the announcement. Two days later Arthur D. Molloy, a researcher from the Hydrographic Office on *Skate*, recorded in his Daily Log the sentiments of most of the submarine's officers and men: "No one mentions the *Nautilus* anymore as we are all confident that what we are doing will have more long term importance."[5]

At 4:00 A.M. GMT on August 10, *Skate* took a radar fix on the northwest tip of Spitzbergen, then dove under the ice pack at 80° North, 10° East. Calvert set course for the pole, traveling at 300 feet and sixteen-and-a-half knots (50 percent power). Experimenting with the NEL and EDO topside echo sounders, he soon determined that the EDO unit was giving a far superior picture of the ice cover. The unit, which transmitted on a frequency of 37.5 kHz and had a beam width of 12°, did have one problem: it did not work well below 230 feet. As the EDO transducer was located in the sail, 47 feet above *Skate*'s keel, Calvert changed his cruising depth to 265 feet. This was, he noted in his patrol report, "shallower than I would have liked, but necessary to the mission."

At 9:45 A.M., with *Skate* now sixty miles inside the ice pack, the EDO echo sounder showed a 500-yard stretch of open water overhead. Calvert made several passes under the area, then came up to 150 feet. Using his UQS-1 mine-detecting sonar, he scanned the opening to determine its dimensions. Satisfied with the information, he pumped up at a rate of 20 feet per minute. Calvert had raised his retractable whip antenna, the tip of which rose 89 feet and 6 inches above the keel. If the tip of the antenna began to bend, indicating ice overhead, he would have enough time to flood down. This time, however, the antenna remained erect. For the last 25 feet, Calvert retracted the antenna and the periscope and came up to a position of sail awash. He then raised the periscope and found that *Skate* was in the center of an oval polynya, 800 yards x 1,000 yards in area. Calvert then brought the submarine to the surface and cracked the hatch. "With typical beginner's luck," he reported, "the first thing we saw on looking around was a full grown polar bear clambering up on the edge of the polynya." Unaware of how rare such a spotting would be, Calvert paid only slight attention to the animal.

Calvert decided to use the opportunity to carry out his orders to experiment with the explosive effects of torpedoes on the ice. Could torpedoes blow holes in the ice that would permit submarines to surface? He fired a Mk 16 torpedo across the polynya and under the surrounding ice. After two minutes and forty-eight seconds, it detonated, throwing up a gray shower of ice, water, and snow. Calvert then fired a second torpedo, set to explode at 2,000 yards. The result, he observed, "proved to be one of the more spectacular torpedo firings I have witnessed." The torpedo ran on the surface of the polynya until it hit a ten-foot-thick ice floe. It then jumped high in the air and went whirling across the ice, throwing snow in all directions. It stopped after 30 yards. It never exploded.

Calvert tried one more shot, again setting the torpedo to detonate at 2,000 yards. This time, the torpedo ducked under the ice floe and disappeared from sight. At fifty-five seconds after firing, Calvert reported feeling a heavy explosion "uncomfortably close to SKATE." There was no surface evidence of the event. "Decided this was enough torpedo firing for today," Calvert noted, "and made preparations to leave."

August 11 found *Skate* heading toward the North Pole along the 10° East meridian. The ice cover overhead was solid, with pressure ridges that averaged fifty feet in thickness. At 8:47 A.M., when 247 miles from the pole, *Skate* slowed and listened for a scheduled explosion from Ice Station Alpha. It heard nothing, nor would it detect other scheduled explosions during the course of the patrol.

*Skate* crossed 87° North at 1300. Navigator Nicholson changed the Mk 23 gyrocompass from the north-seeking to the directional gyro mode. Its drift in this mode could be monitored from drift test data collected at lower latitudes before the shift. Nicholson also took advantage of the North America N6A inertial guidance system, a unit similar to the one that had been carried on *Nautilus*; it had a stable platform kept properly suspended in space with motion monitored and kept current by accelerometers. This system provided him with a heading that was referenced to a polar grid coordinate system centered at the geographic North Pole.

*Skate* reached the top of the world at 1:47 A.M. on August 12. Calvert turned left and headed down the 90° West longitude line at five knots, searching for a polynya or lead that would permit the submarine to surface. He examined three leads in the morning, but they were all too cluttered with block ice to allow surfacing. At 2:30 P.M., however, a small opening appeared some forty miles from the pole. Calvert brought *Skate* up to a position of sail awash. There was not enough room for the entire submarine to surface, so he held position, with the sail gently touching the edge of the lead. He remained in the lead for nearly nineteen hours, ventilating the ship and sending radio messages.

*Skate*'s patrol orders, which had been written by ComSubRon 10, gave Calvert permission to attempt to surface near Drifting Station Alpha if he had managed to reach the North Pole by August 14. This provision, he recalled, had been written into the orders by McWethy "in case we were feeling really ambitious." Such a task, Calvert recognized, would involve a tremendous navigational challenge. Determining one's own position at high latitudes was difficult enough, but to locate a tiny object whose location was not known with certainty would be a monumental test of polar navigation. Calvert, however, was confident that Nicholson was equal to the challenge.[6]

*Skate* dropped out of the tiny lead at 9:30 A.M. on August 13, and Calvert set course for Alpha at twelve knots. "We are now cutting across longitude lines," he noted, "and navigation will become more demanding." Twenty-four hours later, *Skate* surfaced in a polynya at 85°12' North, 135°20' West. Nicholson was sure that he was close to Alpha, at least if the drifting station's reported position, based on sunlines, was reasonably accurate. *Skate* easily made radio contact with Alpha. Calvert learned that there was a polynya only a hundred feet west of the station. He asked Alpha to drop a detonator cap into the water. He then counted the time for the sound to reach the submarine. Translating time into distance showed that Alpha was only 7.7 miles away.

Calvert arranged for the station to drop detonator caps into the water every ten minutes. In addition, the station advised that it would run an outboard motor boat in the polynya. After obtaining a radio bearing from the station, Calvert set course for Alpha at nine knots. One hour later, he located what appeared to be the target polynya. Coming up to periscope depth, Calvert observed "as remarkable a sight as I've seen through a periscope." *Skate* was in the precise center of an 80-yard x 350-yard polynya. A small outboard motor boat circled the perimeter of the opening while its occupant waved enthusiastically. There were huts, radomes, flagpoles, and antennae just off the starboard beam, with twenty-eight station personnel on the ice, waving and taking photographs.

Calvert brought *Skate* to the surface and welcomed on board the occupant of the motor boat, who turned out to be the station commander, Maj. Joseph P. Bilotta of the U.S. Air Force. After locating a suitable mooring site on the west side of the polynya, Calvert put men ashore to drive iron stakes into the ice. "By 2015," he reported, "we were comfortably moored port side to in our private lake, surrounded by all the signs of arctic civilization (radomes, fuel caches, antennae, etc.)—and all this only 294 miles from the center of the famous Arctic Zone of Inaccessibility. A memorable experience."

Alpha, Calvert knew, had been established on a large ice floe in April, 1957, as part of the American contribution to the International Geophysical Year that began on July 1, 1957. One of two drifting stations that were being maintained by the Department of Defense, Alpha supported a team of scientists conducting geophysical research and doing studies in oceanography and meteorology.[7]

Hydrographer Molloy was part of a large group from *Skate* who toured the station on August 15. He visited scientists from the Lamont Geological Laboratory of Columbia University and discussed their experiments in gravity, seismology, geomagnetism, and oceanography. "They have a nice setup," Molloy noted in his log, "but the camp is pretty messy." To his surprise, it started to rain, an unusual occurrence so close to the North Pole, even in summer. Called back to the submarine when the wind shifted and the polynya began to close, Molloy departed without regret. "The stop was nice," he wrote, "but we all felt we would not like to stay on the island [sic] to collect data. Just a little too primitive."[8]

*Skate* departed at 8:00 P.M., having spent twenty-four hours at the drifting station. Calvert set course to explore the Lomonosov Ridge, a prominent undersea range of mountains that divides the Arctic Ocean into two distinct basins. "Information collected during this cruise," he later wrote, "indicates many seamounts, sea knolls, peaks and ridges exist in the Arctic Ocean. Their shape and position can be determined accurately enough to provide valuable navigational aids for future arctic submarine operations."

The next day, August 16, Calvert had the opportunity to try out the polynya search and surfacing procedure that had evolved during the cruise. At 1:23 P.M. the officer of the deck observed on the EDO echo sounder an open area greater than 200 yards. Obeying standing orders, he immediately turned *Skate* sharply to starboard, then came to port and placed the submarine on a reciprocal heading, a maneuver known as a Williamson turn. At the same time, he slowed the ship to nine knots while rising from 265 to 200 feet and calling out the polynya-plotting party.

When the plotting party assembled in the control room, the ship slowed to three knots and the search began for the opening in the ice. *Skate*'s course appeared as a moving pinpoint of light across a piece of chart paper as the Dead Reckoning Tracer (DRT) took input from the gyrocompass and ship's speed to generate an estimated position. On spotting the polynya, the officer of the deck had marked its position on the chart. The plotting party now watched as the pinpoint of light moved toward the mark. It took only twenty minutes to find the polynya. Calvert then brought *Skate* to 150 feet keel depth with a hovering trim as he examined the dimensions of the polynya with the UQS-1 mine-detecting sonar. He found the opening to be irregular in shape, with maximum dimensions of 300 yards x 100 yards.

Raising his attack periscope, Calvert examined the polynya for floating ice. Seeing none, he ordered the diving officer to commence the ascent. Pumping out 5,000 pounds gave a rate of twenty to thirty feet per minute. Calvert used his periscope to observe the tip of the whip antenna as it approached the surface. When it remained straight

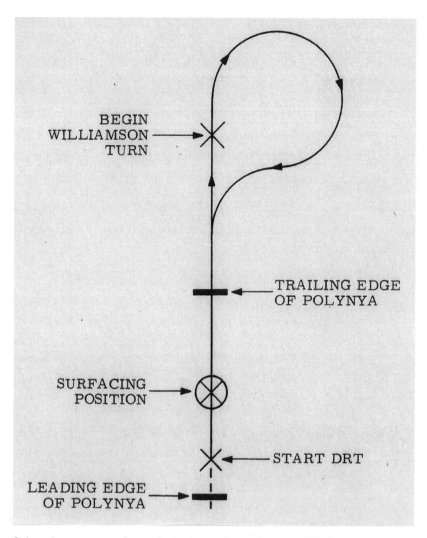

Submarine maneuver for surfacing in a polynya. Courtesy U.S. Navy

at eighty-eight feet, he retracted the antenna and periscope and came up to sail awash at 2:12 P.M. Ten minutes later, after again examining the polynya through his periscope and placing *Skate* in the center of the opening, he brought the submarine fully to the surface. The fifty-nine minutes that it took from first spotting the polynya to surfacing had been about average time for the maneuver.

While on the surface, Calvert attempted to send and receive radio messages. For the first time since he entered the ice pack, he found it nearly impossible to communicate. Only one message managed to get through. Calvert also decided to continue with the torpedo experiment. He fired two torpedoes; both failed to explode. After scientists, led by NEL oceanographer Eugene C. LaFond, collected samples on the ice, Calvert dove at 7:17 P.M. and continued his exploration of the Lomonosov Ridge.[9]

*Skate* surfaced twice on August 17 as it crisscrossed the Lomonosov Ridge en route back to the North Pole. The submarine crossed the North Pole for the second time at 10:58 P.M. Calvert again hoped to surface at the pole, but he could not locate a suitable opening. At 12:57 A.M. on August 18, he crossed the pole for a third time before heading toward Greenland.

Shortly after 6:00 P.M. on August 18, Calvert surfaced in a 120-yard x 300-yard polynya that was located 260 miles from the pole. It took only a record twenty-three minutes from first observation to surfacing. "This is about a minimum in time," Calvert reported, "and would not have been possible had the Williamson [turn] not happened to bring us back in parallel to the major axis [of the polynya]."

Calvert launched the sixth of his icebreaker torpedoes at 7:12 P.M. Fired with zero gyro, depth fifty feet, and set to explode at 1,500 yards, it ran "hot, straight, and normal." The torpedo detonated on time, and Calvert observed the shower of water, ice, and snow rise in the air. When he sent a party to examine the hole, however, they could not find it among the numerous small leads and cracks in the area. As Lyon later observed upon reading the report on the torpedo experiment, the inability to locate the hole created by the explosion was "the basic trouble with this idea—the broken ice falls back into the hole so it 'looks like' the ice sheet before the explosion."[10]

*Skate* submerged at 1:49 A.M. on August 19. Calvert swung the submarine to the southeast in order to cross the top of Greenland en route to the Greenland Sea. The bathymetric record continued over previously uncharted waters. "We have doubtless discovered enough peaks, ridges and canyons in past nine days," Calvert wrote, "to name one after every man in the ship."

As *Skate* neared the edge of the ice pack, Calvert surfaced for the ninth—and last—time in a long, narrow lead. It proved a fitting climax to the richly productive cruise. High mackerel clouds nearly covered the sky, but some patches of blue could be seen, and the sun came in and out. Calvert put the scientists on the ice to take their samples while he "sat back to enjoy the arctic landscape which was breathtaking with this sunlight—the first we have seen."

After three hours on the surface, Calvert took *Skate* down. He cleared the ice pack at 6:30 P.M. on August 20, having steamed 2,405 miles under the ice in ten days and fourteen hours. After surfacing in calm seas to make radio transmissions, he dove again and headed toward Bergen, Norway, at eighteen knots.

Calvert had every right to be pleased with the results of his patrol. Despite the failure of the ice-breaking torpedoes and the inability to detect the sound of explosive charges that were at the heart of a proposed long-range sonar navigation system (known as RAFOS), *Skate* had demonstrated that submarines could operate freely throughout the Arctic, at least in summer. As Calvert pointed out in the conclusions of his patrol report, the feasibility of under-ice operations ultimately depended upon the reliability of the submarine's propulsion plant, atmosphere control equipment, and navigation. *Skate* had experienced no difficulties with its nuclear power plant or its atmospheric control equipment. The most significant development, he pointed out, had been in navigation. The North American N6A inertial navigation system, in combination with the Mk 23 gyrocompass, allowed *Skate* to operate with confidence in its position throughout the polar area. Although Calvert believed that the N6A needed to be modified for extended service and made easier to read, there was no question in his mind that the navigational problems of the under-ice submarine had been solved.

In addition, *Skate* had clearly demonstrated that it was feasible to surface at frequent intervals without fear of damage such as *Nautilus* had suffered. "The ability to search for open stretches of water anywhere in the Arctic Ocean at 16 knots and surface in them in less than an hour after detection," Calvert stressed, "is undeniably a military development of significance. In effect, the surface of a vast new ocean had been opened to the Navy."

As Calvert recognized, however, whether or not surfacing could be accomplished in seasons other than summer "remains to be seen." *Skate*'s sail had to be strengthened for breaking through ice; then, he believed, it might be possible to penetrate ice that was ten to twenty-four inches thick. "Operations to investigate this in-ice surfacing problem in seasons other than summer," he urged, "should be conducted."

He also pointed out that the Davis Strait–Baffin Bay–Canadian Archipelago waterways offered a valuable deep-water entrance to the Arctic Ocean for nuclear submarines and should be explored "as soon as practicable." As Lyon, who was thinking along the same lines would soon learn, Calvert and *Skate* were ready, willing, and eager to continue with the arctic under-ice experiments.

In early October, 1958, Lyon traveled to Washington to discuss the next stage of under-ice research with officials in BuShips and CNO. There was no question in his mind that the most urgent task was to measure sea ice profiles during seasonal periods other than summer. This data was needed to guide the design of hull configuration for submarines that would be able to surface in sea ice on a year-round basis. Although officials in BuShips seemed reluctant to support NEL's work to develop sonar equipment that would provide more precise ice profiles, he found the general atmosphere in Washington to be a far cry from the situation prior to the *Nautilus* cruise. There was, he noted, "a sudden rise of arctic research to a super glamour state with many offices and agencies rushing to get in, or, should we say, to hold out their hands for alms for the love of the Arctic." This was certainly an encouraging development, even if likely to result in some unproductive research.[11]

After receiving his second Distinguished Civilian Service Medal, presented at the Pentagon by Secretary of the Navy Thomas S. Gates, Jr., Lyon went on to New London for a conference at SubRon 10 on a proposal, promoted by Calvert, to send *Skate* back to the Arctic in spring, 1959. NEL, he agreed, would provide *Skate* with an improved model of the NK–Variable Frequency topside echo-sounder equipment that had been used on *Nautilus*. The sonar would be connected by a servo mechanism to the Ship's Depth Detector. Previously, it had been necessary to read the depth of the submarine on one instrument and the ice profile data on another, then mentally subtract one from the other to obtain ice thickness. Now the information would be presented on one instrument with a resolution of plus or minus two inches in thickness of ice at a cruising depth of 400 feet.[12]

By the end of 1958, plans for the *Skate* cruise had matured into a directive from CNO to the commander-in-chief of the Atlantic Fleet. A submarine, CNO wrote on December 30, was to be provided for arctic operations in the late winter or early spring of 1959. The patrol would investigate the problems of under-ice operations in winter conditions and obtain ice data for comparison with information that had been collected during the summer cruises of *Nautilus* and *Skate*. In addition, the patrol would look into "the possibility of surfacing in natural openings in the ice pack during winter."[13]

While an aerial reconnaissance in February revealed the presence of open water in the ice pack, Lyon believed that a submarine should have the capability of surfacing through thinly covered areas of ice. In preparation for the *Skate* cruise, he put together a plan to enable the submarine to break through several inches of thin ice. *Skate's* sail had been hardened with HY-80 steel. Looking at the design, he estimated that the maximum safe working pressure that could be applied against the ice by the sail was sixty pounds per square inch (p.s.i.), with the pressure applied over the area on top of the sail, which measured 6' x 20'. To produce a pressure of 60 p.s.i., the submarine would have to develop a total force of 518 tons by change of momentum on impact against the ice while rising at a controlled rate. As *Skate* had about 3,100 tons of mass, the desired rate of rise had to be small and had to be carefully controlled.

The main problem with these calculations came in estimating the strength of sea ice. Lyon knew from experiments done off Cape Prince of Wales that this figure could vary widely, depending upon salinity, temperature, brine cell/crystal structure, and the presence of lateral compression within the ice sheet. Lyon assumed average values at a surface temperature of minus-20° to minus-30° Fahrenheit. With a rise rate of twenty to twenty-five feet per minute, *Skate* should be able to break through ice fifteen to eighteen inches thick.

While this all sounded nicely scientific, Lyon later confessed that it was "pretty much of a guess." As he explained, "We had to make a number of assumptions, and then much of it was just, I must admit, intuitive; it just seemed right."[14]

In presenting his proposal at BuShips, Lyon kept the intuitive nature of his figures to himself. Although somewhat hesitant to sanction a deliberate *collision* between a submarine and sea ice, BuShips reluctantly gave its approval to the plan. When SubRon 10 issued the operations order for *Skate* on February 18, it authorized Calvert—an enthusiastic supporter of the plan—to surface through "thin, newly frozen ice if this appears feasible."[15]

Lyon had no sooner reported on board *Skate* at New London on March 1, 1959, when a message arrived from BuShips withdrawing permission to surface through the ice. Officials in the bureau obviously had had second thoughts about the procedure and now believed that the planned operation would not be safe. "When you looked at the message," Calvert later wrote to Lyon, "and said, 'It's a cover-your-ass message,' you gave me the courage to see Admiral Warder and ask to go as planned." Fortunately, Calvert noted, ComSubLant "was not a CYA type and gave us permission."[16]

*Skate* got underway from New London at 9:00 A.M. on March 3, 1959. As the submarine headed north at 400 feet and sixteen knots, Lyon and his assistant, W. E. Schatzberg, checked out the NEL sonar equipment. Everything appeared to be working properly. On March 14 Calvert took a final navigational fix on the northern tip of Prins Karls Forland, just off the west coast of Spitzbergen, then proceeded under the ice pack, following the 5° East meridian.[17]

By 8:00 A.M. GMT on March 15, *Skate* was 120 miles inside the ice pack at 81°36′ North. When the trace of the NK–Variable Frequency upward-beamed echo sounder showed a long, flat spot in the ice coverage, Calvert brought the submarine up to investigate. While hovering under the ice, Calvert let out a 160-foot floating wire antenna, which came to rest on the underside of the ice. Lieutenant Kelln had heard about the NEL-designed antenna, obtained one for *Skate*, and managed to connect it without a new hull penetration. Calvert, who had been unable to receive radio transmissions while under the ice in *Skate*'s previous patrol, now heard the scheduled 9:00 A.M. GMT VLF radio broadcast "loud and clear." This gave the first hope that the under-ice communications problem might be on the way to solution.[18]

While listening to the radio transmission, Calvert observed another level area on the UQS-1 mine-detecting sonar. Investigating by periscope, he saw light coming through the ice, which suggested a thinly frozen cover. He decided to attempt to surface. Calvert pumped up slowly, rising at a rate of five to seven feet per minute. The sail touched the bottom of the ice but did not break through. Undaunted, he came up again, this time at a rate of twelve feet per minute.[19]

Lt. Comdr. William H. Layman, who had replaced Nicholson as executive officer, watched on the underwater television camera as the top of the sail approached the ice. "Heavy damage would mean turning back," he recalled, "and some experts had warned us that we should not try to break through ice because of possible damage." Although Lyon's calculations had shown how hard *Skate* could hit the ice without damage, the thought went through Layman's mind: "One misplaced decimal in those calculations, and we've had it."[20]

Layman remained apprehensive until he saw the ice suddenly break apart. Not even a tremble went through the submarine. *Skate*, he remarked, had become "the first nuclear-powered ice pick." Calvert held the ship at forty-two feet keel depth, with only the sail poking through the ice, as he raised number two periscope, turned on the hot air jet to clear the glass, and looked around. Although somewhat difficult to see, he was able to determine that *Skate* was in the middle of a good-sized frozen-over polynya. Calvert blew the main ballast tanks and brought *Skate* slowly to the surface. Opening the hatch, he had to clear ice from the bridge in order to get out. For the first time in history, a submarine had broken through a sheet of sea ice (about six inches thick)—and without suffering any damage.

Despite seeing his calculations vindicated, Lyon had misgivings about his status as senior scientist on *Skate* as the submarine continued northward. "I do not quite understand this ship and the exact position of scientist on board," Lyon wrote in his Scientific Journal. Calvert insisted on relying upon the EDO 255-BM topside echo sounder to profile the overhead ice. He continued to be an enthusiastic supporter of the equipment that had worked so well on *Skate* during the summer, 1958, cruise. Also, Walter Wittmann of the Hydrographic Office, who had been on board in 1958 and was again present, recommended its use. Lyon, the senior scientist, was convinced that NEL's experimental NK–Variable Frequency sonar provided far more accurate data. The difficulty, Lyon suspected, stemmed in part from his association with *Nautilus*. Calvert and his crew had been deeply disappointed when *Nautilus* had become the first submarine to reach the North Pole. The *Skate* summer cruise in 1958 had been a bonding experience, and Lyon was an outsider.[21]

Lyon usually developed a close relationship with a submarine's crew, but this was not happening on *Skate*. "Lyon was withdrawn," Kelln later reflected, "and never jumped into our discussions on how to do better as we summarized our experiments. He just didn't communicate well and we had little time to pull things out of him. He was not exactly open." The crew's relationship with Wittmann, on the other hand, was far more cordial. The wardroom, Kelln noted, "was very comfortable that Walt Wittmann was aboard. He could communicate better than Lyon and seemed to be more helpful in responding to our questions."[22]

Lyon, with quiet insistence, finally persuaded Calvert to run the rest of the night of March 15–16 with the NK–Variable Frequency sonar. "Good profiles obtained," Lyon reported, "which Jim Calvert watched and believe may have convinced him of superiority. There is no comparison."[23]

This incident marked a turning point in the relationship between the senior scientist and the submarine commander, if not with all his officers. As time passed, the two men gained confidence in—and respect for—one another. Lyon would later recall Calvert as "an outstanding commander." Calvert, for his part, remembered Lyon as "a pillar of confidence and a never-failing source of good information. Your name will always have a special place in the submarine forces of our Navy."[24]

As *Skate* continued along the 5° meridian, Lyon observed that the overhead pattern of ice was similar to one he had encountered on *Nautilus* in 1958, although the ridges had a deeper draft. At 2:49 P.M. the officer of the deck spotted a large, thin, flat area and made a Williamson turn. When the polynya-plotting party located the suspected opening, Calvert brought *Skate* to a hover at 200 feet underneath it. Lyon, interpreting the NK–Variable Frequency recorder trace, estimated the ice thickness at nine inches. The ambient light level through what Lyon had labeled a "skylight" confirmed that there was thin ice overhead, as did the picture on the TV screen. When Calvert attempted to penetrate the ice cover, he lost buoyancy control as the submarine neared the surface. *Skate* hit the ice at 10 feet per minute and bounced back to below 100 feet before Calvert regained control. On a second attempt, *Skate* came up at 20 feet per minute. This time, the sail shattered the ice but the submarine started to drop out of the opening. Calvert gently blew the main ballast tank to hold *Skate* in the opening. The ice was found to be seven inches thick, with a surface temperature of minus-20° Fahrenheit.[25]

"Have heard no further questions about using EDO echo sounders for ice profiles," Lyon noted in his Journal as *Skate* approached the North Pole at sixteen knots on March 17. Reaching the top of the world at 10:54 A.M., Calvert brought *Skate* to a hover at 200 feet and looked for a place to surface. Unable to locate a skylight, he began a crisscross search pattern under the pole. Several hours passed without result. Finally, at 2:00 P.M., Calvert noted a possible opening. "At first it was just a faint glimmer of emerald green," he recalled, "visible only through the periscope." Although it looked marginal, he decided to investigate.[26]

*Skate* drifted up to 100 feet. The skylight, Calvert noted, was "dog-legged in shape and treacherously small." Nonetheless, he decided to surface. As *Skate* neared the underside of the ice, Lieutenant Kelln, standing by the ice-detecting sonar, cried out: "Heavy ice overhead—better than twelve feet!" Calvert immediately took *Skate* down.

After another attempt to come up through the small area of thin ice with the same result, Calvert decided on a different tactic. With the skylight's drifting away at the last moment as *Skate* rose to the surface, he factored in an offset that should bring the submarine directly under the thin area as it drifted overhead. This time *Skate*'s sail came within twenty-five feet of the ice; the expected opening, however, failed to appear. "Flood her down—emergency!" Calvert ordered.

At this point a less determined skipper might have given up. But Calvert, who had missed the opportunity to become the first submarine commander to pass underneath the North Pole, was not going to miss the opportunity to become the first one to surface at the pole.

Once again, he tried to locate the small opening, this time allowing for less drift. As *Skate* neared the surface, Kelln shouted: "Thin ice!" Calvert continued his ascent until the top of *Skate*'s sail penetrated the ice. As he held position in the opening, Calvert thought: "If we could break through, we would make history."

"Standby to surface at the Pole," Calvert announced over the ship's public address system. He then blew tanks and brought the entire sail up through the ice. Opening the hatch, Calvert was hit by a blast of icy wind. "It howled and swirled across the open bridge," he reported, "carrying stinging snow particles which cut like flying sand. Heavy gray clouds hung in the sky; the impression was of a dark and stormy twilight about to fade into night." *Skate* was located in a narrow lead with hummocks of ice, some eighteen feet high, on either side. Calvert ordered the tanks blown with high pressure, and *Skate* broke through to the surface.[27]

Calvert had a special mission to perform at the pole. In October, 1958, Sir Hubert Wilkins had spent a day with Calvert on *Skate*. Wilkins had discussed his abortive attempt to reach the North Pole by submarine in 1931, and Calvert had shared with Wilkins his desire to conduct a winter cruise to the Arctic in 1959. Wilkins subsequently sent Calvert a note, thanking him for his courtesy and urging him to push forward for the winter expedition. "Do not be discouraged by apathy and resistance," Wilkins wrote. Less than two months later, on December 1, 1958, Wilkins died of a heart attack.

Prior to *Skate*'s departure in 1959, Admiral Warder told Calvert that he had had a call from friends of Wilkins in New York. Following his October visit to *Skate*, Wilkins had told his wife that perhaps one day his ashes might be taken to the North Pole by a submarine for distribution over the ice. Calvert told Warder that he would be honored to make the effort to carry out Wilkins's wishes. "His heart still belonged to *Nautilus*," Calvert recognized. "It had been his supreme effort of daring imagination."[28]

Lyon, who had talked to Wilkins shortly before the explorer's death, watched from the deck of *Skate* as Calvert made preparations for the service. A small table was fashioned from boxes on the ice and covered with a green baize cloth. Wilkins's ashes were then placed on the table. Thirty crew members left *Skate*, which was flying the flags of Great Britain, Australia, and the United States from masts and periscopes, and assembled on either side of the table. While two men held red flares, Calvert read from the *Book of Common Prayer*, then paid a personal tribute to Wilkins. The shimmering light from the flares that glared through the blowing snow, Calvert recalled, gave the scene an "unearthly atmosphere." The surroundings, Commander Layman agreed, were "dramatic beyond belief."

After Calvert finished his remarks, Lieutenant Boyd picked up the urn, then followed Calvert and the two torchbearers for a distance of some thirty yards from *Skate*. As a rifle squad on the submarine's bow fired three volleys, Boyd sprinkled Wilkins's ashes into the wind. After leaving a flag and marker on the ice, *Skate* slipped below the surface at 8:00 P.M.[29]

For the next three days, *Skate* followed the underwater Lomonosov Ridge toward the New Siberian Islands. At first, the overhead ice was studded with massive ridges. By March 19, however, the pattern changed to large areas of heavy flat ice, with occasional pressure ridges that extended twenty to thirty feet below the surface. Shortly after midnight on the 19th, the NK–Variable Frequency recorder showed an area of ice that was three feet or less. Calvert decided to surface. *Skate*, however, was stopped cold at a keel depth of fifty-two feet. "It was apparently three feet," Calvert reported, "not 'under.'" He flooded down and examined the ice overhead. "We had scarcely left a mark for our effort," he noted, "although the faint outline of the sail was visible."[30]

At 7:09 A.M. *Skate* passed under a large skylight. "It was no trouble to find," Calvert observed, "but the thickness was almost anyone's guess." After some discussion, there was general agreement that the ice likely measured four feet thick. Calvert brought *Skate* up at twenty-five feet per minute. Again, the sail was stopped at fifty-two feet. "Decided to blow main ballast tanks to see if steady pressure would do what momentum failed to do," Calvert wrote in his patrol report. This technique, unfortunately, failed to produce any results.[31]

At 10:40 P.M., with the NK–Variable Frequency echo sounder indicating "three feet or less," Calvert again attempted to surface. *Skate* came up and hit the ice at thirty feet per minute. To Calvert's dismay, however, the submarine made contact with the ice while rising with a 4° up angle. "Not only did we not break through," he noted, "but heard a crunching sound from the sail which we had not heard before."[32]

Between midnight and 8:00 A.M. on March 20, an increasingly concerned Calvert twice more failed to break through the ice and bring *Skate* to the surface. He was reaching the inescapable conclusion that the NK–Variable Frequency echo sounder "cannot tell ice thickness accurately enough for our purpose."[33]

Lyon agreed. The problem, he wrote in his Scientific Journal, lay with the limitations of the ice-detecting sonar. The UQS-1 mine-detecting equipment could only distinguish thickness of ice when sharp contrasts were present. The EDO echo sounder was fine with open water but useless with ice. The NK–Variable Frequency sonar, he pointed out, was designed to be used with open water as a reference. In the absence of open water, depth had to be measured by water pressure. The servo mechanism that connected the Ship's Depth Detector to the sonar was unable to make rapid corrective changes during ascents. As a result, it was difficult to determine with precision ice thickness on surfacing attempts.[34]

Following the failure of his fifth attempt to break through the ice, Calvert called a conference with Lyon, Wittmann, and the ship's officers. "For the first time on either of the *Skate*'s expeditions," Calvert recalled, "I seriously considered turning back." Lyon and Wittmann told Calvert that the chances of finding a suitable area in which to surface should improve at any hour. At 10:53 A.M., as the conference was ending, *Skate* heeled sharply to starboard. The officer of the deck had spotted a flat area and initiated a Williamson turn. This time, *Skate* broke through ice four to six inches thick and surfaced in a long, narrow lead.

"The scene was different from anything we had experienced before," Calvert observed. The sky was clear, with brilliant stars overhead and a three-quarter moon. Small hummocks of ice lined either side of the narrow lead. "The fantastic moonlit landscape and the cloudless star-filled sky," Calvert remarked, "combined to create a scene of fragile beauty." An inspection of the sail showed that the recent collisions with the ice had crushed down the forward lip of the bridge about four inches and had left several four-inch dents in the lower end of the sail. Otherwise, the submarine was undamaged.

Lieutenant Boyd decided to take the opportunity to try out his scuba diving equipment. With Calvert's permission, he and three hearty associates broke a hole in the ice and entered the water. Lyon gave Boyd three water sample bottles and asked him to uncork them at ten, twenty, and thirty feet. With a surface temperature at minus-23°, the water seemed warm to Boyd. After a brief dive under the ice, the swimmers returned without incident—to Calvert's relief.[35]

On March 21 *Skate* reached DeLong Island. Calvert then turned toward Severnya Zemlyna to continue his survey of the Soviet sector of the arctic basin, taking care to remain in international waters. The floating wire antenna again gave excellent reception, and Calvert received a message from CNO congratulating him for surfacing at the North Pole. The communiqué, Lyon noted, also contained a warning "that the sail was not built to be an icebreaker, giving various cautions against what can happen by damage."[36]

CNO's admonition failed to dampen Calvert's enthusiasm for using his "nuclear-powered ice pick" to poke holes in the ice. Twice on March 22, *Skate* broke through to the surface, penetrating ice that was one and nine inches thick. That evening, however, saw an end to *Skate*'s hitherto problem-free cruise.[37]

The previous day Boyd had informed Calvert that a serious leak had developed in the engine room. The leak was occurring at the seal where the drive shaft entered the starboard circulation pump. To repair the leak, *Skate* would have to surface and shut down the condenser, which meant that the submarine would lose half its turbines. A tangle of pipes and wires would then have to be cleared out of the way before the 960-pound circulation pump motor could be lifted by chain hoists. Usually, a repair of this magnitude would be done in the shipyard.

Boyd had placed a canvas cover around the pump to keep the water from spraying while Calvert had searched for a suitable area in which to surface. During the evening, however, as the submarine changed depths, the pressure difference had caused the seal to reseat, and the leak had stopped. When *Skate* surfaced on March 22, Calvert had called his officers together to discuss the problem. In the end, he had decided to continue. "We'll take a chance on it," he told his officers. "No repair, we'll go on. I believe it'll hold."

Calvert soon had cause to regret the decision. At 7:00 P.M. on March 22, as *Skate* cruised below a heavy ice cover, the seal broke loose. "Much worse this time," Boyd reported, "spraying water all over the place." Now there was no option—repairs would have to be made as soon as possible. It took over three hours to find thin ice. Finally, at 10:13 P.M., *Skate* surfaced in a long, straight lead that was fifty yards wide with hummocks on either side. "The sky was gray and depressing," Calvert observed; "the temperature was 31 below, the coldest we had yet experienced." Boyd estimated that the repairs would take about twelve hours. Calvert told him to go ahead.

About an hour after the repairs began, Lieutenant Kelln called Calvert to the bridge. "Look over to port," Kelln advised, "in the ice fields beyond the hummocks." Using binoculars, Calvert focused on a disconcerting vista. Huge ice floes were rising up on end "like giant green billboards," then slipping back into the surrounding ice field. And they were moving toward *Skate!* Suddenly, Calvert heard a loud noise close by—"like the sharp report of a rifle." Looking toward the bow, he noted a new crack in the ice. As the pressure increased, the ice in the lead began creeping up the sides of *Skate,* creating a screeching noise "like banshees" as it scraped against the metal hull.

Called down to the engine room by Executive Officer Layman, Calvert found that the noise of the ice against the hull was much louder. Even more disturbing, however, was the vibration. "The ship fluttered and shook as the ice pressed around her," Calvert recalled. "She seemed to be protesting the agony she felt." When *Skate* abruptly listed to starboard, Calvert determined that he would have to dive. He ordered Boyd to secure the repairs. Moments later, the noise of the ice against the hull abated. Climbing to the bridge, Calvert saw that the ice movement had stopped. It soon started again, but lasted only briefly. As the situation seemed reasonably stable, Calvert concluded that the repairs could go forward. Boyd completed the job in a scant seven hours.[38]

Lyon had been largely nonplused by the situation that had so concerned Calvert. "The ice did move and ice did scrape around the hull," he recalled, "and did make lots of noise, and this was disconcerting to everyone on board." But Lyon had gone through much worse when *Sennet* had been towed through ice in 1947. And he knew that there was no immediate threat to the submarine. As the pressure ridges moved at a predictable rate, it was possible to judge the danger by observing how far away they were and how fast they were moving. On the other hand, Calvert and his crew, never before having experienced the situation, had good reason to be less sanguine about the safety of their ship. The thought that had gone through Calvert's mind was a recollection of the stone cross at Annapolis that honored George Washington DeLong and the crew of *Jeanette*, a ship that had been crushed by the ice in the same area of the Arctic in 1881. He had no desire to cause the navy to erect a similar memorial to James Calvert and the men of *Skate!*[39]

Following repairs to the circulation pump seal, *Skate* continued its cruise through the Franz Josef Archipelago, heading toward Spitzbergen. On March 24 Calvert broke through ice six inches thick and surfaced in a large polynya. The scuba divers again went beneath the ice. Also, *Skate* submerged and resurfaced so a photographic record could be made of the ship's coming up through the ice. Later in the day, Calvert attempted to surface through ice that measured nineteen to twenty-two inches. *Skate*'s sail penetrated the thick cover, but the ship did not come to the surface after the ice damaged the UQS-1 transducer, which was housed in a protective dome on the deck. "Observation of this surfacing," Lyon reported, "left the impression that it was near the maximum breakthrough that we could attain with certainty."[40]

Although the loss of the UQS-1 mine-detecting sonar made difficult the search for thin areas of ice and execution of a Williamson turn, *Skate* managed to surface twice more on March 24. This brought to ten the number of surfacings during the patrol in sixteen attempts.[41]

*Skate* passed along the west side of Spitzbergen on March 26. The submarine cleared the ice pack shortly after midnight. Lyon secured the NK–Variable Frequency echo sounder as *Skate* surfaced, took a radar fix, and made a radio transmission. Calvert then submerged and headed toward New London. Lyon took the opportunity to catch up on his sleep. "Try to shift my sleep habits back to 6 hours in one spell per day," he noted, "instead of 1 ½ hours every 4 hours."[42]

*Skate* reached home port on April 7. At a ceremony held at the Electric Boat Company dock, Calvert received a second Legion of Merit, and *Skate* was awarded a Navy Unit Citation. "We had been successful beyond our fondest hopes," Calvert wrote of the patrol. "We had been pioneers in what we knew would be an ocean of destiny."[43]

Lyon also expressed his delight with the results of the cruise. A great deal had been learned about late winter/early spring operations in the Arctic. Despite the air reconnaissance reports that open-water areas were present in the ice pack, *Skate* had not encountered any during 3,090 miles under the ice. "The probability of finding open water during winter," Lyon wrote in his senior scientist's report, "is negligible." The ability of *Skate* to surface through the ice cover, therefore, had been critical to year-round operations in the far north.

Lyon's calculations of the force required to break through the ice had been proved correct. But as he pointed out, "Accurate determination of ice thickness is the most critical factor in successful surfacing." Although the cruise had demonstrated that current sonar equipment fell short of providing the needed precise information on ice thickness, Lyon now had a good idea of what was required. He planned to reconstruct the NK–Variable Frequency sonar in conjunction with the Ship's Depth Recorder and other topside echo sounders to provide a system that would measure ice thickness with accuracy and guide ascents through frozen-over polynyas and leads.

Several important steps remained in the development of an operational under-ice submarine. Lyon recommended that the next arctic patrol take place during the period of maximum darkness. This would provide an excellent test of the ability of a submarine to surface through ice cover by utilizing exclusively sonar information.

The next winter patrol also would utilize the new scanning sonar, or iceberg detector, which was being developed at NEL. This equipment was critical for detecting ice that projected down to the cruising depth of submarines. "The feasibility of all year submarine transits from Pacific to Atlantic Oceans," Lyon emphasized, "depends on knowledge of the distribution and movement of massive ice floes across the shallow Chukchi–Bering Sea shelf. . . ."

A second patrol, Lyon recommended, should take place in 1960 to complete the developmental phase of the under-ice operations. A submarine, equipped with ice-detecting sonar, should transit the Canadian Archipelago from Atlantic to Pacific. This not only would test the use of ice-covered channels but also would alert the Royal Canadian Navy to the potential danger from Soviet under-ice submarines.

With the completion of this basic under-ice experience in all seasons, Lyon pointed out, NEL would be in a position to specify the equipment for "a proper sonar suit" for arctic submarines and not continue with "a bunch of miscellaneous equipment."

"Our objective," Lyon emphasized, "is military control of the Arctic Ocean and its approaches and is to be obtained through capability to operate submarines in any part of the Arctic Ocean during all seasons of the year." The attainment of this goal was close at hand. Should an emergency arise that required a submarine patrol force in the Arctic, NEL would soon be in a position to specify the necessary equipment and procedures for the successful operation of ComSubArc.[44]

# 7

## Sargo

The widely hailed accomplishments of *Nautilus* and *Skate* gave a visibility and priority to under-ice submarine experiments that Lyon had sought for more than a decade. No longer did he have to plead for limited submarine time for arctic operations, hoping that the submarine force commander in the Pacific or Atlantic might be sympathetic toward his under-ice program. Arctic patrols, once scorned as impractical, dangerous, and useless, now brought decorations, honors, and publicity. It was, Lyon later recognized, "the golden period."[1]

On March 19, 1959, even before *Skate* reached port at the end of her successful winter/spring cruise, CNO authorized BuShips to prepare *Sargo* (SSN 583), sister-ship of *Skate*, for periodic scientific explorations of the arctic basin. Twelve days later, CNO proposed that *Sargo*, assigned to the Pacific Fleet, conduct a winter cruise to the arctic basin via the Bering and Chukchi Seas. As Lyon was well aware, such a patrol would be the most challenging operation of the entire under-ice program. It meant transiting some 1,000 miles of shallow water, about 150 feet deep, at a time of year when it would be covered by massive ice ridges that extended downward for 90 feet or more. And Lyon also knew that to navigate safely through this icy complex, *Sargo* would have to rely on the new iceberg-detecting sonar that was under development at NEL.[2]

The iceberg-detecting sonar was a long-delayed outgrowth of NEL's wartime QLA mine-detecting sonar. As noted previously, Lyon had used the continuous tone, frequency modulated scanning sonar in his early under-ice experiments with diesel-electric submarines. Following the *Boarfish* cruise in 1947, Arthur H. Roshon of NEL's sonar division had proposed modifying the QLA equipment to give it higher resolution for under-ice work. "Wow!" Roshon recalled. "Did we ever get negative reactions." He made several trips to Washington in an effort to persuade BuShips to support the modification program, but to no avail. The bureaucrats, he explained, simply could not understand "what a submarine was doing under the Arctic ice canopy, or why anyone in the Navy would want them to be there." As a result, work on an iceberg-detecting sonar had been put on hold for more than ten years.[3]

The lack of a forward-scanning sonar had created severe problems for *Nautilus* during the submarine's attempt to transit the Chukchi Sea in June, 1958. That experience had emphasized the need for proper sonar equipment if year-round arctic submarine operations were to be possible. In the fall of 1958, following the successful transarctic voyage of *Nautilus*, Lyon and Roshon had gone to Washington to discuss under-ice sonar requirements with BuShips. "We were optimistic enough to believe," Roshon noted, "that the desk jockeys would flip over the new opportunity which had opened." But this did not happen. On the contrary, individuals in the sonar section were "unreasonably rude" to Lyon. They were convinced that the UQS-1D mine-detecting sonar, which was not a continuous tone sonar, was adequate for under-ice work, and they dismissed Lyon's objections.

Following this meeting, Roshon remembered, Lyon had been "unusually quiet, even for him. I know he was hurt, not so much by their diatribe as by the effect which their opposition would have on his program." In the end, as so often happened, Lyon found a way to proceed. Roshon estimated that it would take seven months and $50,000 to develop a proper iceberg-detecting sonar. Lyon managed to obtain the necessary funds from Capt. Robert Watson, who had been a junior officer on *Boarfish* and was now one of the main supporters of the under-ice program at BuShips. Lyon then transferred the money to Roshon.

With funding available, Roshon promptly started on the project. One of the first things he did was recruit NEL employee Frederick Parker to join him. A sonarman on minesweepers during World War II, Parker fortunately had been available for reassignment. Working together, the two men quickly came up with a transducer for the proposed iceberg detector. Similar to the transducer used for the QLA scanning sonar, it projected a repetitive signal that swept the frequency band from 24 to 32 kHz. With a 4° vertical beam pattern and a 90° horizontal pattern, it would be mounted on the forward part of *Sargo*'s lower sail.

Developing a suitable hydrophone proved more challenging. Finally, Glen Liddiard, an engineer in NEL's transducer division, came up with an imaginative design for the hydrophone. The equipment, however, required a large acoustical "window" in the upper leading edge of *Sargo*'s sail that would allow for a forward horizontal scan of 90°. The design division at the Mare Island Navy Yard opposed the modification to the sail, arguing for a fixed mounting for the hydrophone. Again, it was Captain Watson who came to the rescue. Thanks to his support, a suitable window, designed by the rubber laboratory at Mare Island and NEL's transducer division, was cut into *Sargo*'s sail.

While *Sargo* underwent modifications at Mare Island in August, 1959, Lyon flew to Honolulu for a planning session on the winter cruise. He spent August 15 at SubRon Seven, working out the details of the patrol with *Sargo*'s prospective commanding officer, John H. Nicholson. Lyon was pleased to be working with the experienced and talented submariner. A graduate of the U.S. Naval Academy, class of 1947, Nicholson had been one of the first two submarine officers selected by Rickover for the nuclear program. He had put *Nautilus* in commission and served aboard her for three years under Captain Wilkinson, rising from assistant engineer to executive officer. It was during this period that Nicholson had been exposed to Wilkinson's enthusiasm for the Arctic. In 1958, as previously noted, Nicholson had won high marks from Commander Calvert while acting as executive officer and navigator during *Skate*'s summer patrol to the North Pole. He had left *Skate* just before Christmas, 1958, to take command of *Pickerel* (SS 524) at Pearl Harbor. In May, 1959, he had been told that he would be relieving Comdr. Daniel P. Brooks as commanding officer of *Sargo* prior to its scheduled winter cruise to the Arctic.[4]

Lyon and Nicholson proposed an ambitious agenda for *Sargo*. They envisioned a lengthy patrol of thirty days, during which the submarine would transit the shallow waters of the Bering and Chukchi Seas, conduct acoustical exercises with Ice Station Charlie, then follow the 180° line of longitude to the North Pole. *Sargo* next would cross an uncharted portion of the Beaufort Sea to the entrance to McClure Strait. After exploring a portion of the strait, *Sargo* would proceed to the vicinity of Point Barrow before exiting via the Bering Sea. The main purpose of the patrol would be to determine the feasibility of year-round operations from the Pacific, through the Bering and Chukchi Seas, to the arctic basin. The success of the expedition, Lyon recognized, "depends almost entirely on the NEL sonar effort—the iceberg detector and the ice thickness profile sonar."[5]

In early September Lyon, Nicholson, and Capt. William J. Germershausen, Jr., commander of SubRon Seven, traveled to Washington to present the cruise plan to Admiral Daspit and other senior officials at the Pentagon. Daspit, Lyon reported, approved the plan "without slightest argument," even allowing the thirty days in the Arctic, a period that was longer than the combined time spent in the Arctic by all nuclear submarines to date. The patrol would take place in January–February, 1960, "barring only unforeseen critical changes in the world military situation."[6]

The last three months of 1959 were busy for Lyon as preparations went forward for the winter expedition. In addition to his role as senior scientist on *Sargo*, Lyon had been designated by CNO as civilian research director and coordinator for a project that had come to include an icebreaker, a drifting ice station, and a small aircraft from the Arctic Research Laboratory that would land on the ice pack in the Chukchi Sea to collect oceanographic data and measure pressure ridges. "Scientific coordination is paramount," Lyon stressed in his scientific plan, "to realize maximum profit of valid data on the oceanographic, acoustic, and submarine environmental conditions of the Pacific entrance to the Arctic Ocean." It was Lyon's responsibility to insure that all participants and their equipment were ready on time for the patrol, a task that required a great deal of time and effort.[7]

Lyon arrived in Hawaii in early January, 1960, to participate in the sea tests of *Sargo*'s sonar equipment prior to its departure for the Arctic. There was good news and bad news. On the positive side, Roshon's iceberg detector worked well when tested against another submarine that served as a target. All the other sonar gear, however, had problems. The projector cable to the NK–Variable Frequency echo sounder flooded, and Lyon had to request a new transducer from NEL. The UQS-1 mine-detecting sonar was not working, nor was the BQN-4 five-unit upward-beamed fathometer. The problems with the BQN-4 were so serious that Roshon decided that factory repairs would be necessary. On January 8 he flew to the EDO Corporation at Longpoint, New York, carrying the control-indicator console in a suitcase. Four days later, after the equipment received a complete overhaul by its manufacturer, he returned to Pearl Harbor.[8]

By mid-January the problems with the sonar equipment seemed to have been resolved. A broken wire and bad condenser had been found and repaired on the UQS-1, and the unit seemed to be performing well. The refurbished EDO BQN-4 also worked fine during sea trials on January 12, giving good traces on all five transducers. There remained nagging difficulties with the N6A inertial guidance system, but Nicholson believed that they could be remedied at sea. As far as he was concerned, *Sargo* was as ready as it ever would be for the challenging voyage ahead.[9]

Shortly before the submarine departed, Walt Wittmann, HYDRO's ice expert who would be traveling on *Sargo*, returned from a reconnaissance flight over the shallow-water portion of the planned route. He had bad news. Ice conditions, Wittmann reported, were the most severe that he had ever seen, with an extremely heavily ridged area between St. Lawrence Island and the Bering Strait. The picture he painted, Roshon recalled, "seemed rather threatening to all, including Walt himself." Nicholson, however, was nonplused. "His recon flight," *Sargo*'s commander noted in his personal journal, "bore out what we expected."[10]

The naval brass, led by Rear Adm. William E. Ferrall, ComSubPac, were out in force on January 18, 1960, to send *Sargo* on its way. "Wonderful turnout of SubPac," Nicholson wrote. "Well wishes. Beautiful day." The mood of the crew, he reported, was one of "quiet determination" as *Sargo* set course for the Aleutians, cruising at 500 feet and sixteen-and-a-half knots.[11]

It took a week to cover the 2,350 miles between Pearl Harbor and the Aleutian chain. Twice a day, at three o'clock in the morning and three o'clock in the afternoon (*Sargo* kept Hawaiian time on the cruise), Nicholson came up to periscope depth to communicate, navigate, and ventilate. Wittmann gave a talk on ice forecasting, and Executive Officer William K. Yates lectured on navigation. Lyon spoke in the wardroom on January 21. He set forth the purposes of the cruise, terming it the culmination of fourteen years of work, and he pointed out that he would soon need a new objective. "A man of real vision," Nicholson wrote in his journal following Lyon's talk.[12]

As *Sargo* neared the Aleutians, Nicholson observed that "the excitement of the Arctic is increasing amongst everyone." In the hectic months prior to departure, there had been little time to think about what lay ahead. Now, most of the crew were reading about past arctic

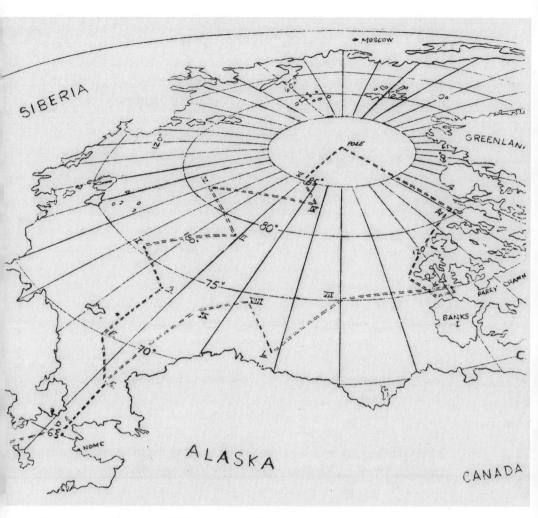

Winter cruise by *Sargo*, 1960. Courtesy U.S. Navy

explorers. "The morale and ability of the crew," he judged, "is truly outstanding."[13]

On the morning of January 25, 1960, *Sargo* surfaced and took a radar bearing on St. Matthew Island, thirty-one miles to the northeast. Radio contact was made with the icebreaker, *Staten Island*, and a rendezvous arranged for late afternoon to the northwest of St. Lawrence Island. *Sargo* made a stationary dive at 11:19 A.M. and entered the edge of the ice pack shortly thereafter. Although the NK–Variable Frequency echo sounder and BQN-4 fathometer were showing scattered block and brash ice overhead, the iceberg detector soon indicated a large target dead ahead. With *Sargo* cruising at 150 feet, the information came as a shock. "Seems impossible that there could be any ice that deep here," Nicholson commented. As he watched the PPI screen, however, the range closed from 1,200 to 400 yards. He swung right with full rudder and slowed to one-third. Shortly thereafter, the iceberg detector made other contacts with what appeared to be deep-draft ice.[14]

"I don't understand it," Nicholson wrote. With the overhead sonars showing only scattered ice, there could not be any deep pressure ridges in the area. Yet the iceberg detector had worked reliably during the trials against a submarine at Pearl Harbor. Roshon soon had the answer. As *Sargo* approached shallow water, he had placed the receiver sensitivity on a higher setting. There were five range scales on the iceberg detector—3,000, 2,000, 1,000, 500, and 250 yards. When shifting scales, it was discovered, the receiver sensitivity, or gain, had to be manually adjusted. If set too high, false targets appeared; if set too low, real targets did not show up on the scope. "Will have to learn by experience where to set gain for each range," Nicholson reported. Also, he discovered that the sound of the contacts could be helpful. Echoes from deep-draft ice produced a sharp sound, whereas shallow-draft ice had a more diffused sound.[15]

It did not take long for Nicholson to regain his confidence in the iceberg detector, and he increased speed to fifteen knots. At 3:00 P.M. he established underwater communication with *Staten Island* at a distance of ten miles. *Sargo* surfaced in block ice near the icebreaker, and an LCVP brought over Comdr. J. B. Larsen, C. O. of *Staten Island*; Gene Bloom of NEL, who was chief scientist on the icebreaker; and Commodore O. C. S. Robertson of the Royal Canadian Navy. Lyon had arranged with CNO for Robertson, his old friend from HMCS *Labrador* and now considered one of Canada's leading arctic experts, to see *Sargo* in action under the ice. Nicholson suggested that Robertson spend the night on *Sargo* and return to *Staten Island* the next day. He was not surprised when the Canadian officer "readily accepted" the invitation, as he had noted that Robertson "just happened" to have brought a small suitcase with him. Larsen and Bloom then departed, and *Sargo* dove at 5:45 P.M.[16]

According to the plan worked out between Nicholson and Larsen, *Sargo* remained within 4,000 yards of *Staten Island* as they worked their way north into the ice pack. As the evening wore on, the ice cover above *Sargo* became two to three feet thick, studded with pressure ridges that extended down to thirty feet. At 3:00 A.M., January 26, Larsen reported that *Staten Island* had become stuck in the ice. It was snowing heavily, he said, with a forty-knot gale blowing and visibility reduced to 500 yards. "It was difficult to imagine these conditions," Nicholson wrote in his patrol report, "as we comfortably orbited her at 120 feet."[17]

Nicholson attempted to surface shortly before 11:00 A.M., but *Sargo*'s sail was unable to penetrate the ice. With a pressure ridge rapidly bearing down on the submarine, Nicholson returned to 120 feet and looked for another opening. Three hours passed without success. "We suspected," Nicholson noted, "Commodore Robertson had high hopes we would be unable to surface so he could complete the cruise in SARGO."[18]

Robertson's hopes, if he indeed had them, were soon dashed. At two o'clock that afternoon *Staten Island* directed *Sargo* toward a frozen polynya, located 300 yards off the icebreaker's starboard quarter. The BQN-4 recorder indicated that the ice was about one foot thick. When Nicholson brought *Sargo* up, however, the sail touched the ice at a keel depth of fifty-one-and-a-half feet. As *Sargo* measured forty-nine feet from keel to top of sail, this meant that the ice overhead was two-and-a-half feet thick. A short blow of the forward ballast tanks enabled the sail to punch through the ice at 2:38 P.M. Minutes later, blowing the after group and all main ballast tanks brought the deck of the submarine through the ice.

When Nicholson cracked the hatch, he found an enormous block of ice on the main deck. Measuring 15' x 20' x 5', it weighed over thirteen tons. An inspection of topside, however, revealed only a few small dents. "This was our first break through," Nicholson reported, "and we had surfaced practically unscathed."[19]

Two men from *Staten Island* came across the ice in snow shoes to retrieve Robertson. The icebreaker then headed north to complete its scientific mission. *Sargo* dove beneath the ice and proceeded toward St. Lawrence Island. By midnight, the water had shoaled to 240 feet. With pressure ridges now extending down to over 50 feet, Nicholson maintained a keel depth of 140 feet. This placed *Sargo* 100 feet from the bottom and 91 feet from the surface.

As Nicholson wanted to take a navigational fix before entering the Bering Strait, he ordered the officer of the deck to execute a Williamson turn on any stretch of thin ice at least 100 yards long. When the NK–Variable Frequency echo sounder, which had not been working during *Sargo*'s previous surfacing, indicated eighteen inches of ice overhead, Nicholson decided to attempt a breakthrough. This time, *Sargo* surfaced easily through fifteen to seventeen inches of snow-covered ice. After *Sargo*'s radar showed her to be forty-one miles southeast of St. Lawrence Island, Nicholson dove and headed north, planning to cross to the west of the island, on the side nearest to Siberia.

As *Sargo* neared the Bering Strait, the water became shallower, and the pressure ridges got deeper. Nicholson decided to modify the concept of command duty officer that he had instituted. There were three teams: Nicholson, Executive Officer Yates, and Engineering Officer Edward O. Dietrich served as command duty officers and were supported by Lyon, Roshon, or Parker. Each team would be on duty for four hours, with eight hours off. The command duty officer would observe the display on the various sonars, consult with his NEL counterpart when necessary, and direct the officer of the deck by interphone (27MC) to maneuver the boat. Faced with the tight situation in the Bering Strait, however, Nicholson changed the procedure. The command duty officer would not only monitor the sonar but also personally conn the ship, giving orders for course and speed changes directly to the helmsman by 27MC. This modification became standard practice during shallow-water transits.

By 2:00 A.M., January 28, the submarine was cruising at 100 feet with the bottom only 26 feet below the keel. Having only 51 feet above the sail to clear downward-projecting ice and with the iceberg detector showing 30-foot pressure ridges in all directions and even deeper ice ahead, Nicholson certainly had his work cut out for him. Just at this point, the stylus of the BQN-4 failed at the same time that the recorder of the NK–Variable Frequency fathometer ran out of paper.

As Nicholson needed both instruments to tell him when he passed beneath deep-draft ice, he ordered a 180° turn while repairs were being made. This was no easy maneuver under the circumstances. "With infinite care," Nicholson recalled, "our planesmen and helmsman brought us about without tilting the boat." The slightest tilt, he knew, "could have resulted in our propellers grinding into the ocean bottom and left us seriously disabled under the pack ice."

The instruments were soon back in working order, and Nicholson was prepared to run the obstacle course of deep-draft pressure ridges.

Diagram of *Sargo*'s shallow-water under-ice transit, 1960.
Courtesy U.S. Navy

For the next twelve hours, *Sargo* underwent what Nicholson described as a "baptism of fire, or rather ice." Maintaining a minimum maneuvering speed of five knots and trying to keep at least twenty feet under the keel, Nicholson and the two other command duty officers found it necessary to twist and turn almost continuously to avoid deep ridges. "Taking violent action," Nicholson reported, "has become a part of our transit routine."

As Lyon observed, the iceberg detector was doing an excellent job of providing the necessary guidance to avoid the deeper pressure ridges. Nicholson, he noted, normally searched on the 1,000-yard scale, occasionally shifting to the 2,000 yards to get "the big picture." If the target remained on the PPI scope into 400 yards, he maneuvered to avoid it, if necessary shifting to the 500- or 250-yard scale. With practice, Nicholson became expert at estimating which ridges *Sargo*'s sail would clear and which ones he had to maneuver to avoid.[20]

By mid-afternoon, the water had become a little deeper, and there were fewer extreme ridges. When a promising stretch of thin ice appeared overhead, registering twelve inches on the NK–Variable Frequency recorder, Nicholson decided to surface. When *Sargo*'s sail touched the ice, he blew the main ballast tanks and punched the sail through to the surface. As it turned out, the sail had penetrated a small pressure ridge, measuring thirty-seven inches. The bow came through twelve inches of snow-covered ice, but the stern remained under heavy ice, four feet thick. "It was a striking sight to see the boat fast with only the bow and sail showing," Nicholson reported. "The sun had just set so we had enough light to enjoy the scene."[21]

*Sargo* spent fourteen hours on the surface, thirty-eight miles from Cape Prince of Wales. Nicholson, after completing communications, dove out of the frozen polynya at 7:32 A.M. on January 29. When *Sargo* reached its cruising depth of 120 feet, he discovered that the bow planes would not rig out. "The pins of the rigging and tilting interlock mechanism were heard to shift part of the way and then hang up," he noted. "We suspected that the holes for the interlock pins were blocked by a coating of ice since the bow had been well out of the water."

Without use of the bow planes, Nicholson recognized, "the boat was in a rather precarious situation." He found it necessary to increase speed to eight knots in order to hold trim. He tried to keep rudder movements to 3°, reserving 5° to 10° "for extremely tight situations where violent maneuvers were necessary." When he used a rudder angle greater than 3°, he had to blow main ballast tanks with the vents open in order to keep *Sargo* off the bottom. Then, as the submarine rose toward the ice, he sometimes had to flood negative to arrest the submarine's upward movement. All-in-all, Nicholson concluded, it was a "nerve-wracking situation."[22]

By 10:00 A.M. *Sargo* was approaching Bering Strait. The water was extremely shallow, only 120 feet. With the overhead ice measuring three-and-a-half feet thick, and pressure ridges 30 to 60 feet deep every 80 yards, Nicholson had to keep *Sargo* at a keel depth of 110 feet. Since the iceberg transducer was mechanically fixed to the sail, making the vertical path of the sonar beam a direct function of the ship's attitude, *Sargo*'s diving officer had to maintain a zero bubble at all times for the proper presentation. The submarine's slow speed and loss of bow planes made this an especially challenging task.

*Sargo* had not gone far into the strait when the depth of the sea bottom suddenly dropped rapidly to 165 feet. Then, just as quickly, it began to shoal. "The video pip on the UQN screen," Nicholson observed, "wheeled around the scope from 55 feet down to 40, 20, 10...." The diving officer on watch, Lt. David A. Phoenix, faced the problem of staying off the bottom, maintaining a zero bubble, and not hitting the ice overhead. "I made a spur of the moment decision," he recalled, "to blow the main ballast tanks with the vents open in order to acquire some upward momentum that I hoped I could control before hitting the ice overhead. As far as I know, this maneuver had never been used before."[23]

As *Sargo* rose ten feet, the pip on the UQN screen reached five feet. "We braced ourselves to bounce off the bottom," Nicholson wrote; however, the soundings began to increase as rapidly as they had shoaled.

Phoenix's ad hoc maneuver had worked. "One bad moment," Lyon commented in his Scientific Journal.[24]

The underwater feature that *Sargo* had so narrowly missed hitting soon had a name. As the sea bottom had dropped away, the fathometer operator had said that it looked like a gopher hole. After *Sargo* averted contact with the bottom, and everyone in the control room had begun to breathe again, Diving Officer Phoenix quipped, "That gopher must have stood up in his hole." One of the men added, "If it is a gopher, his name must be Tall Gonzales."[25]

Word of the encounter with Tall Gonzales soon spread throughout the ship. For several hours, numerous members of the crew found time to stop by the control room and watch the iceberg detector in action as *Sargo* maneuvered around deep pressure ridges. Evidently, all hands became reassured. "The tension which had been building all morning," Nicholson observed, "subsided a bit by evening even though we were still twisting and turning along our track."

*Sargo* completed the transit of the Bering Strait shortly after 5:00 P.M. The water, however, remained shallow as the submarine entered the Chukchi Sea. The pressure ridges, if anything, became more rugged, reaching down to depths of eighty feet. "Iceberg detector is essential," Lyon noted, "and so far is working well." At eight-thirty the morning of January 30, *Sargo* reached the thirty-fathom curve. "Never," Nicholson wrote, "had 30 fathoms looked so deep." Able to cruise at a more comfortable keel depth, the need to avoid deep pressure ridges also became less frequent. "Everyone relaxed a bit," Nicholson reported.[26]

On January 31, as *Sargo* continued toward Herald Island, Nicholson twice attempted to surface but was unable to break through the overhead ice. On a more positive note, the bow planes finally came back into commission. Shortly before 6:00 P.M., the polynya-plotting party was called out, and *Sargo* again attempted to penetrate the ice cover. This time the submarine came up through thirteen inches of ice without difficulty. Radar fixed *Sargo's* position at 17.8 miles from Herald Island.

While Nicholson attempted to establish radio communications, two scuba divers, James G. Tucker and Laurence M. Curtiss, entered the water and cleared a flattened soup can that had jammed the garbage ejector. It took until 2:25 A.M. on February 1 to complete communications. *Sargo* then dove out of the frozen polynya and continued north in search of a trough that Harald Sverdrup had discovered in the 1920s. In 1948 Sverdrup had told Lyon that the trough might afford a good route through the shallow area for submarines. The Norwegian oceanographer was proved correct. *Sargo* located and passed through the trough, reaching deep water shortly after noon. Nicholson increased speed to sixteen-and-a-half knots—"and drew a deep breath"—as the submarine entered the arctic basin. *Sargo*, to both Nicholson's and Lyon's delight, had completed the first-ever winter under-ice transit of the Bering and Chukchi Seas.[27]

Their elation proved short-lived. At 6:30 P.M. the iceberg detector failed. The training motor for the hydrophone had grounded out. If this had happened five hours earlier, Nicholson realized, "we might have been in serious trouble." He could now stay deep enough to avoid ice without reference to the iceberg detector, but he would need the scanning sonar for *Sargo*'s return transit. He decided to surface at the first opportunity and see if the equipment could be repaired.

At 9:38 A.M. on February 2, the NK–Variable Frequency echo sounder recorded a 1,000-yard area of ice that was twelve inches or less in thickness. Nicholson was skeptical about the existence of such a large opening, but he decided to investigate. *Sargo* was now at 78° North, and there was no sunrise at this high latitude in winter. Reaching periscope depth, Nicholson gingerly raised the scope, expecting to encounter ice. Instead, he saw stars overhead. He brought *Sargo* to the surface in a huge polynya, 600 yards x 200 yards, which contained both eight inches of new ice and open water. Inspection of the iceberg detector revealed that no cables had flooded. It would be necessary, therefore, to remove and look into the training mechanism. But as Nicholson noted, in order to do this "the problems which had to be overcome appeared staggering."[28]

The training mechanism, contained in a gear box and attached to the hydrophone, weighed 650 pounds. It had been installed at Mare Island with the help of cranes through an opening in the leading edge of the sail that had been covered by an acoustic window. As the acoustic window could not be removed, the assembly would have to be reached and detached from inside the sail. This meant cutting through the beam assembly that supported the mechanism and lowering it to the 01 deck in the sail. "This would have been a mean feat in any weather because of the weight and shape of the assembly," Nicholson observed. "It was a real feat for two men wearing heavy winter clothing on the cramped 01 level in 20° below zero weather."

Several teams of two men each, working for short periods in the extreme cold under the direction of Chief Robert F. Crowley, labored throughout the remainder of February 2 and into the next day to free the assembly from its mounting. The design of the equipment, as Nicholson noted, was "almost unbelievably inaccessible for maintenance." At noon on February 3, Nicholson held a gloomy meeting in the wardroom with Lyon, Roshon, Parker, and several of the ship's officers. Lieutenant Phoenix, who had been placed in charge of overseeing and coordinating the project, explained in great detail why it was not possible to remove the hydrophone from the training mechanism assembly. After going over the blueprints for the equipment, everyone agreed that the only option—and an undesirable one—would be to cut off the motor cover. Just as this decision had been reached and the blueprints were being folded away, Chief Crowley walked in and said that his team had managed to detach the hydrophone and were anxious to get on to the next step. "What amazing men," Nicholson wrote in his Journal. "Things look better than they have in days. Should know how serious the damage is within a few hours."[29]

As usual, bad news followed good. In the early afternoon, movement of the ice in the unstable polynya began to be felt on *Sargo*. The submarine soon heeled 3° to starboard as the ice pressed against the hull, accompanied by the screeching sound that Lyon had come to know so well. For Nicholson, the situation called to mind the fate of *Jeanette*, which had frozen into the ice not far from *Sargo*'s present position. "It's easy to understand," he noted, "how men in less rugged ships than ours could be terrified by the noise and action of the ice."

It took until two o'clock the morning of February 4 to bring the gear box assembly below and to secure for sea the remaining topside components of the iceberg detector. After spending some forty hours on the surface in temperatures below minus-20°, Nicholson was apprehensive about the effects of the lengthy stay on *Sargo*. As *Sargo* made a stationary dive and planed down to 400 feet, there was unusual vibration and noise, likely caused by ice sticking to the deck and hull. As the submarine built up speed to sixteen-and-a-half knots, however, both soon disappeared. Fortunately, Nicholson wrote, "there was no cause for concern."[30]

As *Sargo* steamed north toward the pole, following a zigzag track along the 180° line of longitude to secure bathymetric data, Roshon examined the training motor. The oil-filled gear box, which contained the training motor, gear train, and transmission, had been fitted with an expansion chamber, consisting of a rubber bellows with sufficient volume to accommodate the expansion and contraction of the oil. The bellows likely had become brittle when exposed to the cold temperatures while *Sargo* had been on the surface on January 28. When *Sargo* dove, the oil had contracted, demanding compensation. The brittle rubber of the bellows had been unable to flex adequately and had ruptured. Eventually, increased pressure had displaced part of the air with salt water. The salt water had reached the training motor and shorted it out.[31]

Roshon and Parker, assisted by the ship's electricians, removed and washed the motor. They found that a number of bearings in the training mechanism had frozen. Although there were no spares on board, machinist's mate William A. Von Thenen came up with a novel but effective remedy for the problem. He soaked the rusted ball bearings in Oil of Wintergreen. They were soon spinning freely.

Meanwhile, the ship's electricians concocted a vacuum oven for baking out the training motor by piping a vacuum pump to the cylindrical body of the spare chlorate candle, then wrapping it with a homemade electric blanket. "Although the motor may not get fixed," Nicholson commented, "it will not be for lack of inventiveness."[32]

While Lyon appreciated the efforts that were being made to repair the iceberg detector, he could not help feeling frustrated, disappointed, and even partially responsible for the failure of the equipment. On February 5 he vented his feelings in a message to NEL.

> PROVERBS 15 CLN 32 ["He who ignores instruction despises himself: but he who heeds admonition gains understanding."] X ICEBERG DETECTOR SINGULARLY RESPONSIBLE FOR FIRST SUCCESSFUL TRANSIT OF BERING/CHUKCHI SEAS YET INEPT MECHANICAL DESIGN CAUSED FAILURE IN ARCTIC BASIN AND ONLY GOOD FORTUNE PREVENTED EARLIER FAILURE IN SHALLOW WATER X PRESSURE COMPENSATION FAILED AND FIELD REPAIR NIGH IMPOSSIBLE X

Lyon urged that every effort be made to correct the problem in the unit that was being prepared for *Seadragon*. A simple hydraulic training mechanism might be preferable to the current electrical system. "ABOVE ALL," he emphasized, "MAKE DESIGN FEASIBLE FOR COMPLETE FIELD REPAIRS WHILE SURFACED IN WINTER ARCTIC OCEAN."[33]

The irony of the situation, Lyon recognized, was that the motor-train system had been the only component of the many devices built by the Submarine Laboratory at NEL that had *not* been tested in the facility's pressure tanks. The incident, he later noted, reemphasized the axiom that all equipment must be tested before being taken to sea, especially to the Arctic. As he again had learned, "Help is not available here."[34]

As repairs on the motor continued, *Sargo* crisscrossed its way toward the North Pole. Nicholson attempted to surface on the afternoon of February 5 when the NK–Variable Frequency echo sounder indicated a flat area overhead. *Sargo*'s sail hit the ice at a keel depth of fifty-six feet, indicating seven feet of ice. Although Nicholson blew the submarine's ballast tanks for five minutes, he could not penetrate the heavy layer of ice. He tried to surface again in a nearby area. This time, the sail made contact at fifty-three feet. He managed to punch through but only the top of the sail was able to come to the surface. "It was an incredible sight," Nicholson reported. The main deck remained five feet below the ice. All that could be seen of the submarine was the sail down to the "583" identifying number on the side. Several crew members climbed down the Jacob's ladder from the bridge and photographed the dramatic scene.

Nicholson tried to communicate for nearly nine hours but without success. Finally, he decided to dive and try again later. He started down shortly after midnight on February 6. The boat, however, was reluctant to leave the surface. With vents open and an extra 4,000 pounds in the auxiliary tanks, the keel depth increased only three feet. Diving Officer Frederick C. Stelter, III, flooded 5,000 pounds into negative. *That* produced results. After some crunching and scraping, the sail slid beneath the ice.

Nicholson attempted to surface again at 6:00 P.M. after the NK–Variable Frequency echo sounder showed overhead ice thickness of one foot or less. Rising at twenty feet per minute, the sail gently contacted the ice at a keel depth of fifty-two feet. Stelter gave the main ballast tanks several short blows in an effort to penetrate the three feet of ice; however, Nicholson reported, "the thing we were particularly anxious to avoid happened." The angle got away from Stelter, and *Sargo* took a down bubble. This caused the stern to swing up and the rudder to slam into the ice. The sail finally broke through what proved to be the thickest ice cover to date, composed of two fourteen-inch layers, two four-inch layers, and two inches of snow.

"The night was beautiful," Nicholson reported. "There were many stars and a bright half moon." The rudder, happily, proved undamaged by its contact with the ice. *Sargo* was in a small polynya, 145 yards x 135 yards. While Nicholson was attempting to communicate, *Sargo* took a sudden lurch as the main deck pushed its way through the ice. "The view of the boat from the ice," Nicholson observed, "was striking." The bow had lifted through thirty-six inches of ice, and the main deck aft was covered with a twenty-eight-inch layer.

After finally managing to get his messages out via Radio San Francisco, Nicholson ordered a stationary dive at 4:30 A.M. on February 7. Again, it took manipulation of the ballast tanks for *Sargo* to tear free of the ice. The submarine then resumed its northern course. The NK–Variable Frequency echo sounder showed that the average ice cover as *Sargo* neared the pole was four feet thick, with pressure ridges forming 60 percent of the canopy. Forty-foot ridges were common, with several ridges extending down to seventy feet.[35]

February 7 also saw the iceberg detector training motor ready to be tested. Shortly after being turned on, it burned up. "It's a shame," Nicholson wrote in his Journal. "The boys really gave it a go."[36]

Nicholson now had two options. He could abandon the attempt to return to Pearl Harbor through the shallow waters of the Chukchi and Bering Seas, exit via the deep waters of the Greenland Sea, then make his way to the Pacific via the Panama Canal. The other option would be to have NEL ship a training motor to Point Barrow. The motor could then be placed on board a light aircraft, which would land on the ice off the coast of Alaska and deliver the motor to *Sargo*, surfaced within the ice pack. Rather than admit defeat, Nicholson leaned toward the second option, but neither choice was especially attractive.

As *Sargo* headed for the pole on February 8, Roshon offered another possibility. The man most responsible for the design of the iceberg detector had been "devastated" by the failure of the equipment. "After some thought," Roshon recalled, "it came to me that SARGO had another large hydrophone assembly." Perhaps it might be possible to receive the signals produced by the iceberg detector transducer on the ship's BQR-2 hydrophone assembly. Although not designed for lobe suppression, as was the iceberg detector hydrophone, the BQR-2 just might work. Nicholson told Roshon to give it a try. "If it works," Nicholson noted, "and it should, this would be an extremely imaginative solution to our problem and would preclude the necessity of further topside repairs."[37]

*Sargo* crossed the Lomonosov Ridge at 5:45 A.M. on February 9. At 9:34 A.M., the submarine reached the North Pole. "Although others had come before," Nicholson wrote, "we had come the hard way—the shallow, dark and cold route." He immediately commenced a cloverleaf search pattern, intently watching the topside echo sounders for an indication of thin ice. Just after completing the first leg of the pattern, a flat area appeared overhead, only twenty-five yards from the spot that *Sargo*'s navigator had decided must be the pole.

*Sargo*'s sail touched the ice at fifty-three-and-a-half feet keel depth. After a short blow forward and aft, the sail broke through thirty-seven inches of ice; the deck remained below the ice cover. The weather, Nicholson reported, was "perfect." It was "clear, calm, and peaceful with a light breeze, a heaven full of stars, and a bright half-moon." The temperature stood at minus-32°.

Harold Meyer, a Honolulu-born Hawaiian crew member, had the honor of raising the Hawaiian state flag right at the Pole, twenty-five yards to the starboard of *Sargo*. "It was a rather Wagnerian scene," Nicholson observed, "with floodlights and red flares illuminating the ceremony."

While surfaced, *Sargo*'s lookout spotted the vapor trail of a high-flying jet. Nicholson ordered *Sargo*'s spotlight shined toward the aircraft. It was obviously seen, as the pilot flashed his landing lights before continuing on his way. Nicholson believed that it was a commercial flight, and he imagined the surprise of the pilot when he saw the light from the North Pole. ComSubRon Seven, however, later told him that the aircraft likely was Russian, as the Soviets had been keeping track of *Sargo*'s progress.[38]

Communications, for a change, were "amazingly good," and *Sargo* managed to clear a backlog of messages. At two o'clock the morning of February 10, Nicholson ordered a stationary dive out of the frozen polynya. Concerned about the possibility of freezing-in due to the long stay on the surface in cold temperatures, he blew the ballast tanks for ten minutes with the low-pressure blower before flooding in 10,000 pounds in the auxiliaries and 20,000 pounds in the negative tank. *Sargo* did not budge. The sail had bonded firmly to the ice. "Little is known about the force necessary to shear this bond," Nicholson noted, "especially at these low temperatures." He again put the low-pressure blower on the ballast tanks, blew them dry, then kept the blower on an extra ten minutes. He opened the vents, producing 30,000 pounds of negative buoyancy. "To our relief," he reported, "the boat quickly tore loose and went right down."[39]

Observing this process, Lyon agreed with Nicholson that far too little was known about the force that might be developed when the submarine structure bonded with sea ice. Was heavy flooding the best way to break free? This obviously was a case for laboratory experiments. Tests in NEL's arctic pool, he believed, should provide the answers to the problem.[40]

Upon leaving the North Pole, *Sargo* headed south along the 105° West meridian toward Nansen Sound. Ice expert Wittmann promised Nicholson that he would soon encounter the most rugged ice in the Arctic. Hardly had Wittmann spoken when *Sargo* passed under a 108-foot ridge.

At 4:30 A.M. on February 12, *Sargo* reached the 100-fathom curve and turned west toward McClure Strait. "Somewhat of a thrill again of surveying in areas without a single sounding," Lyon wrote in his Scientific Journal, "this time with ice overhead and not knowing what islands or shore may lie ahead." The ice cover, as Wittmann had predicted, was extremely rugged, with 90 percent ridges that averaged 75 to 95 feet in depth. Between midnight and noon, *Sargo* passed underneath ten ridges in excess of 100 feet. One giant piece of ice reached down to 122 feet.[41]

Lyon was pleased to report that tests of the jury-rigged iceberg detector "look good." Not only had Roshon, Parker, and Chief Sonarman William M. Page made the connection between the iceberg detector transducer and the BQR-2 hydrophone, but also Chief Electrician Thomas L. Walker had come up with an ingenious system of linkages to provide the necessary 90° of automatic scan. The design and construction of the scanning mechanism, Nicholson noted, had made "this modified sonar equipment into a suitable mechanism." There was, he wrote, "no end of the talent on SARGO." Because of the different location of the BQR-2 hydrophone, it would be necessary to keep a slight up bubble on *Sargo* in order to obtain the proper display on the PPI, but this was not expected to present a major problem.[42]

At 8:00 A.M. on February 14, as *Sargo* approached the entrance to McClure Strait, Nicholson brought the submarine to the surface through twenty-and-one-half inches of ice. "First sun we have seen for many days," Lyon reported. Nicholson decided to use the opportunity to test the "ice destructor" mines that were being carried on *Sargo*. The ice destructors were modified Mk 10 mines that contained 630 pounds of high explosives. They were designed to detonate fifteen minutes after ejection and blow a hole in the ice through which a submarine could surface. Nicholson fired three mines; none exploded.[43]

*Sargo* dropped beneath the ice shortly before 5:00 P.M. By evening, the submarine had penetrated McClure Strait to a distance of 100 miles. Bathymetry showed that the sea bottom was flat and deep, with soundings of 220 to 290 fathoms. "A submarine," Nicholson observed, "should have no difficulty in transiting this portion of the Northwest Passage."[44]

*Sargo* cleared the mouth of McClure Strait outbound at 3:30 A.M. on February 15 and headed toward Ice Island T-3. En route, the submarine surfaced through twenty-two inches of ice at 9:00 A.M. "It was a wonderful day," Nicholson wrote. The sun was a degree above the horizon, there was a light breeze, and the temperature was a balmy minus-13°. When fresh tracks of two polar bears were spotted, a hunting party quickly was organized. The bears, however, could not be found. After communicating, *Sargo* dove at 1:48 P.M. Shortly thereafter, Nicholson gave the iceberg detector its final test. It proved successful. "It was a tremendous relief to have this installation working again," Nicholson noted. "There could be no thought of making the return shallow water transit under the ice without it."[45]

*Sargo* reached T-3 on February 17. A large kidney-shaped piece of ice that had broken off the Ellesmere ice shelf and become embedded in the polar ice pack, the ice island had been occupied by the U.S. Air Force and used as a drifting research station between 1952 and 1954, and briefly in 1955. Designated Station Bravo, it had been reoccupied in 1957 as part of the U.S. contribution to the International Geophysical Year. In January, 1960, following the destruction of Station Charlie, Lt. Comdr. Beaumont M. Buck of the Office of Naval Research and four scientists from the Navy Underwater Sound Laboratory had arrived on T-3/Bravo and established an underwater listening array for tests with *Sargo*.[46]

*Sargo* established underwater communication with T-3 shortly after midnight. As the submarine passed under the 4.5-mile x 9-mile ice island at 390 feet, Lyon watched the recorder of the NK–Variable Frequency echo sounder change abruptly from five feet to 120 feet. The deepest draft of T-3 was registered at 160 feet. *Sargo* then made the requested sound runs between 4:00 and 11:00 A.M. while Buck and his team recorded the noise on their hydrophone array. Nicholson had hoped to surface near T-3, but when this proved impossible he set course for Point Barrow.

On February 19, having passed well clear of Point Barrow, *Sargo* entered the shallow waters of the Chukchi Rise. Nicholson was anxious to communicate. *Sargo* had come to the surface the previous day, but intense aurora borealis activity had made it impossible to send or receive radio messages. Shortly before noon, the upward-beamed fathometer recorded a flat area of three to four feet of ice. The officer of the deck executed a Williamson turn and called out the polynya-plotting party. When *Sargo* returned to the area, Nicholson ordered a vertical ascent. Diving Officer Stelter brought the submarine up at twenty-five feet per minute until the sail hit the ice at a keel depth of fifty-two-and-a-half feet. Although *Sargo* had never before managed to penetrate ice as thick as forty-two inches, Nicholson decided "to give it a real go."

Stelter rocked *Sargo* back and forth in the ice by blowing the forward, then aft ballast tanks several times. Thanks to his skilled manipulation of the tanks, *Sargo*'s sail finally broke through the ice. "We had punched the sail through a record 48 1/2 inches of ice!" Nicholson reported. The view from the bridge was "astounding." *Sargo* had surfaced in the center of a thick ice floe in the midst of open water. "It was not easy," Nicholson quipped, "but somehow we had managed to find a floe instead of a polynya."

*Sargo* remained on the surface for over seven hours as radio operators worked to clear the backlog of messages. Most of the crew took the opportunity to investigate the ice floe. It was another beautiful day, with the temperature 10° and a light wind. When polar bear tracks were discovered, a hunting party set out, but they had no more success than the previous hunters. The pleasant interlude came to an end at 7:00 P.M. when Nicholson ordered a stationary dive. It took four tries to break loose from the thick ice, but *Sargo* finally managed to slip beneath the surface.[47]

Nicholson leveled off at 120 feet and proceeded at seven knots into an area of heavy ridges. The water was extremely shallow. Stelter tried to keep *Sargo* at least 30 feet off the bottom. As it was necessary to maintain a slight up bubble for the proper iceberg detector presentation, this meant that the submarine's stern was only 23 feet off the bottom. The presentation, Nicholson noted, had been causing some problems since the BQR-2 modification. When shifting from the 1,000-yard scale to the shorter scales, side lobes had appeared, making it more difficult to track targets than had been the case with the original equipment. Shortly before 8:00 P.M., as Roshon worked with Nicholson on the best way to operate the equipment, two deep ridges appeared ahead. The detector was operating on the 1,000-yard scale. As it was clear that *Sargo* would not be able to pass beneath the ridges, Nicholson would have to maneuver to avoid them.

Nicholson decided to set a course that would take *Sargo* between the ridges. Shifting to the 500-yard scale of the iceberg detector, he watched the ridges carefully on the PPI display, noting that the ice off the port bow was much deeper. He then changed to the 250-yard scale as *Sargo* neared the ice. When the ridge to starboard could no longer be seen on the PPI scope, he changed course 15° to starboard to give the deep ridge to port more clearance.

Suddenly, *Sargo* was jolted as its sail collided with the ice. The boat heeled slightly to port and took a 6° down bubble. With the bottom only thirty-two feet away, the officer of the deck ordered all stop and sounded the collision alarm. Nicholson, who had been concentrating on the iceberg detector, took the conn and called for engines all-back two-thirds to kill the way. With the submarine rapidly approaching the bottom, he instructed the diving officer to blow tanks with vents open. At the same time, he ordered Maneuvering to put the Emergency Bottoming Bill into effect. Maneuvering promptly shut down the port side of the steam plant in order to preserve one clean condenser in case the other main condenser filled with silt if *Sargo* hit the bottom. Should that happen, the submarine would have propulsion on one shaft when it wriggled free.

*Sargo* narrowly averted contact with the bottom. As the submarine began to rise, Nicholson rang up two-thirds ahead on the starboard shaft and regained depth control. The port side steam plant was then returned to service. A check with all compartments revealed that *Sargo* had suffered no damage from the collision. "This had been a close call," Nicholson reported. "Only the sturdiness of the boat, correct action by each watchstander, and help from above prevented this from being a serious casualty."[48]

"Unnecessary to explain nervous reaction throughout the ship," Lyon wrote in his Scientific Journal. "I am left with nervous reaction for hours afterwards, as everyone is, though none trying to show it, and life on board goes on as usual." He could not help but notice, however, the "intense alertness" of the crew following the collision. "What does the future of this transit out hold for us?" he asked. "We have only begun the long shallow run south."[49]

Roshon, after analyzing the data from the recorders, determined that the problem lay with the strength of the side lobes that had been created by the combined iceberg detector/BQR-2 gear. Although he had realized that the side lobes would have an adverse effect on the presentation on the close-in scales, he had not expected them to mask the targets.[50]

Nicholson and Roshon developed a technique to work within the limitations of the jury-rigged equipment. Henceforth, the selector would be left on the 1,000-yard scale. Course changes would be initiated when large incoming ridges were between 1,000 and 600 yards of *Sargo*. The image of the target then had to be retained in memory while a decision was being made about what to do to avoid the next target. "Fortunately," Lyon recalled, "we never had two ridges in our memory and one coming up, otherwise we might not have been able to do it."[51]

It took several hours before Nicholson and Lyon became comfortable with the new procedure. By midnight, *Sargo* had encountered more than a dozen sixty-foot ridges. This required, Nicholson observed, "a lot of tough maneuvering in 26 fathom water." It gave one "a quivering nerve reaction," Lyon reported, "to watch these massive ridges come in, make the best guess which way to avoid them, and wait for the vertical echo sounder to write out the trace as they pass overhead." After four hours of practice, however, Nicholson was "confident that we could get through safely without another encounter with deep ice."[52]

Between midnight and noon on February 20, *Sargo* continued to dodge around heavy ridges in twenty-five-fathom water. Nicholson was forced to keep a keel depth of only 123 feet, which placed the stern a scant 16 feet from the bottom. There was no choice, he recognized, as "we needed as much clearance from the ridges as possible." *Sargo* reached slightly deeper water in the afternoon, enabling Nicholson to increase keel depth to 145 feet. It was, Lyon remarked, "amazing the more relaxed feeling one gets" from the addition of only a few more feet.

At 8:00 P.M. a solid wall of ice suddenly appeared on the iceberg detector scope at a range of 800 yards. "There was absolutely no opening," Nicholson reported. He changed course to parallel the obstacle, dodging other deep ridges. "For the next six hours," Lyon noted, *Sargo* "had very heavy going." At one point, Executive Officer Yates, who was conning the ship, found himself boxed in by massive ridges. He selected the least menacing target and slipped underneath it. "Nervous tension surely up during this period," Lyon recorded in his Scientific Journal.[53]

Forty- to fifty-foot ridges were common after midnight on February 21, as *Sargo* continued its approach to the Bering Strait. After 5:00 A.M., however, the ice cover became less rugged and the soundings increased to more than twenty-eight fathoms. Nicholson was able to surface in an open-water polynya so that he could inspect the damage caused by the collision with the ice. "The bridge and sail," he reported, "looked no better or worse than expected." The pressure ridge had hit the reinforced top of the sail. It was dished in and shifted aft; the number one periscope, VLF loop antenna, and electronic countermeasure mast could not be raised. The main supporting members, however, were sound. "Although we were all saddened to see the damage to the sail," Nicholson wrote, "each of us realized again how lucky we had been."[54]

On February 22 *Sargo* was traveling twenty feet off the bottom and making eight knots as it entered the Bering Strait. By 11:00 P.M., *Sargo* was south of Fairway Rock. Ahead, in seventeen-and-a-half fathoms of water, lay Tall Gonzales, the place where *Sargo* had nearly hit bottom during the inward transit. "Fortunately," Nicholson noted, "the bow planes were working now." The crew, he reported, "had developed a very active interest in this shoal and its position was carefully plotted on the charts in the Crew's Mess and Maneuvering." This time, Nicholson planned to make a dogleg to the south, then turn west to leave Tall Gonzales at least five miles off the starboard beam.[55]

Nicholson and Lyon had the midnight to 4:00 A.M. watch on February 23 as *Sargo* neared Tall Gonzales. Just before the submarine entered the area of shallow water, however, the N6A inertial guidance system went off line. Although Nicholson would not have precise information on his latitude without the N6A, he was confident that he could avoid Tall Gonzales. By 2:24 A.M., after two hours of maneuvering, *Sargo* appeared to be well clear of the obstacle. As he turned to run past the shoal, the sounding suddenly decreased to 134 feet. With a deep ice ridge dead ahead, Nicholson reversed course and ran back into deeper water. About this time, the N6A came back on line. It showed that *Sargo* was well to the north of its estimated position, placing the submarine on the same latitude as the shoal! In estimating his position, Nicholson had expected a decrease in the current to one-half knot; instead, the current had increased to 1.7 knots.

Creeping toward the southwest, Nicholson encountered the shoal again at 3:45 A.M. and crossed it with a minimum sounding of 140 feet. Tall Gonzales, Lyon recognized, was a ridge rather than a peak. "This area is most critical for exact survey," he noted, "so that we know the best route and can navigate by bathymetry. This is most important and must be done this summer."[56]

*Sargo* continued to encounter heavy ice until 8:30 A.M., when the overhead coverage flattened out with only occasional ridges. Nicholson surfaced off St. Lawrence Island at 9:45 P.M. through eleven inches of soft ice. The temperature, he reported, was "a tropical 37° F." It was, he wrote, "a tremendous relief to know that the toughest part of our transit was behind us and deep water lay ahead."

After completing communications, Nicholson dove at 2:12 A.M. on February 24. Later that morning, he fired an ice destructor mine. It detonated with a tremendous blast, but he was unable to locate the hole that it caused. Nicholson fired another mine later in the day, but it proved a dud. He then headed for Adak at the maximum speed allowed by the damaged sail. *Sargo* managed to work up to fifteen-and-a-half knots. "Noise does not seem too great," Lyon observed.[57]

Lyon again had the midnight to four o'clock watch on February 25. At 12:30 A.M., waves began to appear on the topside sonar recorders, interspersed with ice floes. By 1:30 A.M. *Sargo* had cleared the ice pack, and Lyon secured the NK–Variable Frequency echo sounder.[58]

Nicholson made a routine surfacing shortly after noon. With good communications, the messages of congratulations began to come in.

Admiral Daspit sent his "Congratulations and well done for a tremendous job of skill, perseverance and tenacity in your scientific adventure into one of nature's last frontiers." Adm. Herbert G. Hopwood, CinC-PacFlt, also passed along his congratulations to *Sargo*'s crew and Lyon's scientific group, terming the cruise "another significant milestone in man's conquest of inaccessible areas on earth."[59]

*Sargo* reached Pearl Harbor on March 3. The submarine received a gala welcome. Admiral Ferrall presented Nicholson with the Legion of Merit and announced that *Sargo* had been awarded a Navy Unit Commendation.[60]

After a conference at SubRon Seven on the general results of the patrol and future plans, Lyon departed Honolulu on Pan American at 8:30 P.M. on March 5. After an overnight flight to Los Angeles and a connection to San Diego, he arrived home at nine-thirty the next morning. He wrote in his Scientific Journal on March 6:

> "And so finis SARGO patrol
> The most fabulous patrol of my career."

Thirty-six years later, he jotted next to this entry: "Still true."[61]

Lyon had every right to be pleased with the results of the expedition. *Sargo* had steamed 6,000 miles under the ice, including 2,000 miles in shallow water. It had spent thirty-one days in the ice pack, far exceeding the combined total of all previous submarines. *Sargo* had come to the surface twenty times, at one point breaking through forty-eight-and-a-half inches of ice, a feat that Nicholson termed "one of the major accomplishments of our cruise since it points the way to safe emergency surfacing procedures."[62]

Penetration of the shallow waters of the Bering and Chukchi Seas had been made possible only through the use of NEL-designed sonar equipment. The cruise, Lyon believed, marked the end of the exploratory period of sonar development. Specifications for the first operational under-ice sonar system had now been defined. "The essential point," he argued, "is that now a single, integrated, simple system should be developed immediately for under-ice operations. We should no longer look to the use of many experimental equipments thrown together for each Arctic patrol."[63]

Nicholson agreed with Lyon that the patrol had been an outstanding accomplishment. In his patrol report, he hailed the work of the "capable and dedicated" civilian scientists. In particular, Roshon, by teaching the proper method to interpret the iceberg detector presentation and devising a way to use the BQR-2 hydrophone, had made "a significant contribution to the success of the shallow water transits and is deserving of special praise." He also singled out Lyon's "exemplary" leadership of the scientific group. "The Navy," he wrote, "is fortunate to have such a man leading the way in the Arctic."[64]

Above all, Nicholson took pride in the performance of his crew. He noted the "tenacity and courage" that they had displayed while attempting to remove the iceberg detector training mechanism, as well as their contribution to the successful modification of this critical sonar equipment. "But most of all," he emphasized, "their courage and confidence during and following the tense moments of our near bottoming on 29 January and our encounter with the deep ridge on 19 February was a performance which I will never forget."[65]

Lyon would add to this list of commendations the name of *Sargo*'s commanding officer. His relationship with Nicholson had been exemplary. Never before in his career—and never afterward—had he experienced such a happy and productive partnership between scientist and submariner. Nicholson, he later reflected, had been a "superior" commander, and the success of the patrol had been due in large part to his intelligence and determination.[66]

Executive Officer Yates, who had been Lyon's roommate on the cruise, later pointed out that the scientist had developed "an exceptional rapport" with other members of the crew that contributed to the success of the voyage. "His acceptance by individual officers and crew members throughout the ship," Yates has emphasized, "was an important contribution to confidence and morale. I believe everyone felt Waldo was truly one of us. He spoke our language. He understood our concerns. He was so widely knowledgeable about submarine operations in the Arctic and so easy to talk to, that everyone valued his presence on board. He was always calm, cheerful, assured, and outgoing. It is hard to explain how unusual this relationship was for a civilian on board. I think it is not an exaggeration to say that none of us saw Waldo as a civilian. He was Waldo Lyon, and that meant a lot."[67]

Thanks to the initiative of Admiral Daspit, Lyon's contribution to the development of the under-ice submarine received special recognition. In April, 1960, the American Society of Naval Engineers presented him with their prestigious Gold Medal Award. The *Sargo* patrol, the citation noted, had marked the culmination of pioneering work that had begun in 1946. Its success had been due not only to the equipment that had been developed by Lyon, "but also to the persistence, enthusiasm, leadership, extraordinary professional abilities and devotion with which he pursued his goal against every possible obstacle of technique and personal hardship."[68]

While *Sargo*'s epic voyage represented the pinnacle of the experimental phase of the under-ice program, Lyon recognized that much remained to be done before the arctic submarine could become an operational reality. At the top of his agenda stood preparations for the under-ice submarine transit of the Northwest Passage through the Canadian Archipelago. As this voyage was scheduled to take place in the summer of 1960, he had little time to sit back and enjoy his laurels.

# Closing the Circle

In his senior scientist's report following *Skate*'s winter/spring patrol in 1959, Lyon had recommended that a submarine be scheduled to transit the Canadian Archipelago from Atlantic to Pacific in 1960. Commander Calvert, however, had been eager for additional under-ice work and wanted to advance the date. In April, 1959, less than two weeks after *Skate* returned from the Arctic, Calvert's enthusiasm had been manifested in a letter sent to Lyon by Lt. David S. Boyd, *Skate*'s engineer.

*Skate*, Boyd informed Lyon on April 19, 1959, was already looking forward "to the next cruise north." Why not, he suggested, advance the timetable for the operation through the Canadian Archipelago? *Skate* could depart in early November, 1959, for Baffin Bay, where it could try out the proposed iceberg detector on deep-draft icebergs. From there, *Skate* would head west through Lancaster Sound, Barrow Strait, Viscount Melville Sound, and McClure Strait to the Beaufort Sea. Having completed the Northwest Passage, the submarine would conduct sound transmission studies with scientists on Ice Island T-3 before returning to New London via Robeson Channel.[1]

"I am all in favor of your proposed cruise as well you know," Lyon had responded. NEL, he told Boyd, was building two complete sets of sonar equipment for under-ice operations, one for *Sargo* and the other for *Skate*. Both sets should be ready before the end of the year. Also, he anticipated no problem in securing the required permission from Canada to make the transit. He planned to urge that Commodore Robertson go along on the patrol as an observer and consultant "both to satisfy Canada's national interest and for his personal knowledge of the Barrow Strait–Melville Sound area." Lyon suggested that *Skate* stir up interest at ComSubLant for the cruise—"though as Cdr. Calvert said,

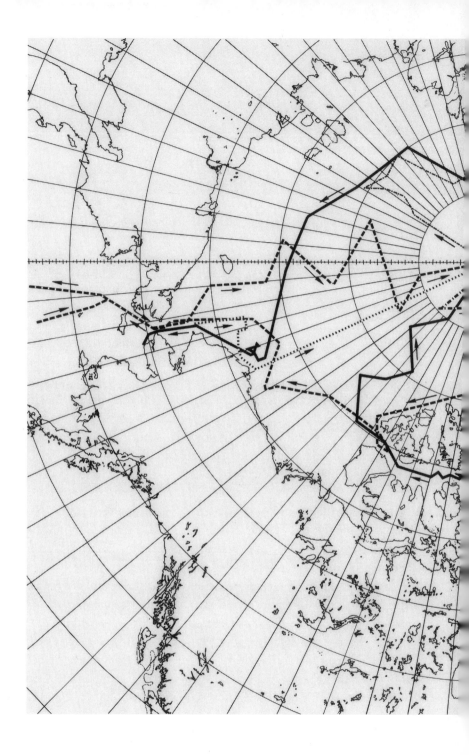

Cruise routes of *Nautilus*, 1957 and 1958; *Sargo*, 1960; *Seadragon*, 1960; and *Skate*, 1958 and 1959. Courtesy U.S. Navy

he cannot appear to be agitating for it." Meanwhile, he would write to Admiral Daspit at Op-31 in an effort to mobilize support for the expedition.²

On May 13 Calvert sent up the chain of command a specific request to conduct a patrol through the Northwest Passage in the fall of 1959. As expected, ComSubRon 10 strongly endorsed the proposal. It received a less favorable reception, however, from ComSubLant. "I cannot understand this," Calvert wrote to Lyon on May 21, "and am frankly getting a little weary of continually having to fight it. I do not intend to give up and realize you have been fighting this sort of thing for many years. But every time we want an Arctic trip the cry about paucity of nuclear submarines and their importance to ASW comes up. I cannot see that any activity in which SSN's have engaged has won the Navy more public support (which it badly needs) than ice work."³

Calvert's pessimistic assessment of the situation was confirmed by Lyon when he visited New London on May 30. ASW exercises and commitments, he learned, would not permit *Skate* to work in the Arctic in the fall. Two weeks later, as Calvert had predicted, ComSubLant appended a negative endorsement to his request. *Skate*, Admiral Warder pointed out, would be one of only two operational nuclear submarines in the Atlantic in the fall. The relative importance of arctic operations had to be weighed against the urgency of using limited submarine time to develop ASW tactics. Furthermore, a transit of the Canadian Archipelago in winter "may be unduly hazardous" as the route through Barrow Strait was poorly charted.⁴

Lyon had not been surprised by the reaction to Calvert's proposal. "I realize and understand full well the overwhelming demand for services of nuclear submarines," he informed Captain Phelps, his superior at NEL. Nonetheless, he continued to lay the groundwork for a voyage through the Canadian Archipelago. In June he visited Montreal for discussions with E. F. Root, coordinator of a project for the ASW defense of northern Canada. Root told Lyon that a nuclear submarine should transit the archipelago as soon as possible "in order to fully alert the RCN to the defense problem and give all concerned an initial feel for submarine operations through ice-covered channels." Lyon suggested that the Royal Canadian Navy send a representative to the U.S. Navy to ask for the patrol. Such a request, he pointed out, also would satisfy Canadian law on the movement of nuclear reactors within the country.⁵

Returning to San Diego, Lyon found waiting for him a letter from

ComSubLant seeking his comments on the hazards of a submarine passage through Barrow Strait. The least known areas, Lyon replied, were the central and southern portions of the strait near Lowther Island. These sectors had never been surveyed, so only opinions were available on the character of the bottom. The very purpose of the transit through the Canadian Archipelago, he reminded ComSubLant, was to examine the feasibility of operating through this area. "We cannot fully visualize the technical requirements of under-ice operations in the channels," he emphasized, "until we have had such experience."[6]

Although the requirement remained for a transit of the Northwest Passage, *Skate* would not have the opportunity to undertake the task. On August 14, 1959, CNO officially killed Calvert's request. "The limited number of nuclear submarines to meet fleet requirements at this time," it pointed out, "precludes an Arctic exploration as proposed [by Calvert]."[7]

As it turned out, CNO's rejection of Calvert's proposal caused only a temporary delay in the desired underwater cruise through the Northwest Passage. While Lyon was dodging ice ridges on *Sargo* in February, 1960, Op-31 was winning approval for a scheme to transfer the newly commissioned *Seadragon* (SSN 584), sister-ship of *Skate*, to the Pacific Fleet by way of the Arctic. "Unless everything falls apart at the last minute," Comdr. M. G. Bayne in Op-31 informed the submarine's skipper, Comdr. George P. Steele, in late February, "*Seadragon* will make the trip."[8]

Lyon learned about the plan shortly after he returned to San Diego from the *Sargo* patrol in March. He hardly had time to unpack his gear before being ordered to a meeting in Room 4E442 of the Pentagon to discuss the details of the operation. At this conference on March 28, the track chart for the cruise received final approval. *Seadragon* would proceed from New London to Pearl Harbor on or about August 1, 1960, via Baffin Bay, Lancaster Sound, Barrow Strait, Viscount Melville Sound, and McClure Strait into the Beaufort Sea. Having completed the Northwest Passage, *Seadragon* would then zigzag to the North Pole. From there, it would head toward the Soviet island of Novaya Zemlya and survey a portion of the Kara and Laptev Seas. This task completed, *Seadragon* would exit the Arctic Ocean via Bering Strait, stopping en route to conduct sound transmission studies with T-3/Bravo.[9]

Lyon hoped to achieve several important objectives on the cruise. Previous under-ice voyages had demonstrated that nuclear submarines could operate in the Arctic in deep water with continuous ice coverage during all seasons of the year. They also could penetrate the shallow-water Pacific entrance to the arctic basin under a winter ice canopy. *Seadragon* would investigate the ability of a submarine to operate safely in the iceberg-infested waters of Baffin Bay and to transit the ice-covered passageways through the Canadian Archipelago. Another major objective of the patrol, Lyon stressed, was "to complete the evaluation of the design of an under-ice sonar system that can be operated by fleet personnel."[10]

*Seadragon* would carry all the components of what was envisioned as an integrated ice suit. NEL would provide the iceberg detector, NK–Variable Frequency fathometer to measure ice thickness, rate-of-rise meter, precision depth recorder, and ambient light meter. Commercially available equipment included the BQN-4 topside upward-beamed echo sounder (five transducers topside and one bottomside), UQN-1 fathometer, and BQS-7 polynya delineator. The latter piece of equipment, produced by the EDO Corporation, would undergo a formal Fleet Operational Investigation during the cruise.[11]

While Lyon prepared the necessary sonar equipment for *Seadragon*, Steele made his plans for the upcoming voyage. The critical area of the Northwest Passage, he recognized, would be the largely uncharted 150 miles of Barrow Strait. Reviewing the few available soundings, Steele concluded that "it seemed deep enough for a submarine here, too shallow there—just a half mile away—without enough soundings to indicate where a place to go through might be. A wild, rugged bottom seemed likely."[12]

In search of information on the area, Steele and his navigator, Lt. Edward A. Burkhalter, Jr., visited the extensive Stefansson Collection of Arctic Literature at Dartmouth College. The naval officers enjoyed the opportunity to discuss their upcoming voyage with Vilhjalmur Stefansson, the famous explorer. "We were enthralled by the still-vigorous old man," Steele wrote, "whose big frame and vitality were so suited to his exploits." Browsing through the books in the library, Steele and Burkhalter discovered "a number of priceless gems," including a copy of Sir Edward Parry's *Journal of a Voyage for the Discovery of a North-West Passage from the Atlantic to the Pacific*. *Seadragon* would be following the same route as Parry had sailed in 1819–20 when he penetrated Barrow Strait and Viscount Melville Sound before being stopped by ice at the entrance to McClure Strait. Steele received permission to take a copy of the book with him on the patrol.[13]

By the end of June, all elements of the upcoming voyage were falling into place. Ottawa had approved *Seadragon*'s transit through Canadian territory, thanks largely to the efforts of Commodore Robertson, who would be going along on the patrol as observer and adviser. The U.S. Navy Underwater Sound Laboratory had agreed to reactivate the acoustical measurement station on T-3/Bravo so that sound tests could be conducted with *Seadragon*. Also, arrangements had been made for the civilian contingent on *Seadragon*. This group included Lyon as senior scientist and expedition coordinator; Art Roshon of NEL and Jonathan C. Schere of the EDO Corporation to look after the sonar equipment; Walt Wittmann and Art Molloy of the Hydrographic Office; and two engineers from the Sperry Corporation to monitor the operation of the newly-installed Ship's Inertial Navigation System (SINS).[14]

Early July saw the sonar gear installed on *Seadragon* and tested during sea trials. One request for equipment, however, had created a major problem for Lyon. Steele wanted a videotape recorder so that he could make a record of the voyage for training purposes. Unlike the later compact VCRs, the recorder of the early 1960s was a bulky piece of gear that occupied several large cabinets and was used only in television studios. Lyon, after a good deal of effort, managed to persuade RCA to build a special unit that could be installed in a torpedo rack on *Seadragon*.[15]

On July 15, 1960, ComSubLant issued the operations order for the cruise. It followed the outline of the March conference except for one important detail. CNO had had second thoughts about the reconnaissance mission to the Kara and Laptev Seas, and had decided to cancel it. Lyon was sorely disappointed. "The glaring default of the SEADRAGON patrol," he later wrote, "was the withdrawal of her track by higher authority away from the continental shelf in the Kara and Laptev Seas." Lyon believed that a survey of this area needed to be conducted "as quickly as possible while we have the advantage of Arctic submarine superiority." The decision not to undertake this mission with *Seadragon*, he predicted, "will be regretted."[16]

*Seadragon* departed New London on August 1, 1960. The submarine made a routine, underwater passage, first heading east to enter the deep water off the continental shelf, then turning north into the Labrador Sea. Wittmann gave his usual talk on ice forecasting, Roshon spoke about the sonar equipment, and Lyon reviewed the history of under-ice submarine operations. A special treat came with Commodore Robertson's well-received talk on the Canadian Arctic and showing of *The Navy Goes Forth*, a film about *Labrador*'s 1954 Northwest Passage.[17]

By August 10 *Seadragon* was proceeding through Davis Strait en route to Baffin Bay. Shortly after midnight, Steele detected the first icebergs on his sonar. At 4:18 A.M. GMT, he surfaced in the vicinity of a medium-sized piece of floating glacial ice, known as a bergy bit. Fifteen miles away, just north of the Arctic Circle, lay a full-sized berg. "Excellent!" Steele wrote in his patrol report. "We shall dive and calibrate our Iceberg Detector gain control on the 26 foot high bergy bit and then go over to the iceberg and evaluate our gear."[18]

Diving to 600 feet, Steele found that both the ship's SQS-4 sonar and the iceberg detector easily registered the bergy bit. Passing underneath the ice, the upward-beamed echo sounder recorded a draft of 82 feet. Steele was then able to adjust the gain on the iceberg detector to give a correct reading. As Roshon observed, Steele was now able to approach the larger target "with confidence."[19]

Steele surfaced for a visual inspection of the full-sized iceberg. It showed that the berg stood 74 feet above the water and had a longest axis of 313 feet. Steele made four underwater sonar runs on the target at a cruising depth of 600 feet to check detection ranges at various speeds. Even at flank speed of 20.5 knots, he found that the ship's SQS-4 active sonar and the iceberg detector made contact at 3,000 yards. Clearly, *Seadragon* would have ample time to avoid icebergs if encountered unexpectedly.

On one of the runs toward the iceberg, *Seadragon*'s sonar operator heard a loud underwater explosion. As ice expert Wittmann had judged the berg to be unstable, it seemed likely that a large piece had broken off. An inspection of the PPI image, however, failed to reveal any change in the dimensions of the target. While the incident gave Steele pause for thought about the potential danger of passing under an unstable iceberg, he nonetheless decided to press on with the experiment. Keenly aware that no submarine had ever passed beneath an iceberg, Steele reported that "I was determined to go underneath—*now*."[20]

Steele approached the iceberg at a speed of seven knots and a depth of 650 feet, making sure to give it ample clearance. It was a dramatic moment, he recalled. The iceberg detector lost contact, as expected, at a distance of 200 yards. It would take another 60 seconds for *Seadragon* to pass underneath the berg. "I bent over the low-mouthed fathometer with the others," Steele remembered. "My hand gripped the stanchion by the periscope; the trace [of the upward-beamed echo sounder] dipped down, down toward our egg-shaped hull—and rose up again." Once clear of the berg, Steele made what Roshon described as an "ecstatic announcement." Picking up the microphone for the ship's public address system, Steele reported to the crew: "*Seadragon* has the honor to be the first submarine to go under an iceberg! Its draft: 108 feet!"[21]

It was soon obvious, Roshon noted, that Steele had become "greatly intrigued" with the notion of running under icebergs. He made five additional passes under the target, sketching the bottom of the ice. The longest underwater axis of the berg turned out to be 822 feet. When added to its above water height of 74 feet and draft of 108 feet, Wittmann calculated its total mass at 600,000 tons.[22]

Steele headed northwest toward Baffin Bay at sixteen knots "with a feeling of accomplishment." He was certain that he would be able to deal with the numerous icebergs that lay ahead. "I had absolutely no doubt," he later wrote, "that our instruments could handle bergs either inside or outside the pack. . . ."[23]

With a position report due to ComSubLant, Steele surfaced amidst block and brash ice at 3:46 A.M. on August 11. In his memoir of the patrol, Steele recounted at some length the procedure employed to bring *Seadragon* to the surface, noting that Lyon carefully observed the process in the control room. Steele's memory, however, failed him. Lyon, in his Scientific Journal, noted that this was the first time in his under-ice career that he had *not* taken an active part in the surfacing process. "I stayed in my rack," he wrote. He did so deliberately because one important purpose of the cruise was to allow the submarine's crew to conduct "routine" operations. Summer surfacing in ice was now routine. "We have come of age," Lyon concluded.[24]

Later in the day, as *Seadragon* proceeded under the middle ice pack in Baffin Bay, the ship's SQS-4 sonar recorded a loud echo on a target at a distance of more than 4,000 yards. An examination of the target with the iceberg detector revealed an extremely deep-draft piece of ice. "The size and depth of this iceberg," Steele wrote, "were impressive."

He wanted to go underneath it and pressed Lyon and Roshon for an estimate of its draft. As best they could tell from the iceberg detector at a distance beyond 1,000 yards, the massive berg appeared to have a draft of 400 to 440 feet. Steele decided to make a pass underneath at a depth that surely would avoid the bottom of the iceberg.[25]

Steele approached the iceberg at seven knots and 700 feet, the maximum peacetime operating depth of *Skate*-class submarines. "Suddenly," Steele reported—in the present tense—in his patrol report, "the fathometer trace rises revealing the sheer side of the berg. Swiftly it rises past our deepest prediction and keeps rising." Just as Steele began to voice the order for an emergency dive to 800 feet, the trace of the fathometer registered a maximum of 138 feet, held for a second, then fell away. The giant iceberg had a draft of 514 feet, more than three times the draft of Ice Island T-3. "The iceberg detector," Steele reported, "was in error by nearly 17% in the unsafe direction." If he had not allowed for an extra safety margin, Steele believed, "there would have been disaster."[26]

Jonathan Schere, the representative of the EDO Corporation, was not surprised at the limited accuracy of the equipment. The iceberg detector projected a 3° vertical beam. The operator noted the range at which the ice overhead faded out of view as the submarine approached, then consulted a chart that related fade distance to clearance. As the fade distance was calibrated for a 1.5° up angle from the top edge of the beam, any angle error—due to beam bending or deviation from a zero bubble—would result in a proportional range error. For example, a 0.5° error in angle would give a 33 percent error in clearance distance. Also, as fade angle depended on target strength, the system had to be calibrated on targets of similar strength. However, there had not been a target of comparable size to the monster berg. "Bottom line," Schere pointed out, "the Iceberg Detector would not give a reliable hit/miss indication, but was just a rough estimator for miss distance. I don't think Steele really understood that."[27]

In his memoir of the voyage, Steele recorded the drama of two additional passes under the iceberg. "The immense block of ice," he wrote, "was instinctively sensed by many of the men, who felt a strong desire to duck as we went under." Actually, Lyon and Robertson had gone to lunch after the first pass under the iceberg and were waiting in the wardroom for Steele to appear. Neither man was especially pleased with Steele's actions in taking the submarine under the large iceberg. Lyon, Schere recalled, "thought it was a stupid publicity stunt and that Steele wasn't thinking about the dangers to the ship." When the captain finally showed up for lunch, Robertson quipped that the group had unanimously decided to honor him with being made "a charter member of Icebergs Anonymous."[28]

Steele certainly had developed a passion for icebergs. With few icebergs in central Baffin Bay, he decided to add two days to the patrol in order to seek out an area of higher concentration. After some discussion with ice expert Wittmann, Steele turned *Seadragon* toward Greenland's Kap York, into an area known as "iceberg alley." He was determined, he wrote, to demonstrate "beyond question the ability of a nuclear submarine to enter an area of high iceberg concentration with safety."[29]

In the early morning hours of August 13, *Seadragon* entered an area rich in icebergs, counting twenty-seven in one ten-mile stretch. The sonar, Lyon reported, had no trouble detecting the icebergs while *Seadragon* passed by at sixteen knots. The submarine surfaced some forty miles off Kap York at 4:00 A.M. and waited for a heavy fog to lift. Lookouts began sighting bergs at 9:45 A.M. At one point, more than thirty-nine towering pieces of glacial ice were in view. *Seadragon* moved on the surface amid the frozen giants, taking photographs and measurements. "The crowd of visitors on the main deck," Steele recalled, "could not seem to get enough of the breath-taking view." At first, he noticed, they were awed by the sight. Their silence, however, soon turned to "exuberant joking," followed by "determined picture-taking."[30]

After completing the surface survey, Steele dove and made runs on selected targets to determine their depth by iceberg detector. In all, he examined five icebergs, passing underneath them a total of eleven times. The point of this exercise, Steele later explained, was to learn how the captain of a nuclear-powered submarine might fight an enemy in Baffin Bay. He discovered that icebergs gave off a loud hissing noise as they melted, raising the possibility that a submarine might avoid them by using passive sonar. This would give the captain an advantage for attack or evasion. An iceberg could be used to mask an attack. If detected by the enemy, the captain could dodge among the icebergs, which would make good targets for the enemy's active sonar and draw off his homing torpedoes.[31]

Finally, at 0300 on August 14, Steele decided that he had sufficient preliminary information on what the iceberg detector/submarine team could do if faced with combat in Baffin Bay. "The weariness of five busy days had us exhausted," Steele noted in his patrol report, "and I swear off bergs in loud tones."[32]

*Seadragon* headed across Baffin Bay toward Lancaster Sound for a scheduled stop at the isolated Royal Canadian Air Force station of Resolute on the southern tip of Cornwallis Island. Steele surfaced shortly before noon on August 16, took a navigational fix, then made radio contact with Resolute. When *Seadragon* arrived off the station at 6:00 P.M., a delegation headed by Squadron Leader S. E. Milikan, commanding officer of the RCAF contingent, came on board for dinner. The stewards, Steele wrote, did "a superlative job," serving shrimp cocktail and chicken cacciatore. There was even "a short snort of Scotch," Lyon reported—but only for the guests and civilian members of *Seadragon*. The ship's officers, Steele noted, could only look on "with a wistful envy."[33]

Squadron Leader Milikan presented Steele with a handsomely carved "seadragon" of walrus tusk ivory. Rendered by an Inuit craftsman, it featured a long and slender creature with numerous fins and a fierce eye. The group departed at 10:30 P.M., carrying mail and publicity material. "This boat is certainly photo conscious," Lyon remarked in his Scientific Journal, "and strong on stories, publicity, releases."[34]

*Seadragon* proceeded toward the western portion of Barrow Strait the next morning. There were four known islands in the area—Griffith, Young, Garrett, and Lowther—and one (Davy) that had been reported by Sir Edward Parry in 1820, marked on the chart as "Existence Doubtful." Soundings in the strait also had been taken by Parry, Steele noted, and were both "unmistakable and rather lonely on the chart we were using." He decided to make a series of survey runs between Lowther and Garrett Islands in the northern part of the strait, then cover the southern portion between Lowther and Young Islands.[35]

After surfacing and taking a navigational fix, *Seadragon* commenced the first survey run, proceeding at 120 feet and seven knots. Steele stood beside hydrographer Molloy as the fathometer stylus traced out the bottom of the sea. Watching the recorder, Steele experienced "a thrill at being in the first ship ever to take extensive soundings of this little-known corner of the world; in a modest fashion we were now entering the great company of explorers."[36]

"The bottom looks rugged as we pass over sharp rises and high hills," Steele noted in his patrol report. "Nothing really menaces us and I begin to relax even though only a few random depths are charted ahead." Fortunately, the waters of the strait remained free of ice, so Steele did not have to worry about dodging downward-projecting ice ridges.[37]

At 4:15 A.M. on August 18, *Seadragon* passed through Barrow Strait and into the 70-fathom water of Lancaster Sound. Steele then reversed course and headed back into the strait, taking a survey line 3,000 yards to the northwest of his original track. At 7:20 A.M., as *Seadragon* approached Garrett Island, intending to pass 2,000 yards abeam, the water shoaled rapidly from 106 to 26 fathoms. When Steele came to the surface in turbulent seas to check his position, radar revealed that the submarine had been swept off course by an unexpected current. "Had we continued on our way," Steele wrote in his memoir of the patrol, "we would have run squarely into an island." As Roshon commented after reading this passage in Steele's book, "Not really of course, since both sonars would soon have displayed it."[38]

After completing his soundings between Garrett and Lowther Islands, Steele concluded that the underwater terrain in the northern portion of Barrow Strait was irregular in character with a least depth of forty-two fathoms. Although a submarine could navigate through this area with proper charts, he decided to continue the survey in hopes of finding a better route.[39]

Shortly after midnight on August 19, Steele commenced an investigation of the area south of Lowther Island. A series of survey lines soon revealed a channel some six miles wide with a least depth of thirty-six fathoms on the southern side and over sixty-five fathoms in the center. The passageway had a regular bottom except for one seamount at forty fathoms that would serve as a handy sonar navigational aid. As a delighted Steele announced to his crew: "*Seadragon* has found a new passage through the Barrow Strait!"[40]

Having accomplished a central objective of the cruise, Steele decided to make a high-speed run across Viscount Melville Sound in order to emphasize the advantages of the Northwest Passage over the Panama Canal route for submarines passing between the Atlantic and Pacific Oceans. Steaming at sixteen knots and 300 feet, *Seadragon* cleared McClure Strait at 2:15 A.M. on August 21. Steele brought the submarine to the surface, established radio contact with Pearl Harbor, announced that *Seadragon* had completed the first submerged transit of the Northwest Passage, and reported for duty with the Pacific Fleet.[41]

Following the track plan, *Seadragon* set course for the North Pole. As the submarine traveled under the ice pack, Steele was treated to a spectacular view through his periscope of the ice ridges passing overhead. "It is beyond belief," Steele noted in his patrol report on August 23, "my power of description is utterly inadequate to do justice to the variety of shape, color and arrangement of heavy ice overhead." When Steele rotated the periscope for maximum elevation, he had an even more stunning sight of the passing ridges. "A procession of boulders seems to roll toward us frightening the senses," he observed, "even though there is at least 30 feet clearance to the deepest ridge."[42]

Confident that the iceberg detector would give ample warning of an especially deep-draft piece of ice, Steele decided to share this impressive view with the crew. For hours thereafter, men came to the control room to use the periscope. "I felt," Steele recalled, "like the proprietor of the greatest show on earth."[43]

The show continued into the afternoon with a junior officer's taking over the conn from Steele. Roshon was at watch at the iceberg detector when a target beyond the 2,000-yard range produced a loud sound. "The echo strength," he noted, "was a certain clue that it was larger and had a deeper draft than the other ridges." The sonarman on duty reported the contact to the conning officer. "Very well," he replied, continuing to supervise the use of the periscope by members of the ship's crew. As the target appeared on the edge of the 2,000-yard scale, Roshon observed that it seemed "even more foreboding." Again, the conn was informed, and again the information was acknowledged.

At *Seadragon*'s current speed of seven knots, Roshon knew that they would reach the target in nine minutes. "The sonarman and I were becoming anxious," he recalled. The sonarman alerted the conning officer with an increased sense of urgency as the target approached the 1,000-, then 500-yard scale. Each time, the young officer replied, "Very well." Finally, when the target reached 200 yards, the sonarman shouted, "Mr. X, we are going to strike that big target if you stay at this depth; it is less than 200 yards ahead!"

This announcement finally caught the conning officer's attention. He looked quickly at the PPI scope of the iceberg detector, then ordered an immediate dive while lowering the periscope. Roshon, the sonarman, and the conning officer held their breath while awaiting the possible collision. In less than a minute, the ridge passed overhead with only a few feet of clearance. It had been a close call, and a needless one.[44]

*Seadragon* arrived at the North Pole shortly after 10:00 A.M. on August 25, the fourth submarine to reach the top of the world. Steele badly wanted to surface at the pole, but he found no suitable openings in the thick ice cover overhead. He immediately initiated a search for a polynya, but the hours passed without success. Finally, at 10:20 P.M., he brought *Seadragon* through two inches of ice into a small, circular polynya some twelve miles from the pole.[45]

*Seadragon* remained on the surface for the next twelve hours. As there was no place to moor, the submarine had to stay underway with a full-steaming watch, although the engines rarely had to be used to hold the ship in position. With good weather, Steele decided to organize a softball game on the ice. "Playing softball," Steele later pointed out, "was intended to show friends and potential enemies that our nuclear submarines now had things under control in the Arctic." In any event, everyone had a good time—the crew defeated the officers and chiefs by a score of thirteen to ten—and the story later generated front-page coverage in the *New York Times*. Communication was excellent, as it had been throughout the cruise, and Steele was able to talk not only to ComSubPac, but also to hold a press conference.[46]

Before leaving the polynya, Steele wanted a photograph of *Seadragon*'s surfacing through the thin ice cover. Lt. Glenn M. Brewer, head of the underwater development section of the Naval Photographic Center, set up his camera on the ice. He would remain on the surface, along with two other volunteer photographers (the ship's doctor, Lt. Lewis A. Seaton, and Sonarman Robert E. Hammon) as *Seadragon* dropped down to 150 feet, then reemerged. It would be a "very simple" maneuver, Steele believed. "Twist to port, move ahead a little, and come up in a few minutes to recover the party."[47]

Steele initiated the vertical descent at 10:20 A.M. on August 26. The diving officer, however, was unable to stop *Seadragon* at 150 feet, and the submarine continued down to 235 feet. As the ship began to rise, Steele applied full left rudder while backing the port engine at one-third speed and going ahead one-third on the starboard engine. Reaching 150 feet, he raised the periscope. To Steele's great dismay, he found that *Seadragon* had drifted toward the edge of the small polynya. He tried desperately to keep the submarine in the opening. "Everyone in the control room," he recalled, "was frozen in horror as I continued the losing struggle." Finally, he realized that he had no choice except to make a quick Williamson turn and come back to the polynya.[48]

On the surface, Brewer and his companions watched *Seadragon* disappear, then heard the pinging of the submarine's sonar begin to grow distant. They were alone on the ice, without survival gear. What was planned as a simple operation suddenly had become far more complex—and threatened three lives.

Finding a small polynya amidst the ice pack was never an easy task, as Steele had learned during his recent lengthy and frustrating search for an opening near the North Pole. "Many long and anxious moments," Lyon noted in his Scientific Journal, "knowing the difficulty to locate a polynya." Finally, after the longest fifteen minutes of his life, Steele found the opening. "Needless to say," Lyon wrote, "men on ice were most thankful." "I pause here," Steele wrote in his patrol report, "to give this advice with all earnestness to future captains of submarines in the Arctic: until equipment is vastly improved, do not under any circumstances leave men outside your ship and dive in the polar pack."[49]

Just how close the disaster had come was emphasized later in the day when Steele fired an ice destructor mine, one of two that were tested during the patrol. This one, unlike the previous mine, exploded. He searched for the hole for two hours but could not find it, bringing home how easily he might have failed to locate the polynya with the photographic party. "Death would have been easier to bear," Steele later wrote, "than the responsibility for leaving those men to die at the North Pole."[50]

*Seadragon* set course for T-3/Bravo, located off Point Barrow, running at 360 feet and sixteen knots. Everything was routine until shortly after midnight on September 2. As Steele worked on his patrol report, the general emergency alarm sounded. "Fire in the torpedo room! Fire in the torpedo room!"

Steele immediately went to the control room and ordered *Seadragon* turned in the direction of the last polynya that had been detected by the ice-profiling sonar, located some two miles away. As smoke began to circulate throughout the submarine, the crew plugged their emergency breathing masks into the outlets of the clean-air manifold. By the time *Seadragon* reached the polynya, however, the fire had been extinguished and the smoke cleared.[51]

Steele later learned that a crew member who had been charging a torpedo battery had placed the heavy battery clamps used for the charge on top of an oily rag on the deck. The clamps "bit" through the rag to the deck, causing sparks that ignited the rag. As clean air from the "scrubber" system on *Skate*-class submarines went first to the torpedo room before being routed by piping to all other compartments, the smoke from the fire quickly spread throughout the ship. Thanks to the experience of *Nautilus* in 1958, however, *Seadragon* was equipped with a multi-outlet, clean-air manifold, connected to the air supply used to blow the submarine's ballast tanks. Each crew member had an emergency breathing mask that could be plugged into the outlets.[52]

"The fire alarm under the ice," Steele later wrote, "would live in men's memories as the most dreadful sound of the cruise." In retrospect, however, "our most severe trial of all was just ahead." This trial took place two days later while *Seadragon* was conducting sound transmission studies in shallow, ice-filled waters near T-3/Bravo.[53]

Early on September 3, Steele was called to the control room by Lieutenant Burkhalter, who had the conn, and informed that the iceberg detector had failed. "It was probably the most dangerous moment of the trip," Steele wrote in his memoir of the cruise. "Ahead lay certain danger; we could not surface since there was no polynya. Even if we turned around we could not go far before the drifting ice would throw an obstacle in our way. We could not hover with enough accuracy in these cramped quarters to prevent hitting bottom or ice."

Steele, relying on the "instincts" developed during his sixteen years of naval service, quickly determined that a sharp thermal layer might be distorting the beam from the iceberg detector. He ordered the depth increased from 110 to 120 feet, placing *Seadragon* 30 feet from the bottom. The technique worked. The submarine moved out of the thermal layer, and the iceberg detector showed the proper display. Steele describes the scene: "'Whew!' exploded the heretofore imperturbable Waldo Lyon. 'That was too close.'"[54]

Lyon's Scientific Journal and Roshon's memoirs give a somewhat different version of this incident. Both Lyon and Roshon had been long aware of the potential problem that might be caused by a sharp temperature gradient. The hydrophone for the iceberg detector was mounted in the upper portion of the sail's leading edge, while the transducer was located six feet further down on the sail. It was theoretically possible for the hydrophone to be in colder water than the transducer. This would cause the sound beam from the transducer to bend down; the sound reaching the hydrophone would then come from the water stratum above the temperature gradient. Roshon, in fact, had alerted the crews of both *Sargo* and *Seadragon* to this possibility.

When Roshon arrived in the control room on the morning of September 3, he recalled that "faces were truly white." Targets on the PPI screen were not appearing until about 500 yards, then vanishing after holding for 100 yards. After a quick check of the equipment, Roshon concluded that the hydrophone was not hearing what the transducer was sending. "Voilà! We were in that improbable situation of a sharp gradient." As he told this to Steele, Roshon recalled, "Waldo just smiled and nodded his head."[55]

"Believe trouble was upward refraction in confused temperature gradient," Lyon noted in his Scientific Journal, "such as observed in previous submarine and icebreaker cruises in this area." He recommended to Steele that depth be increased by ten feet both to afford a greater margin of safety from the overhead ice and to improve target discrimination. "New experience for SEADRAGON," he noted, "which became old hat for SARGO. Always takes time to entice submarine [commander] closer to safety by running closer to bottom but for which there is much more reliability of flatness than that of ice canopy overhead."[56]

Having "survived" the encounter with the temperature gradient, *Seadragon* completed the sound transmission study with T-3/Bravo and headed for Nome. The submarine passed through the Bering Strait on the surface in calm seas on the evening of September 5. It arrived off Nome the next afternoon and moored alongside the icebreaker *Northwind*.[57]

Rear Adm. Henry H. Caldwell, commander of the Alaskan Sea Frontier, led a welcoming delegation that included the mayor and other prominent citizens of the city, pleased to celebrate the arrival of the first nuclear submarine to call at Nome. During a press conference on *Northwind*, a "Well Done" message from President Eisenhower was read to the cheering crew. Following a reception at the North Star Hotel that was hosted by the local chamber of commerce, Steele gave his men liberty to tour the town. This marked the first time that a submarine crew had been granted liberty in Nome since a fight between navy and army personnel in 1946. The men, Steele reported, lived up to the challenge "and no incident of any kind was reported ashore." Lyon, who had been to Nome many times, used the opportunity to take a three-hour walk along the beach.[58]

*Seadragon* got underway at 8:00 A.M. local time (3:00 P.M. GMT) on September 6. Eight days later, following a routine submerged passage, the submarine reached Pearl Harbor. *Seadragon* was led to its anchorage by water-spouting fire boats as helicopters dropped flower petals. Rear Adm. Roy S. Benson, ComSubPac, expressed his congratulations, then awarded Steele the Legion of Merit and announced that the crew would receive Navy Commendation Medals. Following a press conference, Lyon went over to the nearby *Sargo* for lunch with Nicholson. "Most enjoyable," he recorded in his Scientific Journal; "seemed like old home."[59]

Steele believed that *Seadragon* had accomplished its mission objectives in exemplary fashion. The patrol had not only demonstrated the feasibility of an underwater Northwest Passage, but it also had discovered an important deep-water channel through Barrow Strait that could be used to advantage by submarines in the future. In addition, the cruise had shown in a convincing manner that submarines could proceed at high speed through areas with high concentrations of icebergs. "Going under icebergs was not just a publicity stunt," he later emphasized. "My task in the Baffin Bay was to see if properly outfitted nuclear powered submarines could operate effectively in that iceberg rich environment. If so they could defend the Atlantic shipping lanes against Soviet nuclear submarines coming down from the Arctic Ocean between Greenland and the Canadian Archipelago and through Baffin Bay bypassing the well defended Greenland–Iceland–United Kingdom Gap. There the SOSUS detection system was backed up by surface, air, and submarines, but in theory only submarines could work under and around Baffin Bay ice and icebergs and *if this could be proven,* they could constitute the only barrier force possible there."[60]

Steele reported himself "very pleased with the performance and reliability" of the ice suit. The iceberg detector, polynya delineator, and upward-beamed echo sounders had all lived up to expectations. Only the NK–Variable Frequency echo sounder gave him cause for concern as it had failed to measure ice thickness with sufficient precision.

Communications, unlike the situation experienced by some earlier arctic submarines, had been excellent. It had been possible to talk, loudly and clearly, on single sideband radio to at least one station during the entire patrol. The high point had come while *Seadragon* was moored near the North Pole and Steele was able not only to talk to his superiors in Pearl Harbor but also hold a press conference.

The performance of the Sperry SINS (MK I), Steele noted, had proved "a pleasant surprise throughout the cruise." There was no question in his mind that the equipment, when used in conjunction with other information available, would allow nuclear submarines to navigate at high latitudes with confidence. Steele believed, however, that ship's personnel required additional training in the operation of the SINS.

Perhaps the most disappointing development concerned the ice destructor. Two mines were tested, but only one had detonated. "What is it," Steele asked, "about Arctic operations that defeats our firing circuits?" He stressed the need for "utmost reliability" in these circuits. The failure of the ice destructors in tests by *Seadragon* and other submarines, he concluded, called into question the suitability of all submarine weapons for arctic use.[61]

Lyon agreed with Steele on the valuable contributions that *Seadragon* had made. The patrol, he wrote in his senior scientist's report, had "closed the circle" on experimental arctic cruises that had begun in 1947 with *Boarfish*. Submarines could now patrol safely in the ice-covered waters of the Arctic Ocean and its approaches during all seasons of the year. The necessary sonar equipment, operated by ships' crews, was ready to shift from experimental prototypes to production models.

The next step in the development of the arctic submarine as an effective combat weapon, Lyon argued, would involve a research program to produce a submarine that could take maximum tactical advantage of the ice canopy. He proposed that the Submarine Research Facility at NEL undertake a study of the momentum-impact method of breaking through the ice cover. The techniques used by *Skate* and *Sargo*, he pointed out, had been based "on rather questionable broad assumptions." A controlled laboratory experiment, on the other hand, followed by a field study, would provide the necessary data for the design of an arctic submarine with "decisive tactical advantage" over opponents.

Lyon also agreed with Steele that a study of weapons was necessary. War-fighting exercises should be conducted to determine if torpedoes could be used effectively in ice-covered seas. In addition, submarine versus submarine war games would yield valuable information on the tactical aspects of arctic operations.[62]

"The Arctic Ocean is the submariner's private sea," Lyon emphasized; "hence, his sole capability to exploit and control." The *Seadragon* patrol had marked "an end of a chapter of all experimental cruises." It was now necessary to move on to the next phase—military control of the Arctic Ocean and its approaches.[63]

*Nautilus* welcomed in New York, August 25, 1959. Courtesy U.S. Navy

Lyon and Hyman G. Rickover on deck of *Nautilus* in New York harbor, August 25, 1958. Courtesy U.S. Navy

*Skate* surfaces through frozen lead, March, 1959. Courtesy U.S. Navy

Lyon (lying on deck) and Art Roshon work on iceberg-dectecting sonar on *Sargo*, February, 1960.
Courtesy U.S. Navy

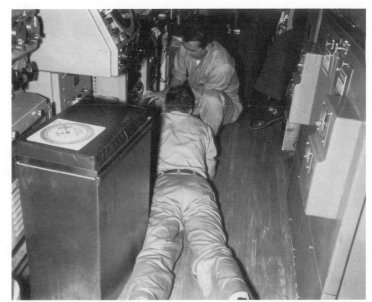

Under-ice sonar equipment on *Sargo,* February, 1960.
Courtesy U.S. Navy

*Sargo* surfaces through forty-eight-and-a-half inches of ice, February 19, 1960.
Courtesy U.S. Navy

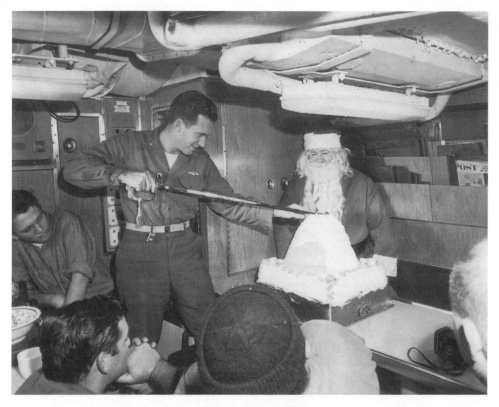

Commander John H. Nicholson cuts cake to celebrate attainment of North Pole as Santa Claus looks on, February 9, 1960. Courtesy U.S. Navy

Crew inspects damage to *Sargo* caused by collision with ice on February 19, 1960. Courtesy U.S. Navy

*Seadragon* is dwarfed by iceberg in Baffin Bay, August 11, 1960. Courtesy U.S. Navy

Lyon, Commodore O. C. S. Robertson, and Art Malloy on *Seadragon,* August, 1960. Courtesy U.S. Navy

*Skate* and *Seadragon* moored at the North Pole, August 2, 1962. Courtesy U.S. Navy

Scuba team from *Skate* prepares to dive into polynya, August, 1962. Courtesy U.S. Navy

President John F. Kennedy congratulates Virginia Lyon, who received her husband's Presidential Medal for Distinguished Federal Service, August 7, 1962. Courtesy U.S. Navy

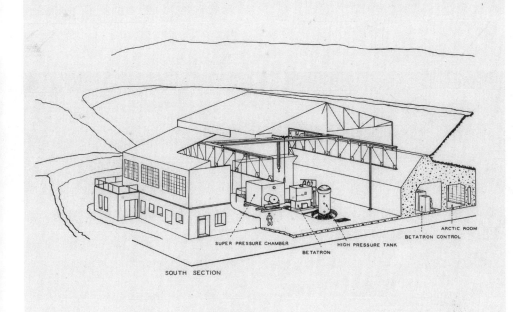
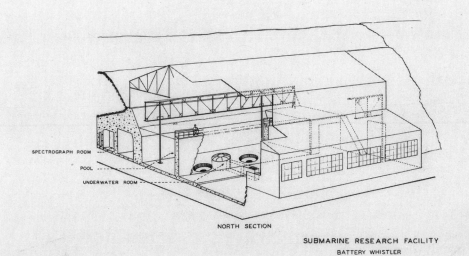

Arctic Submarine Laboratory. Courtesy U.S. Navy

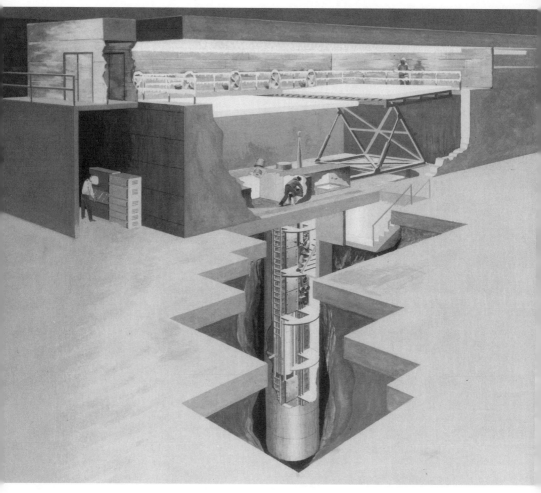

Sketch of preparation for ice break-through tests at Arctic Submarine Laboratory, 1965.
Courtesy U.S. Navy

Staff of Arctic Submarine Laboratory receive award from Capt. Harry C. Mason, *standing far left,* commander of NEL, August 1, 1963. *Standing left to right:* Mason, Donald H. Stephens, Gene L. Bloom, Clem Walton, A. Wayne Medlin; *seated left to right:* M. Allan Beal, Daniel R. Leonard, Walter E. Schatzberg, Bruce A. Grenfell, William O. Carroll. Courtesy U.S. Navy

USS *Whale,* first 637-class submarine to surface at the North Pole, April 6, 1969. Captain William M. Wolff, Jr., in center, with hand on the state flag of South Carolina (for *Whale*'s new home port of Charleston). Courtesy William M. Wolff, Jr.

Recipients of the Lowell Thomas Award from the Explorers Club on USS *Growler*, New York City, November 12, 1997. *Left to right:* Vice Adm. George P. Steele, Vice Adm. John H. Nicholson, Vice Adm. James F. Calvert, Dr. Waldo K. Lyon, Capt. William R. Anderson. Courtesy Eve Josephson

# 9
## Tactics and Weapons

When Lyon returned from the *Seadragon* patrol, he was ready to move on to the next phase in developing an operational arctic submarine. The recommendations set forth in his senior scientist's report, Lyon believed, would require the Submarine Research Facility at NEL to focus on five areas. Top priority, requiring 60 percent of available manpower and funds, should go toward modifying the arctic pool to grow sea ice. Controlled impact studies could then determine how submarines broke through sea ice, leading to design modifications. Second priority should be given to under-ice war games to evaluate weapons, tactics, and the prototype sonar ice suit that was being developed by the EDO Corporation. Also, there remained an urgent requirement to conduct the oceanographic survey of the Soviet sector of the Arctic Ocean that had been scheduled for but not done by *Seadragon*. Additional projects included establishing a sonar surveillance array at Point Barrow, continuing icebreaker studies of sea ice, and completing the oceanographic survey of the Bering and Chukchi Seas.[1]

While work proceeded on the arctic pool, Lyon turned his attention to the other facets of the program. In November, 1960, he raised with Calvert at SubDiv 102 the possibility of scheduling *Nautilus* and *Skate* for an arctic patrol in spring or summer, 1961. Calvert replied in December that no Atlantic Fleet submarine would be available for a spring cruise; however, prospects for the summer looked good. Not only would *Nautilus* and *Skate* be able to participate in summer war games, but there also was a possibility that the Pacific Fleet would send a submarine through the Bering Strait to rendezvous with an Atlantic Fleet boat near the North Pole in August. "This would be a fine idea so far as I am concerned," Calvert noted, "and would be a particularly good idea if anything should fall through with one of our two [submarines]." In any event, Calvert emphasized, "I think it is imperative that we get on with the project of finding out who can hear whom up there."[2]

ComSubLant, although initially in favor of submarine versus submarine under-ice war games during the summer of 1961, soon had a change of heart. In late February, 1961, ComSubLant pointed out to Calvert that the prototype ice suit for *Skate* would not be available until June; it would be installed during a scheduled overhaul that would last into mid-summer. In addition, the ice suit for *Nautilus* was not expected to be ready before September. In view of "the uncertainty of completing sonar installation in a timely fashion," ComSubLant advised, the scheduled arctic submarine operation should be postponed until the summer of 1962 so that "installation can be completed and plans formulated in an orderly manner."[3]

This decision was confirmed in March at a meeting between ComSubLant and ComSubPac in San Francisco. "From what I hear," Calvert informed Lyon, "the arctic point of view was poorly represented." Equipment problems, as usual, had been "in the air." Also, Nicholson's impending reassignment from command of *Sargo* to Op-311 in the Pentagon likely had helped to "remove the steam" from ComSubPac's enthusiasm for 1961 summer operations. Looking on the bright side, Calvert wrote, "The admirals are all a little sheepish about turning down this year and we should capitalize on it!"[4]

Lyon also received disappointing news about NEL's proposed arctic sonar surveillance array. "The capability of submarines to fight in Arctic areas, the ability to conduct ASW operations and the methods to provide submarine detection and surveillance," NEL had advised CNO, via BuShips, in January, 1961, "must now be determined or solved before the U.S. Navy can determine the nature of the threat of submarine warfare in the Arctic, how to neutralize it and how best to use the Arctic Ocean to the advantage of the United States." NEL wanted to install a sonar surveillance station at Point Barrow. If successful, a second station would be built on Gore Island, located off the northwestern coast of Banks Island. The initial cost of the network would amount to $826,000.[5]

Although ComSubPac had strongly endorsed NEL's plan, pointing out the need "to make a decision that we want to use the Arctic for our purposes or let it go by default," BuShips was not impressed. The surveillance installation, it informed NEL, could not be funded as part of the current arctic research program. As additional money would be required, BuShips recommended that the proposal be submitted for consideration by the *Trident* Strategic Ballistic Missile Submarine Steering Committee. "This killed it," Lyon soon learned. "Trident had no interest in the Arctic."[6]

The string of bad news continued into the end of March when Lyon went to Long Beach to play in his first U.S. National's Badminton Tournament. Competing in the men's senior doubles (players over forty), Lyon and partner Wally Kinnear lost in the semifinals by a score of fifteen to fourteen. "I blew our winning point," Lyon recalled many years later, "right next to the net—a simple shot."[7]

Fortunately, things began to look up in the spring. In February, Calvert had informed Lyon that Richard J. Boyle, who had served as a junior officer under Calvert on *Skate* in 1959, was leaving the navy. "I think," Calvert had written, "he could be of value to you." Lyon agreed that the nuclear-trained Boyle would be "a most useful person to the arctic business in view of his background in engineering and submarining." Although NEL's technical director objected to hiring Boyle, Lyon managed to overcome his opposition and place Boyle on the payroll of the Arctic Submarine Facility as of May 15, 1961. Boyle would prove a significant addition to Lyon's staff, as he quickly became the scientist's strong right arm in matters dealing with nuclear submarines.[8]

June brought word that ComSubLant and ComSubPac had agreed to conduct a joint operation in the summer of 1962. Two months later, CNO approved the proposal for two submarines, one from the Pacific and one from the Atlantic, to meet in the arctic basin and conduct war games. A planning conference to work out the details of the operation would be held at the Pentagon in December.[9]

With no polar submarine operations scheduled for the remainder of 1961, Lyon focused on the modifications necessary to grow sea ice in the arctic pool. This was a major project that obviously would require several years to complete at the current level of funding. Not until the early months of 1962 did his attention once again turn to under-ice warfare.

Since 1959 Atlantic Fleet diesel-electric submarines had been conducting annual late winter/early spring exercises in the fringe ice areas of Cabot Strait. Conditions in the strait closely simulated those that could be expected in potential submarine operating areas in the Denmark Strait, Norwegian Sea, and in the northern and eastern Barents Sea during winter. Unlike the previous years, however, Submarine Ice Exercise 1–62 (SUBICEX 1–62) scheduled for March, would include a nuclear submarine. In January, 1962, *Skate* had been equipped with the first production model of the EDO Corporation's integrated sonar ice suit, the AN/BQS-8. Unlike earlier equipment, the integrated sonar suit featured an electronically steered hydrophone array, replacing the mechanically trained array that had so plagued *Sargo* in 1960. The exercise would afford valuable training in under-ice operations for the crew, only two of whom remained on board from previous arctic cruises. In particular, it would enable watchstanders to qualify in the new sonar equipment.[10]

SUBICEX 1–62 got underway on March 5, 1962. Comdr. Shannon D. Cramer, who had replaced Calvert as ComSubDiv 102, led a three-submarine force that included *Skate* and diesel-electric boats *Entemedor* (SS 340) and *Tusk* (SS 426). Lyon had sent Boyle to ride on *Skate* and observe the operation of the BQS-8. In place of separate and scattered units, the BQS-8 featured a control indicator console with a recorder for the upward-beamed echo sounders and a PPI screen for the iceberg detector and polynya delineator.

*Skate* spent twenty hours under the ice during the exercise, which lasted until March 28, and encountered ridges with drafts of up to twenty-seven feet. Preliminary results indicated that the BQS-8 had proved both reliable and effective. The summer patrol, of course, would afford a more rigorous test of the equipment.

SUBICEX 1–62 also featured war games. *Tusk* fired one exercise torpedo at *Entemedor*; it surfaced and became trapped on top of an ice floe. A second torpedo, launched by *Entemedor* at *Skate*, malfunctioned and went deep. The third and final torpedo, aimed at *Skate* by *Tusk*, failed to acquire the target. "This exercise," ComSubDiv 102's final report emphasized, "demonstrated that at present the submarine force had no weapon that can kill an enemy submarine lying-to in a polynya surrounded by ice of appreciable thickness."[11]

While SUBICEX 1–62 was taking place, Lyon was making preparations for the major summertime operation. In addition to evaluating the BQS-8 ice suit, the exercise also would offer the opportunity for a high-latitude test of the NEL-developed OMEGA navigational system. OMEGA involved a network of ground stations that transmitted low-frequency radio signals over long distances that could be received by submerged submarines equipped with floating wire antennas. Experts at NEL believed that the geographical area of the patrol would be beyond the normal operating range of the current OMEGA network, which consisted of only three stations; crossing angles of lines of position would not permit accurate fixes. Lyon, nonetheless, was persuaded that the cruise would afford an important test of the limitations of the system, if nothing else.[12]

Another piece of special equipment that would be tested on the cruise was the Gertrude Alpha, a new long-range, low-frequency directional sonar transducer that could be used with the UQC-1 ship-to-ship underwater communication system. BuShips believed that the Gertrude Alpha modification would provide "a markedly increased communication capability basically by lowering the frequency to 2 kc and employing a directional reflector."[13]

A good omen in the midst of these preparations came in early April when Lyon competed in his second badminton national seniors' tournament. This time, Lyon and his partner won the senior men's doubles championship, the first in what would become a series of national titles for the badminton enthusiast. Lyon enjoyed not only the physical exercise of badminton but also the mental relaxation. "You can't be thinking about anything else when you're playing a badminton game," he once remarked. "It's just a complete break from whatever one is working on."[14]

By early June, 1962, following several trips to Pearl Harbor, Norfolk, Washington, and New London, Lyon believed that everything was in place for the cruise. All special equipment had been installed on both *Skate* and the Pacific Fleet submarine, *Seadragon*. The track plan for both ships had been approved. To Lyon's delight, the plan included forays into the Kara and Laptev Seas. As usual, there was a last-minute addition to the program. When intelligence sources indicated that the Soviets were likely to conduct atmospheric nuclear tests during the summer, CNO ordered a radioactive monitor installed on *Skate*'s deck.[15]

As Lyon soon would discover, however, all was not well with *Skate*. Lt. Comdr. Frank L. Wadsworth, who had joined the ship during an overhaul in the winter of 1961–62, recalled that the submarine had "two significant and special problems . . . and both concerned safety." One involved the lack of arctic experience by the ship's crew, especially the officers. *Skate*'s skipper, Comdr. Joseph L. Skoog, Jr., had only limited experience in nuclear submarines and none in the Arctic. His junior officers also had never sailed under the ice. Partly to offset the inexperience of *Skate*'s wardroom, the Bureau of Personnel had ordered Wadsworth, chief engineer on *Sargo*, to relieve the executive officer, Lt. Comdr. William Matson, who had been scheduled to leave shortly before the beginning of SUBICEX 1–62.[16]

Wadsworth certainly was a welcome addition to *Skate*. As a junior officer, he had served on *Nautilus* during its arctic cruise in 1957. He later participated in *Sargo*'s epochal winter patrol in 1960. When he reported to *Skate*, Wadsworth had been "shocked to see how shallow the experience level in the wardroom was." He decided to take on the demanding role of navigator in addition to his duties as X. O. rather than assign a less qualified junior officer to the task. Captain Skoog concurred. Although SUBICEX 1–62 had provided "a modest exposure to the ice," Wadsworth remained concerned over the crew's lack of experience as *Skate* prepared to depart for the demanding—and potentially dangerous—arctic waters.

An even more troubling problem involved the weakness in the integrity of *Skate*'s salt water piping systems. The submarine had suffered numerous failures of silver-soldered joints prior to overhaul. Although sister-ships *Sargo*, *Seadragon*, and *Swordfish* had encountered similar problems, *Skate*'s were much more severe. "During overhaul," Wadsworth noted, "special tests of *Skate*'s piping joints showed many had little or no bonding of the silver braze to the metal of the pipes." Some of the joints were rebrazed. When sea trials and subsequent tests showed that many of the joints remained vulnerable to failure, special measures were taken. "Some of the critical pipes," Wadsworth recalled, "were braced so that a joint failure would not lead to a complete separation of the two pipe sections and wide-open, double ended flooding." It was not the most satisfactory solution for the problem but was the best that could be done under the circumstances.

SUBICEX 2–62 got underway on July 7, 1962, when *Skate* departed New London for the Arctic via Baffin Bay. Five days later, *Seadragon*, commanded by Comdr. Charles D. Summitt and carrying Walt Wittmann as senior scientist, left Pearl Harbor for a scheduled rendezvous with *Skate* on July 31 at 82° North, 150° East. Plans called for *Sea-*

*dragon* to pass through Bering Strait and Chukchi Sea into the Arctic Ocean, then make a penetration of the Laptev Sea before proceeding to its meeting with *Skate*.[17]

As *Skate* made its way toward Baffin Bay, Jonathan Schere of the EDO Corporation worked with the ship's crew to insure that the BQS-8 ice suit would be working at top efficiency when the submarine entered the ice pack. As one of the objectives of the cruise was to determine if a navy crew could operate and maintain the equipment, Schere would not be allowed to touch the BQS-8 or even talk to the operator once the ship entered the ice pack. If the equipment broke down, he would not be allowed to fix it unless the sonar chief informed the captain that the crew could not make the repair. The captain then would advise Lyon that the navy was not capable of maintaining the equipment. Lyon, in turn, could allow Schere to become involved.[18]

The transit northward was routine until shortly after midnight (GMT) on July 14. After coming up to periscope depth to receive radio traffic, Skoog was taking the submarine down to a cruising depth of 600 feet when the ship experienced a major flooding casualty. Just as *Skate* reached 570 feet, a three-inch line in the auxiliary cooling sea water system failed at a silver-brazed joint. A geyser of sea water sprayed over the lower engine room, shorting out all exposed electrical equipment in the compartment, including a critical motor-generating set. Large quantities of water then began shooting along the curved hull of the engine room before cascading down in a way that led watchstanders to believe that the leak was overhead. As they searched for the source of the leak, Skoog initiated an emergency surfacing. By the time *Skate* reached 200 feet, however, the men in the engine room had located the problem and secured the broken line.[19]

Examining the failed joint with Skoog, Lyon noted that little of the brazing had taken hold. Fortunately, when the joint had pulled apart, an adjacent pipe had prevented a complete separation. As a result, only a slot around the periphery of the failed pipe had leaked water. "Brazed piping," Lyon concluded, "should be replaced by welded pipe to assure a thorough and complete connection of metal at all joints and fittings." There was, he pointed out, no method to insure a foolproof inspection and testing of a brazed fitting. "It all depends," he noted, "on workman skill and honesty."[20]

It had been a close call. "Needless to say," Boyle speculated, "had this casualty occurred under the ice with the pipe sections completely separated, the results could have been disastrous."[21]

*Skate* crossed the Arctic Circle on July 15 and proceeded through the middle ice pack into Baffin Bay. Skoog encountered numerous icebergs but had no trouble identifying and avoiding them. He selected one large piece of glacial ice to calibrate the iceberg detector. He made five passes underneath it, determined its depth at 240 feet by the upward-beamed echo sounder, then set the proper gain on the iceberg detector.[22]

Lyon, as he had on *Seadragon* in 1960, acted as an observer and allowed the ship's crew to conduct all operations under the ice. At one point, as Skoog was surfacing in a polynya, Lyon noticed that the upward-beamed echo sounder system had been shut down at a keel depth of eighty feet. He called Skoog's attention to the possibility that a block of ice could drift over the ship during the final stages of the ascent. It would be wise, he counseled, to keep the sonar turned on until the submarine had reached the surface. Skoog agreed to change the surfacing procedures.[23]

In the early morning hours of July 16, *Skate* passed through Kane Basin. The challenging navigational problem of negotiating Kennedy and Lincoln Channels into Lincoln Sea lay ahead. Both channels were narrow and permanently frozen. Neither had ever been transited by an under-ice submarine. Soundings on the charts were not reliable, nor were the adjacent land masses accurately plotted. *Skate*, X.O and navigator Wadsworth pointed out, would have to travel "blind" through a "tunnel" of ice, relying on active sonar to locate both sides of the tunnel.[24]

To Wadsworth's dismay, Captain Skoog decided to make the transit at the relatively high speed of fifteen knots. Recalling the damage that *Sargo* had suffered after hitting the ice at eight knots, Wadsworth suggested a more moderate speed. Skoog, however, "was determined to make up some of the lost time resulting from the flooding casualty and insisted on the high speed." At the same time, Skoog said that he was leaving the control room. "I advised Skoog," Wadsworth recalled, "that I would relieve the OOD of the conn and guide the ship myself. As navigator I was responsible for safe navigation and I was also the most experienced officer aboard for this type of operation."

As Skoog became involved in a poker game, an activity to which he seemed addicted, Wadsworth took *Skate* through the channels. At one point, the Ship's Inertial Navigation System showed positions wandering off to the west of the channel. When the SINS reached a position indicating that the submarine was a mile *inland*, Wadsworth obtained permission from Skoog to surface in a rare opening in the ice to check the ship's position with visual bearings and sun lines. He then was able to reset the SINS and proceed with more confidence.

"In my judgment," Wadsworth concluded, "transiting these 'tunnels in the ice' with only an occasional piece of open water was far more challenging and dangerous (because of our speed) than the subsequent probing of shallow waters of the Kara and Barents Sea which *did* concern Skoog enough to give up poker and take the conn himself."

By evening, *Skate* had entered the Lincoln Sea, where few soundings were available. The bottom proved to be rough, with hills and plateaus. At one point, *Skate* passed over unexpected shoal water with a least depth of 270 feet. Plans for a complete survey of the area, Boyle noted, would be given a high priority.[25]

*Skate* finally crossed over a steep underwater slope and into the 800-fathom waters of the arctic basin. With heavy ice ridges overhead, Skoog set course toward Novaya Zemlya. During a routine radio contact with Norfolk, he received word that the Soviet Union had announced that it would be conducting military exercises in the Barents and Kara Seas to test "various kinds of modern weapons." Ships and aircraft were warned to stay out of this area, which included the frequently used nuclear testing site on Novaya Zemlya, from August 5 to October 20. Shortly after receiving this message, *Skate*'s radio transmissions were jammed.[26]

This combination of circumstances—the flooding incident, the Soviet announcement of impending nuclear tests, and the jamming of *Skate*'s radio transmissions—caused a good deal of apprehension in the usually unflappable Lyon. "One gets the feeling," he wrote in his Scientific Journal, "that the USSR could use our appearance for a shot, using propaganda that we came to spy." A bit self-conscious about recording his concerns, Lyon added: "If this is read later, it will appear very foolish; if it is never seen again, I suppose it might be considered real."[27]

Actually, Lyon was reflecting the tense state of the Cold War in 1962. In August, 1961, Soviet Prime Minister Nikita Khrushchev had initiated a war of nerves with the Kennedy administration by constructing a wall between east and west Berlin. The following month, the mercurial Russian leader had announced the resumption of atmospheric nuclear testing, ending a voluntary moratorium that had existed since 1958. Even as Lyon wrote in his Journal, Soviet medium-range missiles were secretly en route to Cuba. Three months later the superpowers would approach the brink of war. As historian Michael R. Beschloss has emphasized, the early 1960s were "the crisis years" in U.S.-Soviet relations.[28]

*Skate* entered the Kara Sea on July 27. Shortly after noon, Skoog gave the command to surface in a small polynya so that the scientists could check the radioactivity equipment on deck. When he reached periscope depth, Skoog examined the opening. As it looked like it would be "tight at the stern," he adjusted trim to stern down, then twisted the stern away from the ice to port as the ship approached the surface. With the stern approaching the ice to starboard, he ordered the screws held stopped. Just at this moment, a watchstander in the stern reported hearing a loud "clunk." Believing that the submarine likely hit a piece of ice with its rudder, Skoog was not overly concerned. After a short time on the surface, *Skate* dove and resumed course.

As the submarine built up to cruising speed, however, excessive vibration began when the starboard shaft exceeded 180 revolutions per minute. Keeping 180 rpm on the starboard shaft and 260 rpm on the port shaft, Skoog found that he was able to make thirteen-and-a-half knots. Locating a large polynya, he brought *Skate* to the surface to inspect the damage. Scuba divers reported that one blade had been bent aft about one-and-one-half inches, beginning five inches from the tip. Otherwise, there was no metal missing and no nicks.

As nothing could be done immediately to effect a repair, Skoog decided to continue on his survey mission. That evening, the commissary department laid on a special meal—fillets of beef wrapped in bacon, cake, cherries jubilee, et al.—in honor of Skoog's birthday. "I try to be good company," he noted in his patrol report, "but it's not been altogether a dandy day."[29]

For the next two days, *Skate* worked in the shallow waters of the Kara Sea off Novaya Zemlya, making a bathymetric survey of the area. In general, Skoog found that the depths being recorded by *Skate* were checking reasonably well with the charted data. Early in the morning of July 29, as his mission neared completion, Skoog faced a decision. He could run directly north to deep water or head northeast and attempt to pass through a charted slot near Ostrov Uyedineniya into an area that had been sparsely sounded. "We are here to verify the accuracy of charted data," Skoog wrote in his patrol report. The avid poker player opted for the slot.[30]

By noon, *Skate* had passed through the slot and entered the "desert" of a blank chart. Skoog proceeded cautiously at 110 feet, with 90 feet under the keel and heavy ice overhead. "It's going to be tight," he noted. Between 12:30 P.M. and 5:00 P.M., Skoog reported, "we spend some hairy hours groping our way by soundings through a complex bottom area, varying course and depth almost continuously." Skoog at first tried to follow valleys through an area of softly rounded hills, but this technique frequently diverted the submarine east or west of course. In one hour, *Skate* made only three-and-a-half miles to the north. Skoog then began to steer northerly courses whenever the water was at least thirty-five fathoms deep. This tactic produced much better progress.

At 5:00 P.M. *Skate* reached fifty fathoms. "It never sounded like so much to me before," Skoog reported. The submarine was again in charted waters. Skoog, who had had the conn since 2:30 A.M., could now relax. It had been "a long, sometimes tense day." A major task, however, had been completed. "In no case," he wrote, "have we found sounding discrepancies that could not be attributed to small errors in position determination. I think the validity of the published data has been proven."

To everyone's relief, *Skate* left the Soviet section of the arctic basin at thirteen-and-a-half knots and headed for the rendezvous with *Seadragon*. A few minutes after midnight on July 31, *Skate* made contact with *Seadragon* on its UQC underwater communications gear at a range of 4,800 yards. Skoog and Summitt had an extended UQC conversation, primarily concerning the failure of *Seadragon*'s gyrocompasses. The rendezvous had been fortunate, as *Seadragon* had arrived close to the meeting point before the problem had become severe. Now, the submarine was unable to determine accurate directions. Skoog and Summitt agreed that *Seadragon* would use *Skate*'s reported course headings to help settle its compasses while maintaining station on *Skate*'s beam by relative sound bearings as the two submarines headed for the North Pole. "We snapped into formation 4,000 yards beam to beam," Summitt reported, "and watched our station with all the pride of a destroyer task force."[31]

*Skate* and *Seadragon* reached the North Pole at 6:30 A.M. on August

2. Two-and-a-half hours later, they surfaced in a small polynya three miles from the Pole and moored alongside an ice floe. "The weather is glorious," Skoog observed in his patrol report, "unlimited visibility, bright sun, strong white and blue colors, 3-knot wind, and air temperature 45." Accompanied by color guards, both commanding officers, executive officers, and senior scientists marched across the ice and exchanged plaques and letters from ComSubLant and ComSubPac. Despite the warm weather, Skoog noted, "I hypocritically prescribe Arctic dress for PIO [Public Information Office] reasons."[32]

Lyon, for one, was not impressed. He believed that it would have been more in keeping with the traditions of the naval service had the two ships tied up alongside one another and conducted the exchange. After all, as he had been emphasizing for years, they were in the middle of an *ocean*, albeit a partially frozen one.[33]

Skoog took advantage of the good weather to attempt repairs on the damaged screw. Several non-critical I-beams were removed from the lower-level engine room and welded together to form a clamp that would fit over the twisted blade. Bolts holding the beams together could then be tightened in an effort to flatten the blade. Scuba divers worked in half-hour relays to place the clamp in four successive positions on the damaged blade. The procedure worked. Following "a careful muster," Skoog dove out of the polynya and worked *Skate* up to fifteen knots without vibration.[34]

The next day, while proceeding toward a planned meeting with *Burton Island*, the two submarines carried out tactical exercises. *Skate* attempted to maintain sonar contact with an evading *Seadragon* as the two submarines moved under the ice. As soon as *Seadragon* rose into ice clutter, *Skate* lost the target due to the refraction and geometry of sonar rays. This came as no surprise to Lyon, as he had seen the effects of ice clutter on sonar during the trials with *Redfish* in 1952. "Hiding in ice cover appears to be the method of operating," he noted in his Journal. "Quieting is obviously important. It is ice jungle warfare."[35]

On the evening of August 4, *Skate* received a message that Lyon had been selected to receive the President's Medal for Distinguished Federal Service, the highest honor accorded to civilian employees of the government. "It is heartwarming to observe the wholesale pleasure that the crew and officers derive from this word," Skoog noted in his patrol report. "There is a flood of congratulations of obvious sincerity. Dr. Lyon reacts with characteristic quiet modesty."[36]

Although Lyon knew that he had been nominated for the prestigious award, he also had realized that the competition would be intense. "Most appropriate and happy to receive medal while on a cruise," he remarked in his Journal, "since submarine forces have been so instrumental in making it possible."[37]

*Skate* made UQC voice contact with *Burton Island* shortly after midnight on August 7. Three hours later, Skoog and Lyon boarded the icebreaker for a meeting with Summitt and Comdr. George H. Lewis, *Burton Island*'s commanding officer, to discuss procedures for torpedo tests. *Skate* and *Seadragon* would fire a series of Mark 37 Modification 0 practice torpedoes. A sonar-homing torpedo that could be set in either active or passive mode, the Mk 37 was designed to run out toward a target, acquire it by either pinging (active) or listening (passive), then attack at a speed of twenty-five knots. The navy considered the torpedo to be "quite effective" against both shallow- and deep-running submarines.[38]

Exercises began in the afternoon with *Seadragon* as the shooter and *Skate* as the target. *Seadragon* fired the first practice Mk 37 on passive mode while *Skate* was making a hard, twisting turn at periscope depth. Running at 125 feet, the torpedo appeared to attack the target successfully. From this point on, however, the tests produced far less favorable results.

A second torpedo, fired under similar conditions and again using passive mode, was drawn into an ice ridge, likely by the reflected sound of its own noise. The third shot, set on active homing, acquired an ice ridge, which caused it to turn and attack *Seadragon!* The Mk 37 slammed into the submarine but did not cause any damage. The fourth and final Mk 37, set passive, acquired *Skate* while the submarine was running at 400 feet but passed by without attacking.[39]

The next day was devoted to active and passive sonar tactical exercises between *Skate* and *Seadragon*. When the submarines returned to *Burton Island* at the end of the day, a message arrived from CinCPac-Flt that the Soviet Union on August 5 had exploded a forty-megaton nuclear weapon in a testing area on the west side of Novaya Zemlya.[40]

Lyon shifted to *Seadragon* on August 9 for a continuation of the torpedo tests. He enjoyed the change, as there seemed a more relaxed atmosphere on the Pacific Fleet boat. "Easier living on SEADRAGON," he noted in his Journal, "but ship is sharp and the officers are sharp. Ship works very well and easily." Also, in comparison with *Skate*, *Seadragon*'s cook made excellent biscuits.[41]

When the torpedo trials resumed in the afternoon, *Skate* fired four Mk 37s at *Seadragon*. The first shot, launched by *Skate* in active mode while *Seadragon* circled at periscope depth, failed to acquire the target. The second torpedo, fired in passive mode under similar conditions, acquired and attacked *Seadragon*. After the third torpedo failed to perform because of a gyro problem, the final Mk 37, set active, acquired and attacked an ice ridge.[42]

After the exercise ended at 3:00 A.M. on August 10, Lyon managed to get a couple of hours of sleep. He arose to an excellent meal. "Enjoyed beautiful sunny side up eggs for breakfast; quite a contrast to SKATE." Shortly thereafter, he received a message from CinCLantFlt that briefly described the award ceremony for his presidential medal that had been held on August 7.[43]

It had been a gala occasion, as Lyon later learned in full detail. President John F. Kennedy presented six awards in a ceremony held in the Rose Garden of the White House. Recipients included Dr. Frances O. Kelsey of the Food and Drug Administration who had blocked the marketing of thalidomide in the United States, a drug that was later found to produce severe birth defects when used in Europe, and Llewellyn E. Thompson, Jr., a distinguished diplomat who had recently returned to Washington after five years as ambassador to the Soviet Union.[44]

One recipient, of course, had not been present. The navy explained to Kennedy that Lyon was on "a confidential mission." In his place, Virginia Lyon would receive the medal. Kennedy expressed his admiration for Lyon's work and quipped, "Those of us who had difficulty navigating at sea are astonished at the ability to navigate under ice." The citation accompanying the medal read:

> "He has been singularly responsible for the pioneering development of the knowledge, techniques and instruments that made it possible for a submarine to navigate under the ice cap [sic] in the Arctic. In the face of formidable obstacles he persevered in believing that transarctic submarine navigation could become a reality and directed his efforts toward this objective. His achievement represents a highly important contribution to the Nation's security."

In addition to his wife, Lyon's two children and his mother also attended the ceremony (his father had died in 1960). While Secretary of the Navy Fred Korth led a high-ranking contingent of naval officers in attendance, Lyon was more pleased to note the presence of such old supporters as Calvert, McWethy, and Steele. As Lyon wrote in his Journal on August 10, "I could not have asked for a more happy or appropriate arrangement for receipt of the medal."

Following a conference on *Burton Island* on August 10 to review the results of the torpedo test and to plan the next shots, Lyon transferred back to *Skate*. When the tests resumed, *Seadragon* fired two torpedoes in active mode while *Skate* maneuvered close to the ice canopy. Both Mk 37s acquired ice. "Cannot seem to get active torpedoes past ice," Lyon reported, "unless at 400 feet." Of three further shots by *Seadragon*, two missed the target and one ran erratically.[45]

August 11 and 12 saw another series of tactical exercises between the two submarines. They again demonstrated that active sonar was useless against a submarine that remained close to the overhead ice canopy. The final phase of the torpedo tests came the next day. *Skate* surfaced near *Burton Island* and took on board six expendable warshots with dummy heads. This time, *Skate* would be firing against *Seadragon* "to hit."

During the course of the final torpedo tests, *Skate* fired three Mk 37s in passive mode and three in active mode as *Seadragon* proceeded at a known depth, speed, and course. Of the three passive shots, the first two were erratic while the third ran true but did not acquire the target. The active mode torpedoes did somewhat better. The first ran true but did not attack; the second failed to acquire the target; and the third, running at 550 feet, acquired the target, attacked, and hit *Seadragon*'s sail.[46]

"Well," Skoog wrote in his patrol report, "it's not impossible to hit a deep target under ice—only damned unlikely." With regard to attacks on submarines surfaced in polynyas, he noted, "We had better start drilling our landing party." Skoog continued, "My recently acquired belief that the Mark 37-0 is a decent torpedo is gone, at least for the case of Arctic environment." The exercise had demonstrated that the Mk 37, set on active homing, would be captured by ice while running at depths of less than 400 feet. Furthermore, he concluded, "We have no torpedo capable of hitting a submarine which is shallow under ice. . . ." *Seadragon*'s commander agreed. "We definitely determined," Summitt reported, "the capabilities and limitations of the Mk 37 torpedo in the Arctic—but it was all disheartening."[47]

*Skate* and *Seadragon* separated on August 15. While *Seadragon* returned to the Pacific via Bering Strait, *Skate* retraced *Seadragon*'s 1960

Northwest Passage, this time from west to east. *Skate* proceeded without incident through McClure Strait, Viscount Melville Sound, Barrow Strait, and Lancaster Sound into Baffin Bay. "Simple, fast run," Lyon noted in his Journal.[48]

While passing out of Davis Strait and into the Labrador Sea on August 22, *Skate* surfaced and monitored a news broadcast that President Kennedy had just announced the "historic rendezvous" at the North Pole. "This is the first time that two of our submarines have worked together in this manner under the Arctic ice pack," Kennedy stated at the beginning of a news conference, "and I want to congratulate all those involved in this exceptional technical feat."[49]

*Skate* completed a routine underwater passage to New London and arrived dockside at 2:30 P.M. local time, August 28. Ironically, the submarine encountered the worst weather of the entire patrol upon reaching home port, with heavy rain pelting down as a delegation from ComSubLant welcomed home Skoog and his crew.[50]

In his patrol report and in subsequent speeches, Skoog emphasized that under-ice operations could now be considered routine deployments. X.O. Wadsworth, however, did not see the patrol in this light. "I strongly disagreed with [Skoog's conclusion] then and still do today," he wrote in 1997. "The presence of thick ice overhead," he explained, "adds significant dangers to submarining and could easily lead to the loss of the ship and all hands if a fire, flooding or plane control casualty occurred while trapped under the ice." Arctic submarine operations, he stressed, would *never* be routine.[51]

SUBICEX 2–62, Skoog and Wadsworth agreed, had pointed out a number of shortcomings in the development of an operational arctic submarine. Tests of the OMEGA navigation system, for example, had produced mixed results. It was found that the low frequencies of OMEGA successfully passed through arctic ice up to seven or eight inches thick, and that a navigational line of position could be obtained. This line, however, was defined as a percentage of the distance from one line to another parallel line on an OMEGA chart. "The catch," Wadsworth noted, "was that I needed to know my approximate position in order to know which two lines we were between. Hence, OMEGA was helpful in refining an approximate position and, with more stations, could have been used for accurate fixes."[52]

Skoog agreed. While *Skate* heard each of the three existing stations, the submarine was not able to obtain a useful line of position. Nonetheless, Skoog believed that the potential advantages of the system had been demonstrated. "It seems clear," he noted, "that an SSN could verify SINS positions by OMEGA without surfacing amid or through ice." He suggested that two high-latitude stations be added to the system and that an improved trailing wire antenna coupling be devised. "This important operational capability," Skoog concluded, "is worthy of the necessary effort and funds."[53]

Tests of the Gertrude Alpha modification to the UQC underwater communications system also were less than satisfactory. Although the signal from Gertrude Alpha could be heard at a distance of 50,000 yards, as compared to 18,000 yards for the unmodified UQC sound system, the range of readable two-way voice communications was 12,000 yards for both devices.[54]

The BQS-8 prototype ice suit, on the other hand, had fared much better during the exercises. Operated and maintained by ship's personnel, the BQS-8, Lyon noted, had "performed well" throughout the patrol. Only one major problem remained to be solved. Precise measurement of the thickness of ice skylights had proved unreliable. It would be necessary, Lyon believed, to design a special depth detector for use with the ice suit sonar.[55]

Conclusions from the tests of tactics and weapons, Lyon pointed out in his senior scientist's report, had been "strikingly similar" to those derived from the exercise between *Redfish* and *Burton Island* in 1952. It had again been shown that the sea ice canopy provided a unique protection against detection and attack. As to weapons, Lyon agreed with Skoog and Summitt. "The capability of torpedoes to operate in an arctic environment," he stressed, "is zero." This lack of effective weapons remained "the most serious deficit in our knowledge of arctic submarine warfare."[56]

The next major research and engineering period, Lyon argued, should be aimed at giving the arctic submarine "the capability to use the ice canopy to maximum advantage." Sound transmission studies should be "a first priority task" because of the unique sound characteristics of the area created by the ice canopy. Also, the laboratory study of the momentum-impact method of ice breakthrough should continue, followed by a critical field study. In addition, Lyon emphasized, increased knowledge of the oceanography, ocean-cryology (that is, the freezing process of seawater), and geophysics of the Arctic Ocean "is basic to any development in arctic submarine warfare." As special submarine expeditions could provide "a flexible yet powerful method to conduct research studies in the Arctic Ocean," he suggested that USS *Triton* (SSN 586) might be assigned to conduct scientific studies.[57]

With ship's personnel now operating all equipment and making all decisions pertaining to under-ice operations, the polar submarine clearly had passed from an experimental status to become part of the regular fleet. On the other hand, what the operating units of the navy intended to do with this capability remained to be seen as the tactical and strategic importance of the Arctic Ocean had not been determined. Lyon was concerned over the possibility that further development of the under-ice submarine as an effective weapon might languish if the Arctic was deemed an area of low military priority. Should the region take on crucial strategic significance at some point in the future and the United States had neglected to develop the war-fighting capability of the arctic submarine, the results could well be calamitous. It would then be much too late, Lyon pointed out, "to initiate the research and development necessary to meet the challenge."[58]

While Lyon was participating in the war games of SUBICEX 2–62, Commander Nicholson was fighting a bureaucratic battle in the Pentagon to insure that the next generation of attack submarines would have an arctic capability. As Nicholson learned shortly after his assignment to Op-311, sea trials of the new attack submarine *Thresher* (SSN 593) had revealed flaws in its design. The submarine's commander, Comdr. Dean L. Axene, was full of praise for the overall performance of *Thresher*, a deep-diving, highly maneuverable, quiet submarine with reduced and streamlined external appurtenances. He complained to Nicholson, however, that *Thresher* was difficult to control at periscope depth, particularly in heavy sea states, due to its small sail. Also, Axene thought that the submarine's single periscope and lack of surveillance equipment were mistakes. After accepting Axene's invitation to ride on *Thresher* and see the problems for himself, Nicholson agreed that modifications would be necessary.[59]

Nicholson discussed *Thresher*'s deficiencies with the staffs of the Type Commanders; they agreed on the need for a redesign. He also secured the support of Comdr. John Potter, on the staff of the Ships Characteristics Board in CNO. Potter and Nicholson then contacted BuShips, who assigned John Wakefield to the problem. The talented designer came up with plans to extend the length of the boat by thirteen feet eight inches, enlarge and harden the sail, and add a periscope and surveillance mast. Wakefield also redesigned the sail planes to permit 90° of travel, which would facilitate vertical ascents and descents through the ice. With the addition of an ice suit, the modifications would not only solve the depth control and surveillance problems but also would make the submarine capable of operating under ice. The total cost for all the changes came to $35 million.[60]

Nicholson and Potter took the proposal to the Future Characteristics Board. While the board agreed that the modifications were necessary, it said that no money was available. After a good deal of discussion, Potter managed to get the board to make a commitment for one-third of the necessary funds. Nicholson then persuaded his superiors in Op-311 to find another third by canceling or delaying some items in the submarine-building budget that already had been approved. With two-thirds of the money in hand, Nicholson next approached Rear Adm. Ralph K. James, head of BuShips. Impressed with Nicholson's initiative, James agreed to cover the remaining third.

A delighted Nicholson drew up a document that set forth the funding arrangements. The paper then went to Rear Adm. Lawson P. Ramage, Op-03B, for his signature or "chop." Called into Ramage's office, Nicholson expected to be congratulated for his hard work and innovative approach to the problem of funding the modifications. Instead, he found the wartime skipper of *Parche* (SS 384) and holder of the Congressional Medal of Honor, to be "absolutely livid." Ramage told Nicholson that if there was one thing worse than a "Nuke," it was an "Arctic Nuke." "So, you want my chop," he said. "Well, here it is!" Ramage then threw the document on the floor and began to jump up and down on it. Rarely had a superior communicated his wishes so clearly to a subordinate. Nicholson managed to retrieve his paper and flee the office.

Taking the document back to his desk, Nicholson read through it carefully to see what had so offended Ramage. But the only reference to the Arctic in the paper had been the term "iceberg detecting sonar." Nicholson, always willing to take prudent risks, decided on a simple approach to the problem. He took out the offending reference and replaced it with "mine detecting sonar"—that is, the same thing by a different name. After obtaining the Type Commanders' support for the change, he sent the paper back to Ramage. The tactic satisfied the feisty admiral, and he signed the document. In October, 1962, the modified design was redesignated a new class, the *637/Sturgeon*.[61]

As Lyon later pointed out, Nicholson had been "singularly responsible" for incorporating the under-ice capability into the 637-class attack submarine. In doing so, he had assured the future of the arctic submarine.[62]

# The Arctic Submarine

While awaiting the appearance of the 637-class arctic-capable attack submarine, Lyon planned a series of under-ice cruises that would achieve both operational and research objectives. Additional war games were needed to build a base of operational knowledge on the capabilities of the arctic submarine. Also, research cruises should be made to continue survey work in the arctic basin and its approaches, conduct sound propagation studies, and collect oceanographic data. In October, 1962, Lyon was delighted to learn from ComSubLant that *Triton*, a large, twin-reactor submarine that had made an around-the-world submerged voyage in 1960, would be converted for arctic research. "We hope to schedule TRITON for a special Arctic cruise," Admiral Grenfell advised Lyon, "and plan to equip her with an ice suit during her current availability. In addition, we are improving her habitability, communications and command spaces."[1]

Following discussions between Lyon, Nicholson, and Calvert in the Pentagon in early October, Comdr. Jon L. Boyes, commander of Submarine Division 71, drafted a seven-year plan for under-ice operations. Although concerned about "the fading interest in the Arctic now that the publicity angle seems to be dying out and the vagueness that seems to mark the Navy's interest in the submarine arctic program," Boyes nonetheless remained optimistic that his plan would secure favorable endorsements. A coherent and rational approach to the development of the navy's arctic capability, the seven-year program envisioned a series of thirteen operational and five research cruises between fiscal 1965 and 1971. As Boyes predicted, the plan won the approval of higher authorities and was issued as OpNav Instruction 03470.4 on May 27, 1963.[2]

As the seven-year arctic program wound its way through the naval bureaucracy, an event took place that no doubt helped assure its favorable reception. On January 27, 1963, the Soviet press reported that the nuclear submarine *Leninsky Komsomol*, under the command of Captain 2nd Rank Lev Zhil'tzov, had reached the North Pole "sometime before last July." During this training exercise, Captain Zhil'tzov stated, the November-class submarine had passed under the arctic ice and taken station near the Pole "with the aim of stopping rocket-carrying enemy submarines from using the Arctic area for striking a rocket blow." His submarine, Zhil'tzov warned, had the capability "to detect and destroy the enemy nuclear submarines that were trying to approach Soviet shores."[3]

Clearly, the arctic basin was no longer the private preserve of the U.S. Navy's under-ice nuclear submarines. Under ordinary circumstances, Soviet competition could be expected to act as a spur to American naval interest in the Arctic. But this did not happen. Three months after the Soviet announcement, a major disaster took place that would have a profound impact on the entire U.S. submarine program, including the planned under-ice operations.

On April 9, 1963, the nuclear attack submarine *Thresher* left Portsmouth Navy Yard for a two-day post-overhaul cruise. During a deep-dive test the next morning, sound contact with the submarine was lost. It soon became clear that the U.S. Navy had suffered its worst-ever submarine casualty. One hundred and twenty-nine men went down with *Thresher*, including seventeen civilian technicians. A court of inquiry concluded that *Thresher* had likely experienced a situation similar to the one that endangered *Skate* in 1962. A sea water leak in the engine room, probably from the failure of a silver-brazed joint, had short-circuited electrical equipment and caused the reactor to shut down. The commanding officer, John W. Harvey, had been unable to blow ballast because ice had formed in the valves from the high-pressure air tanks, causing them to freeze.

Although the findings of the court were in large part speculative, there was no question that major deficiencies existed in the navy's nuclear submarines. On June 3, 1963, Rear Adm. William A. Brockett, chief of BuShips, established a program that was designed to insure the safety of the fleet's submarines. Known as SubSafe, it included a redesign of sea water piping in all nuclear submarines, and a modification of the high-pressure blowing system in *Thresher*-class ships that were under construction.[4]

Although SubSafe was destined to severely curtail submarine availability over the next five years, its effect was not immediately felt. Lyon continued to prepare for the 1964 scientific cruise with *Triton* that had been scheduled under the new seven-year program. His laboratory designed and built a large winch that would be mounted in the submarine's sail. Operated by remote control, the hydraulically-driven winch would enable scientists to take precise temperature, salinity and sound profiles, and sea bottom cores to 15,000 feet while *Triton* hovered beneath the ice. Planned by Allan Beal and based on Sverdrup's idea for Wilkins's 1931 *Nautilus* expedition, the winch was envisioned as an important scientific tool for studying the arctic basin.[5]

NEL also worked on the development of a receiving/recording sonar buoy for sound propagation research. *Triton* would plant the buoy, with its hydrophone array, on the ice. The buoy then would record sound transmission from the submarine as it maneuvered under the ice. A transponder beacon would help to locate and recover the buoy. This system, Lyon believed, would provide valuable sound data in a much more convenient and expeditious manner than previous operations with ice stations and ships.[6]

Plans for the 1964 scientific program on *Triton*, however, took second place to the long-delayed sea ice break-through tests in the arctic pool at Battery Whistler. Between November, 1963, and March, 1964, Lyon devoted most of his time to the sea ice experiments. "It was necessary for me to live at the laboratory continuously," he observed, "every day except three hours in the morning and evening in order to control pool freezing processes and assure the safety of plant procedures."

"It sounds simple," he later recalled, "but we found it was very complicated in controlling the sea water to grow sea ice in an enclosed space." There were a number of major problems that had to be overcome. For example, during the freezing process in the open ocean, some of the salt in the water does not freeze; it simply trickles out into the ocean depths. In the pool, however, controlling salinity presented a difficult chemical challenge. A quarter of a million gallons of sea water had to be held at the correct temperature and correct salinity level. And this could not be done automatically. "You continually had to monitor," Lyon pointed out, "and then you've got to make computations and change."

Air bubbles in the sea ice also caused difficulties. It proved impossible to control the amount of air bubbles that came through the system's piping and into the pool. As a result, the pool's sea ice had a density of only .87 percent as compared to the .92 percent of open ocean ice. In other words, the home-grown sea ice did not have the strength of open water ice.

It took about a month to grow a sheet of sea ice in the pool that was thick enough for break-through tests. The ice grew an inch on the first day in a temperature of minus-40°. At the end of the first week, there was one foot of ice in the pool. Finally, after a month of careful monitoring and adjustments, Lyon had three feet of sea ice. At this point, a hydraulic ram rose from the bottom of the pool at a controlled rate and smashed through the ice while cameras recorded the process and instruments on the ice measured the amount of force applied. The shattered ice then had to be pulled out of the pool and hauled away before a new sheet could be started.

Despite the imperfections of the laboratory sea ice, the arctic pool proved "a powerful tool" in understanding the dynamics of the breakthrough process. Lyon's experiments yielded a mathematical model that was used to design the sail of the 637-class submarines. All the work that he had done over the years to give the navy an arctic test facility was paying rich dividends for the submarine force.[7]

In the midst of the break-through tests, Lyon received word of the first in what was to prove a series of delays in implementing the seven-year under-ice program. On January 15, 1964, CNO canceled the 1964 *Triton* research cruise. "This decision," ComSubLant advised, "was necessitated by the serious need for submarine services to the fleet and our recent delays in completion of new SSNs at building yards."[8]

Despite the disappointing news, Lyon remained optimistic that under-ice cruises would soon resume. By July, 1964, he had managed to secure approval for a research cruise by *Triton* in the fall of 1965. Also, Comdr. J. M. Snyder, who had replaced Boyes as ComSubDiv 71, worked up a program of under-ice war games with *Seadragon* and *Sargo* for the summer of 1965.[9]

Unfortunately, it did not take long for these plans to fall apart. By the summer of 1965, both the *Seadragon-Skate* war games and the *Triton* research cruise had been postponed until 1966. But even this schedule soon came into question. In September, 1965, Commander Burkhalter, the navigator on *Seadragon* in 1960 and now C. O. of *Skate*, explained to Lyon that the submarine's modifications would not be completed before August, 1966, at the earliest. This would leave insufficient time to work up for an arctic cruise. The SubSafe program, he noted, had created a problem in availability for all nuclear submarines. Also, Burkhalter reported, a recent visit to the submarine desk at CNO to mobilize support for arctic operations had proved disappointing. "Frankly," he wrote, "I did not find too much enthusiasm for an Ice Trip."[10]

One way to give priority to arctic operations did exist. A group of arctic enthusiasts in the submarine service wanted to explore the possibility of stationing ballistic missile submarines (SSBNs) under the ice. Commander Boyes had included SSBN operations in his seven-year plan but higher authority had deleted them. This did not, however, end the debate.[11]

In August, 1963, Comdr. James T. Strong, who had served as executive officer on *Seadragon* in 1960, wrote to Lyon and noted the absence of SSBN patrols in the seven-year plan. Now in command of *Lafayette* (SSBN 616), a newly commissioned "boomer," Strong believed that the "desirability of developing an SSBN arctic capability is obvious." He envisioned a strategic scenario in which an SSBN would take station in a polynya and await launch orders. "On receipt of the order to launch," Strong continued, "she should launch surfaced or submerged, through the polynya."[12]

Strong prepared a plan for arctic operations in the spring of 1965, which he submitted to his superiors. Previously, he pointed out, missiles could not be launched north of 67° N due to the limitations of the ship's navigation system. *Lafayette*, however, was scheduled to be equipped with the new TRANSIT satellite system and the Mk 84 missile fire control system, which continuously computed missile corrections as the submarine moved through the water. As a result of these improvements, missiles could now be launched accurately from latitudes as high as 75° N. Also, expected equipment enhancements soon would extend the launch position to 90° N.

Strong wanted to launch two A-2 Polaris missiles—one from the surface and one from underwater—from the edge of the permanent ice pack east of Greenland toward the CAESAR underwater surveillance network southeast of Newfoundland. "Missile tests for the near-arctic would appear to be especially valuable," Strong argued. "Not only would they anticipate a future arctic missile capability but they would confirm or deny that missiles presently on station in the Norwegian Sea are not adversely affected by their near-arctic environment."[13]

Strong's request went nowhere. Both ComSubRon 16 and ComSubLant, he wrote to Lyon, gave "faint but negative endorsements" to the proposal. Nor did the outlook seem any brighter. "The climate at the Fleet and LANT staff," he noted, "is also that of polite disinterest so I do not expect their action to be any different." Nonetheless, Strong concluded, he expected to spend four more years in the SSBN command bracket, "and I plan to keep plugging for it as long as I can."[14]

The issue of SSBN operations in the Arctic came to a head in the fall of 1967 when Dr. Edward Teller, "father of the hydrogen bomb" and an individual who could not easily be ignored by the naval hierarchy, proposed that missile submarines be deployed "under the Arctic ice." Their proximity to Soviet targets, he wrote to the director of the Office of Research and Engineering, would allow each missile "to utilize its throw-weight in a more effective way." Also, Teller warned that the Russian presence in the far north represented a menace to the United States. "The Arctic Ocean," he emphasized, "is a strategic territory which we cannot continue to ignore."[15]

The response to Teller's inquiry was drawn up by Capt. James B. Wilson, head of the Polaris/Poseidon Branch in CNO. Sent out over the signature of Rear Adm. Philip A. Beshany, director of the Submarine Warfare Division, it represented—Wilson informed Lyon—not "a Navy position" on the subject of submarine arctic operations, but "an indirect reply to Teller." While it may not have been "official," it spelled out with rare clarity the dominant thinking in the submarine branch of CNO on under-ice operations.[16]

"There is no significant present advantage to SSBN under-ice deployment," Beshany wrote. "Target coverage of heavily populated, industrially significant regions of the USSR is not improved over that provided by present operating areas." Furthermore, the navy did not foresee any near-term source of SSBN vulnerability that would cause it "to accept the greater operational difficulties posed by such [under-ice] deployments." Soviet ability to search for, detect, and identify submarines in the open ocean was "extremely limited," and it was expected to remain so "for the next several years." Should the Soviet ASW threat increase, deployment of SSBNs to the Arabian Sea and Indian Ocean would be preferable to the Arctic.

The navy recognized that the Arctic offered unique protection against ASW surface ships and aircraft, Beshany acknowledged. The ice cover neutralized optical, infrared, and surface effect detection.

Also, SSBNs could hide from active sonar among the ice pinnacles. However, these advantages were outweighed by the disadvantages of arctic operations. "Other than the obvious hazard of operations under the ice where it would be impossible to surface in case of a major casualty," Beshany pointed out, "precision navigation, weapons launch and communications would be more difficult."

Beshany went on to review the navy's under-ice program since 1957. "No major strategic doctrinal changes governing SSN deployment have resulted from these operations," he noted. On the other hand, the navy must continue to assess the potential of the Arctic for use by U.S. and enemy forces. It would be necessary to develop tactics for attack submarines, to gather data on ASW weapons performance, and to collect intelligence on Soviet activity in the far north. In short, although the navy had a continuing interest in the Arctic, this was not an area of high strategic priority.

Beshany's position came as no surprise to Lyon. He recognized that neither the Central Intelligence Agency nor military intelligence sources had ever uncovered a naval threat to the United States via the Arctic Ocean. "We who deal in Arctic submarines," he pointed out in 1966, "really deal in futures. . . ." Strategic planners, he noted, believed that opposing submarine forces would reach a situation of balance in the next ten to twenty years. In this event, it was likely that submarine operations would focus on the marginal sea ice zones—the area where the ice pack meets the open sea. The navy had to be ready to fight in these waters. Preparations had to be made *now* for the submarine wars of the future.[17]

The summer of 1966 saw the first of the 637-class submarines nearing completion. Although *Sturgeon* (SSN 637) was the lead ship in the class and had been laid down at the Electric Boat Company in August, 1963, the submarine closest to launching was *Queenfish* (SSN 651), which was being built by the Newport News Drydock & Shipbuilding Company. In late July Lt. Comdr. Alfred S. McLaren, executive officer on *Queenfish* and formerly a junior watch and photographic officer on *Seadragon* in 1960, alerted NEL to a potential problem with the BQS-8 sonar. The equipment had been installed on the ship, but there were no checkout instructions. With the builder's trials scheduled to begin in September, McLaren wanted to insure that the ice suit sonar was operating properly.[18]

Boyle, who had become Lyon's closest associate in matters concerning nuclear submarines, was dumbfounded by the news. "The complete lack of written instructions as to check-out procedure," he informed sonar-expert Roshon, "seems a little incredible to me." Boyle pointed out that the lack of a proper checkout could well be disastrous. "We seem to have a really dangerous combination on our hands," he emphasized, with the first of the 637-class submarines and the first production model of the under-ice sonar.[19]

In late August Comdr. Jackson B. Richard, prospective commanding officer of *Queenfish*, wrote to Lyon and reaffirmed his interest, as conveyed by McLaren during a visit to NEL earlier in the month, in testing *Queenfish*'s arctic capability as soon as possible. Richard said that he had submitted a revised schedule for the submarine's shakedown cruise to include a trip to Cabot Strait in January, 1967. He also wanted to lay the groundwork for a full-scale arctic cruise in the fall of 1967.[20]

Lyon followed up Richard's request for a trial of *Queenfish*'s under-ice capability by writing to Admiral Wilkinson in the Pentagon and strongly recommending a seven- to ten-day cruise to the edge of the polar ice pack. Wilkinson, a long-time proponent of under-ice operations, expressed his strong support for the plan. "I would certainly like to see an operational test of the first production AN/BQS-8 equipment in QUEENFISH," he informed Lyon on September 20. He said that he would write to ComSubLant and ComSubPac about the desirability of giving *Queenfish* sufficient time to check out the sonar gear in Cabot or Davis Strait before departing for its assignment to the Pacific Fleet. Also, Wilkinson added, "Hope that we may be able to mount some submarine Arctic operations one of these years."[21]

With Wilkinson's endorsement, ComSubPac agreed to a transfer or "chop" date of March 1, 1967, for *Queenfish*. This would permit three days of under-ice operations in February. At long last, the arctic submarine program was getting back to sea.[22]

Lyon and Boyle departed from New London on *Queenfish* at 8:00 A.M. on January 30, 1967. As the submarine steamed east at twenty-one knots and 300 feet, they held discussions about the upcoming tests and gave lectures on procedures. The polynya-plotting party conducted exercises, and Richard practiced vertical ascents. Early on February 2, as *Queenfish* turned northeast around Grand Banks, the BQS-8 made contact with a group of whales at 3,500 yards. Approaching these strong submarine-like targets, it was possible to identify individual whales at 800 yards on the PPI scope.[23]

Two days later, as *Queenfish* headed north in the Labrador Sea toward Davis Strait, an unfortunate accident took place. Lt. Ronald E. Burdge, the engineering officer, had the conn as *Queenfish* came up to periscope depth to copy radio transmissions at 5:30 P.M. His right hand became caught in the dashpot cups of the rising number one periscope, and he suffered a traumatic amputation of his thumb and a portion of his forefinger. Richard changed course at once toward St. John's, Newfoundland, and requested the evacuation of Burdge by helicopter.[24]

*Queenfish* reached a point some twenty-five miles northeast of the entrance to St. John's harbor at midnight on February 6. A gale was passing through the area, making it impossible for a helicopter to make the pickup. It took fifteen hours for Richard to negotiate the high seas and enter the port. After Burdge was transferred to a Canadian pilot boat, Richard set course for Davis Strait.[25]

*Queenfish* entered an area of scattered ice the next day. The sea state remained extremely high, with the BQS-8 topside recorder registering waves of more than 20 feet. *Queenfish* rolled a good deal at a depth of 200 feet, and there was even some roll at 300 feet. In the afternoon, *Queenfish*'s sonar operators began to report iceberg contacts. On closer examination, however, they turned out to be surface effect caused by a positive temperature gradient. The contacts, Richard noted in his patrol report, "began to vanish as we gained experience."[26]

Not until February 9 did the iceberg detector have its first real operational trial. At 3:37 P.M. the BQS-8 picked up a strong target at 1,350 yards. "The scope presentation," Richard noted, "is sharp and bright, the echo clear. Evaluated iceberg." The ship's sonar operator tracked the berg as *Queenfish* closed at two knots to investigate. He watched as the target approached on the 800-yard, then 400-yard scale. Expecting the iceberg to drift left, Richard held his course. At this point, Neil Kelly, an electronic engineer from NEL's High Resolution Sonar Division, came into the control room. He stared in amazement at the PPI screen as the submarine continued to head for the iceberg. *Queenfish* cleared the overhead ice by seventeen feet. The submarine, however, likely had passed under a spur of the berg, narrowly missing deeper draft ice to port.[27]

"Cannot help but note that perhaps once again luck has played so great a part," Lyon wrote in his Scientific Journal after talking to Kelly. "Again a close situation that could have been most severe. Each cruise seems to have one and you wonder how your luck holds out."[28]

This incident, Lyon concluded, emphasized the need for better training procedures. Tapes should be prepared that could be played into the BQS-8 console and generate realistic presentations of the various ice piloting situations that might arise. The conning officers could become familiar with iceberg contacts, various sound refraction conditions, biological targets, and other aspects of under-ice navigation. They would then know what to expect and be prepared to take appropriate action.[29]

Richard and his crew soon became expert at identifying and avoiding icebergs. On February 10 *Queenfish* encountered twelve bergs. The majority were initially picked up by the ship's AN/BQS-6B sonar at ranges of up to 6,000 yards. As the submarine closed on the icebergs, they were acquired and confirmed by the BQS-8 equipment. "Both sonars," Richard noted, "are providing good clear indications and performing outstandingly in concert with each other when the sound velocity profile is isothermal." Although the BQS-6B provided early indications of icebergs, it could not give confirmation as to clearance. Only the BQS-8, Richard pointed out, could provide this "most essential information as the iceberg is closed."[30]

As *Queenfish* continued north under the ice, Electric Boat Company technicians conducted recordings of the ice conditions as detected by the sonar on the Mk 48 prototype torpedo, a wire-guided, fast (fifty-plus knots) homing weapon. Richard had to secure the BQS-8 sonar while the tests were being run because it interfered with the torpedo's sonar. There was only 2 kHz separation between the units in their signal processing. "This is not satisfactory," Lyon noted, "if MK48s are to be used under ice."[31]

By February 11 *Queenfish* was traveling under ridges with thirty- to forty-foot drafts. Richard had been looking during the previous two days for a suitable polynya in which to surface, but he had not found one. At 12:30 P.M., however, just as the submarine passed the entrance to Hudson Strait, he located a promising stretch of thin ice. Richard brought *Queenfish* to the surface through two inches of new ice in a large, oval-shaped polynya. "It is a beautiful clear windless day," Richard reported, "and the air temperature is measured at minus 14°." He allowed all hands to come to the bridge and look around or view the scene through the periscope. "Morale couldn't be higher," he noted, "as we accomplished a real first with QUEENFISH."[32]

After two hours on the surface, Richard dropped out of the polynya, simulating a shallow-water dive. As the *Queenfish* submerged, Richard spotted a large polar bear observing the process from the edge of the polynya. A photographer attempted to take a picture of the bear through the periscope, but in his excitement he used the wrong camera settings![33]

Although the descent was accomplished without incident, it pointed out the need for a quick locking/unlocking mechanism for the sail planes of the 637-class ships. The planes required six minutes to rig in a vertical position, then ten minutes to re-rig for normal operations. "This is a real weakness of the present design," Lyon recognized.[34]

On a more positive note, Richard expressed his satisfaction with the performance of the new secondary propulsion motor. Recessed into the hull during cruising operations, the small electric motor could be deployed to position the ship in and under polynyas. Trainable through 360°, it added a great deal to the maneuverability of the single-screw 637-class submarine. Although only capable of two or three knots, Richard also noted that it could be used to propel the submarine out of the marginal ice zone in the event that the main propulsion system failed.[35]

On February 12, after ninety-six hours and 1,075 miles under the ice, *Queenfish* reached open water and headed for New London. "The QUEENFISH and follow ships of the SSN 637 Class submarine," Richard concluded, "are suitably equipped for Arctic operations. Only minor modifications and refinements of installed systems and state of the art are required." He was satisfied with the performance of the BQS-8 sonar except for the topside ice profiler. This equipment, he reported, was "unstable and unreliable." He recommended that the vertical profile recorder and depth sensor be redesigned to provide reliable indications of ice thickness of plus or minus six inches for sustained periods of time.[36]

Lyon also was pleased with the first under-ice trials of the 637-class submarine. With the *Queenfish* patrol, he wrote in his senior scientist's report, the development of the arctic submarine had now reached a third phase. The initial phase had involved a series of patrols by diesel-electric boats, beginning with *Boarfish* in 1947. This had been followed by the nuclear submarine cruises of *Nautilus* and the *Skate*-class ships between 1957 and 1962. *Queenfish*, he pointed out, "Signifies the start of a Submarine Force capability to control the ice covered seas, twenty years after the first experiment under ice."[37]

Lyon agreed with Richard that the BQS-8 sonar, except for the topside profiler, had performed as expected. The problem with determining ice thickness had frustrated Lyon for years. The flaw in the system lay not with the topside echo sounder but with the Sperry Depth Detector. Air that became trapped in the ship's piping while at periscope depth or submerged could cause an error in the indicated depth as large as six feet over a twenty-four-hour period. Frequent venting was required, but this could be conveniently accomplished only at a keel depth of less than 150 feet. While adequate for normal submarine operations, the Sperry unit was "unsatisfactory for use as a depth reference for ice thickness determination." NEL technicians had been working since 1959 to understand and correct the problem, but they had not had much success.[38]

In many ways, the most valuable lesson that Lyon had learned—or relearned—centered on the importance of that area where the open sea and the ice came together. The cruise, he emphasized, had "struck home for me that there is a very real danger to the Navy in this interface between the open sea and the ice cover, which we call the marginal sea ice zone." He had first become alert to this problem during the patrol with *Carp* in 1947, but later voyages with nuclear submarines to the central arctic basin had pushed the concern out of his mind. *Queenfish*, however, had again demonstrated that the first hundred miles or so of the ice cover presented "a mixed up, muddled, unpredictable ocean environment." Contrasting temperatures and salinity resulted in confused sonar conditions. "It's a perfect place to hide and move," Lyon pointed out, "where detection is extremely difficult." Future research should be directed at acquiring a greater understanding of conditions in this area of potential conflict between opposing submarine forces.[39]

Having taken a major stride in establishing the ability of the 637-class submarine to operate successfully in the Arctic, the next step should involve an extended under-ice cruise. It was vital to test the ability of the sail planes of the 637-class submarine to break through sea ice skylights. Lyon hoped that this trial could take place later in 1967.

Commander Richard certainly was eager to conduct arctic opera-

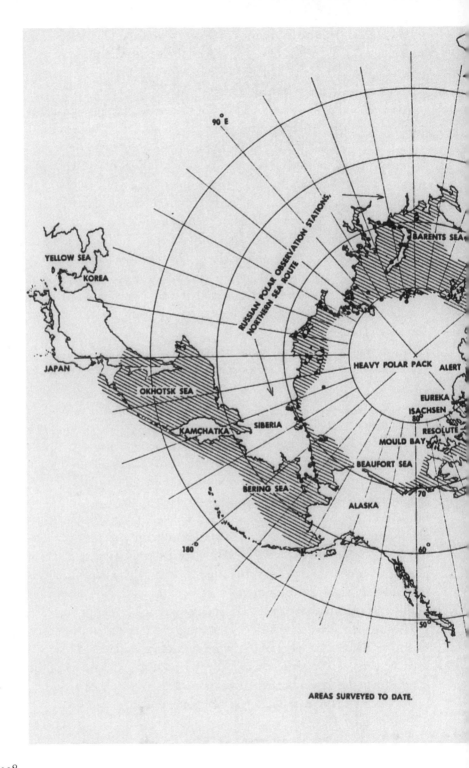

AREAS SURVEYED TO DATE.

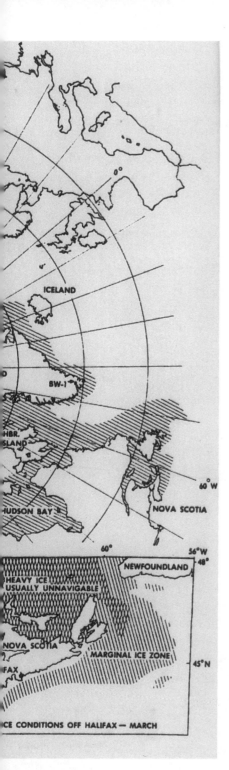

Areas surveyed to date. Arctic area showing regions of marginal ice (shaded areas) and consolidated ice (white area in center). In the sector between about 50° and 190° E, dots indicate Russian observation stations along their Northern Sea Route. The inset is an enlargement of the area between Nova Scotia and Newfoundland. Courtesy U.S. Navy

tions. On April 28, 1967, he pointed out to ComSubPac that there were at present three operational 637-class submarines, with twenty-three more to follow. "It is considered imperative," Richard wrote, "that the Submarine Force vigorously continue Arctic operations." Accordingly, he requested that *Queenfish* be scheduled for an under-ice cruise in the fall.[40]

Richard was optimistic that his request would be approved. "I believe that the local 'climate' remains receptive to the proposed operation," he informed Lyon, "but will directly hinge on the scheduling aspects. Everyone here seems to have a pet project for the QUEENFISH to do." Less than two months later, however, Richard learned something about the priorities of arctic operations.[41]

While ComSubPac "appreciates fully" the rationale for an arctic patrol, higher headquarters responded on June 13, the problem was one of priorities. "The paucity of nuclear submarines in the Navy and in the Pacific Fleet in particular," ComSubPac pointed out, "has caused CINCPACFLT to settle upon and issue a binding list of priorities, some ten in number, which must determine the utilization of these submarines." Under-ice operations ranked tenth on this list. Richard's proposed cruise, therefore, "must, regrettably, be disapproved at this time."[42]

There would be no under-ice operations in 1967. And, as it turned out, no 637-class submarine would be available for an arctic cruise in 1968. Only in 1969 would Lyon be able to conduct the ice breakthrough trials necessary to prove the capability of the new arctic submarine.

The impetus for an arctic submarine operation in 1969 came from Captain Strong. The former executive officer on *Seadragon* in 1960 and commander of *Lafayette* was now in charge of tactics and training at ComSubLant. In September, 1967, Boyle visited Norfolk with a proposal that Lyon had prepared for an under-ice patrol to test the capabilities of the 637-class submarine. In forwarding this plan to his superiors, Strong gave it an enthusiastic endorsement. He pointed out that all under-ice cruises that had been scheduled under CNO's seven-year program had been canceled. The last significant arctic operation, in fact, had taken place in 1962. Strong recommended that preparations begin for an under-ice exercise in March, 1969, which would involve a 637-class submarine and *Skate*. "Go ahead with the planning," Com-

SubLant replied; "this too may have to abort in light of the situation in 1969."[43]

In order to enhance the prospects that his superiors would approve the under-ice cruise, Strong hit upon an innovative ploy. The arctic operation, he suggested, should include a test of the capabilities of the SOSUS underwater detection network in the North Atlantic. Strong knew that he could count on the support of the commander of Ocean Systems, Atlantic, who happened to be none other than Capt. Robert McWethy, a long-time under-ice enthusiast. Objectives connected with the SOSUS network, Strong informed Lyon, "can bring the highest priority."[44]

Strong's tactics worked. In March, 1968, ComSubLant favorably endorsed the plan for SUBICEX 1–69. CNO gave its approval on June 17, 1968, for a full-scale arctic exercise that would include *two* 637-class submarines and *Skate*. A 107-page operations order, issued by ComSubLant on March 7, 1969, set forth the details of the largest under-ice submarine operation ever conducted by the U.S. Navy.[45]

The first major task of SUBICEX 1–69 would be to test the ability of 637-class submarines to break through ice-covered polynyas. *Whale* (SSN 638), commissioned on October 12, 1968, would evaluate the BQS-8 ice-detecting sonar, hardened sail, and fairwater (or sail-mounted) dive planes of the new class of submarine during "carefully controlled ascents" against ice skylights that measured about six to twelve inches thick. "Great care must be taken," the operation order cautioned, "to avoid shaft or screw damage."

The second phase of the elaborate exercise would involve an assessment of the SOSUS detection capability in the marginal sea ice area of the Norwegian and Greenland Seas. *Whale* would take up a barrier patrol near the northeastern end of the Denmark Strait while *Skate* attempted to make an undetected westward transit through the area. To assist in the interception, ComSubLant would radio SOSUS contact information to *Whale*.

Following this exercise, *Whale* would attempt to trail *Skate* as it took evasive maneuvers under the ice. The two submarines then would rendezvous for a weapons capability test. During four days of war games, *Whale* and *Skate* would fire twenty-four modified Mk 37 torpedoes. Thirteen not-to-hit units would be fired from under the ice to open water; eleven torpedoes, intended to hit, would be launched while both attacker and target were under the ice.

Another phase of SUBICEX 1–69 involved *Pargo* (SSN 650), a 637-

class submarine that had been commissioned on January 5, 1968. Under the command of Comdr. Steven A. White, *Pargo* would rendezvous with *Skate* northwest of Iceland, prior to *Skate*'s attempt to make the undetected transit through the Denmark Strait, in order to conduct reverberation tests of the Mk 48 torpedo. *Pargo* then would proceed to Holy Loch, Scotland. After three days in port, the submarine would return to Portsmouth via the Barents Sea, collecting detailed oceanographic and bathymetric data "to aid in a thorough evaluation of this route from the standpoint of submerged navigation."

All-in-all, SUBICEX 1–69 involved a tremendously ambitious agenda. Beyond question, the arctic capability of the 637-class submarine would be given a full test.

*Whale*, commanded by Comdr. William M. Wolff, Jr., and with Lyon and Boyle on board, weighed anchor from Charleston, South Carolina, at ten o'clock the morning of March 18, 1969. Seven days later, while approaching Iceland, Lyon received some sad news. "REGRET TO INFORM DR LYON," a message from ComSubLant read, "THAT HIS MOTHER PASSED AWAY 2100 22 MARCH." This came as no surprise, as his mother had been in ill health since Christmas. "She faded away," Lyon wrote in his Journal, "a shadow into the night." He recalled the time on *Skate* in 1962 when he had received word about the presidential award and how much his mother had enjoyed the visit to the White House. "This cruise has no such happy award," he commented, "just this message."[46]

Over the next five days, Lyon had cause to wonder if the cruise might be ill-fated. On the evening of March 27, an explosion destroyed the aft oxygen generator. While there were no injuries from the fire, a torpedoman broke his arm in falling down a ladder while hurrying to reach the emergency. Three days later, a faulty diode caused the ship's nuclear reactor to shut down. "A FIRST UNDER-ICE SCRAM," Lyon wrote in capital letters in his Journal. It was the first time in all his years of under-ice operations with nuclear submarines that he had experienced such a potentially disastrous event. Fortunately, the rapid restart procedures that had been implemented in the wake of the *Thresher* calamity brought the plant quickly back on line.[47]

The crucial ice break-through tests began on April 1. There was, Commander Wolff recognized, "a large body of opinion within the submarine community that single-screw submarines were inappropriate for Arctic Ops, due to the possibility of shaft/screw damage leaving the ship crippled." This was the reason why the operations order for SUBICEX 1–69 had cautioned against ascents through ice that was more than twelve inches thick.[48]

While proceeding along the Greenwich meridian into the Arctic Ocean en route to T-3, Wolff spotted an area of thin ice. He came to a hover at 120 feet, placed the sail planes in the vertical position, and trimmed the submarine for a 5° up angle to protect the screw. He then ordered the ship surfaced. The diving officer pumped sufficient ballast to give *Whale* a rate of rise of 12 to 16 feet per minute. The submarine easily penetrated five-and-three-quarter inches of ice. "Classic break through," Lyon commented in his Journal. The temperature was minus-26°, with a bright sun and no wind. After a brief stay on the surface, *Whale* dropped out of the polynya without difficulty.[49]

Later in the day, Wolff located a thicker section of flat ice. Confident that *Whale* could handle the situation, he decided to surface through thirty-six inches of ice. The top of the sail came through but the sail planes remained imbedded in the ice. The rudder post also broke through; however, the deck stayed beneath the ice. Concerned that the sail planes might bond with the ice, Wolff ordered a vertical descent after only a few minutes on the surface.

*Whale* had less success on its third attempt of the day. The ship hung up upon striking an estimated thirty-six inches of ice. Wolff rocked the submarine by blowing tanks in groups, but *Whale* failed to break through. The problem, Lyon suspected, was that the initial impact against the ice had failed to produce the expected crack pattern.

As *Whale* continued across the Greenland Sea, the overhead ice became more rugged, with frequent, heavy ridges and few skylights. On April 2 the number two oxygen generator became inoperative, reducing the oxygen content in the submarine. Lyon, who had been running for twenty-five minutes each day in the nuclear tunnel, had to stop his exercise regimen. Even when the number two generator came back on line, the quality of the air in *Whale* remained a problem.[50]

The next morning, Wolff surfaced through nineteen inches of ice into a large polynya. "Did not feel impact," Lyon noted. The sail planes came completely out of the ice sheet. After one hour on the surface, *Whale* made a vertical descent. Although blocks of ice lay under the sail planes, they failed to impede the submarine's progress. "Ice probably well fractured from ascent," Lyon commented. A longer period on the surface at low temperatures, he believed, would have been required to re-cement the fractured ice.[51]

At 7:00 P.M. GMT on April 3, *Whale* surfaced through two inches of new ice at a distance of 14,000 yards from T-3. Plans called for *Whale* to make a triangular run toward the Queen Elizabeth Islands to provide a sound source for the hydrophone array on T-3. Wolff promptly began the first 200-mile leg. The test continued over the next two days, with *Whale* making two successful breakthroughs while running between T-3 and the Canadian Archipelago.

*Whale* completed the exercise on April 5 and headed toward the North Pole. It reached the top of the world at 10:15 the next morning. Wolff, who had done a great deal of reading about arctic exploration, was aware that Robert Peary had purportedly reached the pole on April 6, 1909, and he had adjusted *Whale*'s schedule to coincide with the sixtieth anniversary of that event. He wanted to surface at the Pole and searched for a skylight, but ten feet of ice lay overhead. At 8:00 P.M. he attempted to break through an estimated two to three feet of ice, but he lost speed control on the ascent, causing the sail to strike with little momentum. Air blows failed to break the ice. A second effort also came to naught. Although *Whale* hit the ice at a higher rate of rise, the submarine was unable to break through what the BQS-8 registered as thirty inches of ice.[52]

Finally, at 9:50 P.M., Wolff brought *Whale* to the surface through twenty-four inches of ice; the ship's Sperry Mk 3 SINS showed that the submarine was 600 yards from the North Pole. As the hull remained underneath the ice cover, a rope ladder was rigged from the sail to permit the crew to take turns on the ice. "A classical polynya," Lyon noted, "surrounded by ridges." It was a lovely day, with bright sunshine and a temperature of minus-22°.[53]

*Whale* departed the Pole at two o'clock the morning of April 7 and headed south for the barrier exercise with *Skate*. By April 10, having successfully surfaced through the ice eleven times, the submarine reached a rendezvous point at 80° North to conduct a sonar buoy test with an aircraft from Patrol Squadron 11. Wolff surfaced through thirteen inches of ice at 9:10 A.M. Again, the weather was bright and sunny. The rendezvous went off smoothly, with *Whale* easily recovering an airdrop on the ice. After some two hours of exercises, *Whale* bid the aircraft farewell and slipped beneath the ice.[54]

By April 14 *Whale* was holding station in the fringe ice zone just northeast of Jan Mayen. A SOSUS message placed *Skate* to the north of *Whale*. Although *Whale* kept a careful sonar watch, *Skate* passed through the area during the early hours of April 15 without detection. Wolff then opened the sealed orders that gave the rendezvous point with *Skate*, should the interception fail. By the afternoon, he established listening contact with the elusive submarine.[55]

The next phase of the operation called for *Whale* to maintain contact with *Skate* as it attempted to elude under the ice cover of Denmark Strait. All went well at first, with *Whale* holding listening contact with *Skate* at 8,000 yards throughout April 16. "SKATE has such a recognizable whine of gear train," Lyon noted, "that she is easy to recognize and hold." At seven-thirty the next morning, however, *Skate*'s captain, Comdr. David Phoenix, who had been a diving officer on *Sargo* in 1960, dumped a false target that *Whale* followed all morning. The trailing exercise resumed in the afternoon, this time with *Whale*'s maintaining contact by use of its active sonar.[56]

On April 18 the two submarines reached the area that had been designated for torpedo tests. The firing exercise began at 9:00 P.M. with *Whale*'s firing modified Mk 37 torpedoes to hit *Skate*, which maneuvered under the ice canopy. The first shot ran straight and true but the torpedo's acoustics failed. The next nine Mk 37s, however, acquired and hit *Skate*. The last torpedo failed to start. "A most successful torpedo shoot," Lyon wrote. In contrast to the "fiasco" of 1962, none of the torpedoes had been captured by ice. On the other hand, Lyon was quick to point out that the target had not attempted to use the overhead ice to mask its movements.[57]

On April 20 *Whale* and *Skate* made a rendezvous with the tug that would attempt to recover the not-to-hit torpedoes. This phase of the exercise began in the evening and continued throughout the next day. In all, *Skate* fired eleven torpedoes and *Whale* fired two. Tug personnel recovered nine of the Mk 37s so that the recording data could later be analyzed.[58]

As *Whale* left the area and steamed eastward toward Scotland, *Pargo* headed northwest from Holy Loch to conduct its under-ice portion of SUBICEX 1–69. On April 22 *Pargo* entered the ice pack and surfaced through thin pancake ice. The next day brought a stiffer challenge when White brought his boat up through eight inches of ice in a refrozen polynya.

White's operations order and pre-departure briefings had emphasized the many dangers to the submarine when surfacing through ice. *Pargo*'s single screw easily could be put out of action; the sail planes and top of the sail were vulnerable; the submarine could become frozen in the ice pack. After these first two surfacings, White concluded that the operations order was overly cautious. "I came to believe," he recalled, "that if we could develop the proper techniques, the ship had a far greater capacity to break through ice than the '6 to 12 inches' that the OpOrder mandated."[59]

*Pargo* came up through the ice twice on April 24. On one surfacing, White used a "healthy" rate of rise and broke through a re-frozen polynya 36 inches thick. "I admit," White wrote in 1998, "even after all these many years, that I still can vividly remember the tremendous and frightening noise that accompanied that first break through of three feet of ice." Hydrographer Arthur E. Molloy, assistant to senior scientist Walt Wittmann, was impressed. Although the BQS-8 sonar was giving imprecise readings on ice thickness, he wrote in his journal,"The captain is working out his [surfacing] technique very well. . . ."[60]

*Pargo* reached the North Pole on April 25 and White broke through heavy, clear ice. He remained on the surface only thirty minutes, however, as he feared that *Pargo*'s sail planes might become frozen in the ice. The next morning he found a large polynya with a thin cover of ice at the Pole and again surfaced the boat. It was a beautiful day, with the sun shining, wind calm, and a temperature of 0°. White left a photographic party on the ice to take a picture of *Pargo* as it resurfaced in the polynya. "We submerged with the intent to go about one hundred feet," he recalled, "and then resurface through the 1½ feet or so of ice." As the submarine slipped below the surface, however, White lost depth control and *Pargo* began to move out from under the opening. "There ensued an agonizing period as we tried to 'refind' the polynya where we had left the party. How far away could we be and still be able to find them? How long would it take? What if . . . ?" Only George Steele, who nearly lost the photographic party from *Seadragon* in 1960, could fully understand White's apprehension. Fortunately for the two men on the ice, White quickly located the opening and resurfaced.[61]

Over the next seven days, as *Pargo* conducted a classified shallow-water reconnaissance mission, White brought the submarine up through the ice numerous times, including one penetration of ice more than forty inches thick. In all, *Pargo* steamed a total of six thousand miles under the ice and surfaced twenty-one times through the ice cover. "I definitely considered the ice break through tests a success," White reported. But there was the matter of exceeding the limits of his operations order. When he returned to Norfolk and briefed Vice Adm. Arnold F. Schade, ComSubLant, and his staff on the patrol, White's superior interrupted him at one point and asked, "I thought the Orders said six to twelve inches of ice." At this point, the room grew very quiet. White paused, frowned, and replied, "I must have misread the OP-ORDER; I thought it said 'six to twelve feet.'" The room broke out in laughter. Admiral Schade smiled and said that he would discuss the matter later with White. "We never did," White recalled.[62]

SUBICEX 1–69, Lyon wrote in his senior scientist's report, "completes the task of evaluating under-ice operations by SSN-637 class submarines." The ice break-through tests with *Whale* and *Pargo* had laid to rest any doubts about the ability of single-screw submarines to operate in the Arctic. The submarines had suffered only minor damage: *Whale* had a crushed overtaking light atop the rudder and a partially misaligned clamshell cover over the bridge, while *Pargo* showed a few tears to the rubber covering on the BSQ-8 sonar at the front of the sail.[63]

The BQS-8 sonar, Lyon noted, satisfied the requirements for under-ice navigation. As Commander White pointed out, the sonar "was superb in alerting to deep ridges as well as defining the area of the polynya in preparation for surfacing." Also, Lyon continued, the hovering system, with the use of tiltable sail planes and special propulsion motor, provided excellent under-ice maneuverability. Finally, it appeared that the ice capture problem with the Mk 37 torpedo seemed close to solution.[64]

A number of problems remained, however. In particular, the difficulties with the upward profile recorder had not been resolved. As White and Wolff had discovered, it simply did not measure ice thickness with any degree of precision. "The recorder," Lyon agreed, "is unsatisfactory with regard to basic design, precision and reliability." This deficiency in the ice suit had to be corrected.

"The 637-class submarine," Lyon concluded, "is the culmination of twenty years of experimentation." Research should now turn to the creation of the next generation of arctic submarines. Three factors suggested the direction of future work: the tactical disadvantage of slow, vertical ascents in the ice canopy, the dominant upward refraction of sound, and the apparent success of acoustic weapons against targets transiting below the ice canopy. If it had taken twenty years to develop a true arctic submarine, it likely would take another twenty years of laboratory experiments and field trials to come up with a submarine superior to the 637-class. This would be Lyon's new challenge. He was eager to begin.[65]

# Epilogue

Arctic patrols by 637-class submarines continued during the 1970s, demonstrating the capability of the U.S. Navy in ice-covered seas. In the summer of 1970, *Queenfish* (SSN 651) retraced the path that *Nautilus* had covered in 1958 en route to the North Pole. A comparative analysis of under-ice measurements was taken in hopes of providing clues to changes in the world heat balance. The submarine, under the command of Alfred McLaren and carrying Allan Beal and Boyle from Lyon's staff, then surveyed the thirty- and forty-fathom curves along the 2,600 miles of the Siberian Shelf from Severnaya Zemlya to the Vikitsky Straits, through the Laptev Sea, around the New Siberian Islands, and through the East Siberian and Chukchi Seas. *Queenfish*, frequently working in shallow waters with irregular bottoms, encountered heavy ice with drafts in excess of 70 feet for considerable periods of time, and a deepest draft of 165 feet. The survey, Commander McLaren noted, turned out to be "a very challenging task."[1]

Lyon went back to sea on *Hammerhead* (SSN 663) in the late fall of 1970. The submarine, commanded by Powell F. Carter, first conducted trailing and torpedo-firing exercises with *Skate* in Baffin Bay. *Hammerhead* then proceeded north, through Smith Sound, Kane Basin, Kennedy and Robeson Channels, and into the Lincoln Sea. On November 13, as *Hammerhead* entered the arctic basin, the computer memory of the ship's inertial navigation system failed. As this computer also was an essential component of *Hammerhead*'s satellite navigational system, Commander Carter would have to rely on star sights to determine his position. The Mk 19 gyrocompass would provide directional information, while propeller turn-count would be used to determine distance traveled. "This run to the Pole and back," Lyon wrote in his Journal, "will be most interesting because it will be the first without use of SINS."[2]

Fortunately, *Hammerhead* enjoyed clear skies during the remainder of the patrol. The ship's navigators used the submarine's Type III periscope sextant to take 336 celestial lines of position. Not only did *Hammerhead* reach the North Pole without difficulty, but it also located a new deep-water submarine passage along the eastern side of the Lincoln Sea, connecting the arctic basin and Robeson Channel. This discovery, Lyon pointed out, represented "an all important accomplishment" of what he considered "a highly successful patrol."[3]

After a two-year hiatus, Lyon returned to the Arctic in the spring of 1973 on board *Hawkbill* (SSN 666). SUBICEX 1–73 focused on the operating problems of trailing, surveillance, and under-ice piloting. The under-ice navigational phase of the patrol went extremely well. Lyon refrained from offering advice to the ship's crew during the first shallow-water transit under winter ice by a 637-class submarine. Under-ice deployments, Lyon recognized, no longer required arctic specialists.

The war-fighting phase of SUBICEX 1–73, however, produced less happy results. Working with *Seadragon*, *Hawkbill* conducted a series of trailing and torpedo-firing exercises in the Bering Sea that confirmed earlier problems of detection and attack in the marginal sea ice zone. On a more positive note, an acoustic array that had been developed by the Applied Physics Laboratory of the University of Washington was placed on an ice floe. It successfully tracked six torpedoes under the ice. As Lyon noted, this represented "a very significant first."[4]

Following the patrol, Lyon attended a briefing at Pearl Harbor for ComSubPac and his staff. Lyon used the occasion to raise the financial problems that were plaguing arctic research and development efforts. Recent funding difficulties, he informed the assembled submariners, were about "to obliterate the Navy's small arctic R&D community that works for you." Inflation and overhead had caused costs to skyrocket. At the Arctic Submarine Laboratory, he pointed out, costs per man-year had increased from $25,000 in 1969 to $54,000 in fiscal 1974. During the past year, he had been forced to cut his staff from twenty-six to sixteen. Further reductions would put the laboratory below a threshold from which recovery would not be possible.

In his twenty-six years of serving the navy, Lyon continued, "I have never asked what I must now ask. If you, the Fleet, feel there will be a future concern about your back door to the north, then you must impress that concern on the Chief of Naval Material." Either sufficient funds—about $2.8 million a year, adjusted for inflation—must be provided to the arctic group to continue their research and development work, or they must be relieved of crippling overhead costs. "If the answer by the Navy's R&D management continues to be NO," Lyon warned, "then, Gentlemen, I say not aloha but good bye."[5]

Over the next five years, Lyon led a dual existence. During part of the year, he was at sea for annual arctic exercises. He devoted the other portion of the year to the struggle to save the navy's arctic research and development program. As it turned out, the arctic patrols went far better than the administrative battles.

SUBICEX 1–75 saw *Bluefish* (SSN 675), commanded by Frank Kelso II, conduct a special intelligence gathering operation in the Barents and Greenland Seas, followed by an under-ice transit of the marginal sea ice zone off the east coast of Greenland. The next year, *Gurnard* (SSN 662) completed a successful winter crossing of the Bering and Chukchi Seas, duplicating the feat of *Sargo* in 1960—but without the excitement of the earlier voyage. In 1977 *Flying Fish* (SSN 673) investigated the marginal sea ice zone of the northern Greenland Sea, one of the Arctic's most turbulent areas. *Pintado* (SSN 672) surveyed the bathymetry and environmental conditions of the Northern Sea Route in the eastern Kara Sea in 1978; it also made the first survey of McClintock Channel. Finally, in 1979 *Archerfish* (SSN 678) participated in trailing exercises in Baffin Bay and fired nine of the new Mk 48 torpedoes.[6]

During these patrols, submarine commanders encountered constant problems with their under-ice sonar array. This equipment, unlike earlier sets, had been manufactured by the Hazeltine Company. After the successful performance of the prototype BQS-8 on *Skate* in 1962, BuShips had approved the equipment for commercial production. Roshon and his team at NEL had worked closely with Jonathan Schere of the EDO Corporation to write specifications so that the production equipment would be similar to the prototype. When the bids came in, Hazeltine made the lowest offer. Roshon, who was a member of the committee to review the proposals, had been suspicious about the low bid. Hazeltine had never before built sonar equipment, and it had submitted a figure that was close to EDO's estimated cost of materials. Although his committee had rejected the Hazeltine bid in favor of the more experienced EDO Corporation, BuShips had overridden the committee's recommendation. Hazeltine's low bid, BuShips believed, simply reflected the company's desire to break into the sonar business. NEL had protested, pointing out that Hazeltine had failed to deliver satisfactory components for the Polaris program. As a gesture to EDO for its work on the system, BuShips had awarded the company a contract for five sets. Hazeltine, however, would manufacture all subsequent under-ice sonar equipment for 637-class submarines.[7]

As it turned out, Roshon had had a right to be suspicious. Operators of the Hazeltine-produced equipment soon discovered that the company had cut corners. For example, the prototype BQS-8 had had a target simulator with an elaborate and expensive circuit that adjusted target strength with range to closely match what happens in sea water. The specifications, Schere recalled, included a provision that target strength had to vary with range but did not say how closely the target simulator had to match real life. Hazeltine had bid a simple, inexpensive circuit that did not come close to matching actual propagation loss but had met specifications. "This was repeated over and over," Schere pointed out, "including the more critical parts of the system, such as the transducers." Having helped prepare the specifications, Schere noted, "it never occurred to us to look for shortcuts and dodges to get around its intent."[8]

The EDO-manufactured equipment, which had been installed on *Queenfish*, *Whale*, *Sturgeon*, and *Tautog* (SSN 639) (the fifth set went to the sonar school at Key West), had caused few difficulties. But the Hazeltine-produced sets, carried by subsequent 637-class submarines, were a different story. The problems with this sonar gear, Boyle recalled, "were horrendous." Projectors failed on nearly every patrol during the 1970s, as did scanning switches. The ice-profiling overhead sonar proved so inadequate that it was necessary to go back to equipment that had been used by *Skate*-class submarines in the late 1950s and early 1960s. These deficiencies in the under-ice sonar, Lyon emphasized in a series of reports, had been "designed into the system by awarding the procurement contract to the low bidder irrespective of the bidder's qualifications." In a slap at the way the navy awarded production contracts, he concluded: "Neither documentation nor legal process of procurement will substitute for or assure good design engineering."[9]

Roshon and Boyle documented the many problems with the under-ice sonar and reported to BuShips. "The list of needed corrections," Roshon recalled, "became at least one-half inch thick." Although the submarine section of the sonar branch supported the changes, higher officials not only turned them down on the grounds that no funds were available, but they also acted to reign in Roshon's independence. Henceforth, a high-level BuShips committee informed him in 1970, all future work would be assigned to his staff by the bureau's sonar branch. Roshon would be required to use an elaborate R&D format when submitting requests for research; these requests then would receive careful scrutiny. No longer would Roshon and his staff be able to accept a problem from the Fleet, evolve a simple unit, then test the equipment at sea. Rather than work in such a restricted environment, Roshon chose to resign.[10]

Lyon also used his senior scientist's report for each patrol as a forum to discuss a variety of issues dealing with the navy's arctic program. In 1976, for example, he pointed out that the defense establishment had failed to define the relative importance of ice-covered areas. "The Arctic," he wrote, "has a clouded presence, lurking in the wings of the Navy's main stage of action." In contrast, the Arctic held "center stage" for the Soviets. The Russians, he stressed, had "overwhelming world leadership" in arctic science, engineering, laboratory facilities, construction of icebreakers, and in other areas. Only in the realm of under-ice submarining did the United States have a clear advantage. He argued that it was vitally important for the U.S. to maintain this superiority as a counter to the Soviets in the world political arena. The presence of American submarines in the Arctic could have a sobering effect on Soviet dreams of world power.[11]

The organizational battles over arctic research and development became a constant theme in Lyon's reports. As director of the Arctic Submarine Laboratory, he pointed out, he occupied a dual position. On the one hand, he was responsible to the Fleet for supporting arctic operations. Lyon worked closely with the staff of the appropriate force commander in formulating the objectives of each SUBICEX. He then participated in the necessary planning, directed and funded the contractors of arctic field installations that might be required by the exercise plan, negotiated with Canadian scientists when necessary, and arranged for any special equipment. Lyon then took part in the field operation, providing technical direction, evaluating results, correcting deficiencies, and defining future research directions.

At the same time, Lyon served the navy's Material Command in conducting arctic research and development. The Arctic Submarine Laboratory was a part of the Naval Ocean Systems Center. In terms of administrative structure, the laboratory was an anomaly, with Lyon having unusual independence in the formal chain of command. The scientist constantly had to remind his bureaucratic superiors that the system might be administratively untidy, but it *worked*. "This dual role is an unusual organizational arrangement," he acknowledged, "but it is essential to making the right, quick response" to the demands of the Fleet.

The main problem, Lyon recognized as he grew older, was that the under-ice program depended almost entirely on his personal role in it. Program continuity and responsibility had been executed "by a personal tradition; not by a conventional line management authority, and therefore will disappear with my departure from the program." He urged that the position of director of the Arctic Submarine Laboratory be formally recognized within the Material Command's administrative structure as "the technical executive agent for CNO in the conduct of the Submarine Arctic Warfare Program."[12]

Lyon participated in what proved to be his final arctic patrol in November, 1981, when he rode on *Silversides* (SSN 751) during a SUBICEX to Baffin Bay, Nares Strait, and the Kara Sea. As it happened, just as his active role in arctic submarine operations was ending, the navy was formulating a new Maritime Strategy that placed increased emphasis on the far north. Prior to the 1980s, a central mission of the navy in the event of a non-nuclear war with the Soviet Union was to prevent Russian attack submarines from entering the Atlantic Ocean and disrupting Allied shipping. U.S. attack submarines and other ASW forces, aided by the underwater surveillance network (SOSUS), would attempt to seal the gap between Greenland, Iceland, and Great Britain, thereby preventing Soviet submarines from reaching Allied convoys. In this scenario, naval analyst Norman Polmar has pointed out, the Arctic was looked upon "as an interesting place to explore, but few hard military requirements could be found."[13]

The administration of President Ronald Reagan, however, adopted a more aggressive posture. As part of a general military buildup, naval strength was to be increased to 600 ships, including 100 attack submarines. Together with the larger navy went a new Maritime Strategy, which was outlined by Admiral James D. Watkins, chief of naval operations, in 1986.[14]

In the event of a threatened war with the Soviet Union, Watkins wrote, "The Navy will seize the initiative as far forward as possible." The main Soviet target was the Northern Fleet. Based on the Kola Peninsula, this largest of four Soviet fleets included 65 percent of Russia's submarine ballistic missile force and 75 percent of its attack submarines. Under the new Maritime Strategy, U.S. carrier battle groups, preceded and accompanied by some fifty attack submarines, would be sent into the Norwegian Sea. If the presence of this large naval force failed to deter Moscow from a decision for war, the United States would be in a position to engage the enemy. "As the battle groups move forward," Watkins stated, "we will wage an aggressive campaign against all Soviet submarines, including ballistic missile submarines."

The battle against Soviet submarines would be fought primarily by American submarines. As Adm. Kinnard McKee, director of the Nuclear Propulsion Program, told a congressional committee, "There is nothing that can work against a submarine under the ice except another submarine." Although the 637-class attack submarine was capable of under-ice operations, the newer 688/*Los Angeles*-class was not. Intended as a high-speed escort for carrier battle groups, the 688-class had a small sail with dive planes that could not be rotated for ice breakthroughs. In conformity with the new Maritime Strategy, the navy decided to modify the 688-class for under-ice operations. Beginning in fiscal year 1983, all new 688 construction would be modified to include bow-mounted, retractable dive planes, a hardened sail, steel covering over sonars, improved ice-detection sonar, and enhanced navigational and communication equipment. Although these changes enabled the 688I-class to operate in the deep ice-covered waters of the arctic basin, the submarine was less effective in shallow, ice-covered waters than the 637-class boats due to its larger size and inferior depth control.[15]

Lyon, who stepped down as director of the Arctic Submarine Laboratory in February, 1984, and became its chief research scientist, watched these developments with growing concern. Executing the arctic phase of the Maritime Strategy, he believed, raised the possibility of "a catastrophe" for the U.S. Navy if sent into ice-covered waters to attack Soviet submarines. The Bering Sea, the entire North American coast, Hudson Bay, and the Gulf of St. Lawrence, he argued, formed a marginal sea ice zone that was "ideally suited for . . . hostile deployment and missile attack."

While there would be no problem in locating and attacking an enemy submarine that was moving under the ice pack in the arctic basin, this marginal sea ice zone, with its confused sonar conditions, afforded a hiding submarine a clear tactical advantage. U.S. submarines, he pointed out, did not have the necessary equipment or knowledge to fight effectively in the marginal sea ice zone.

Lyon advocated that a laboratory study be made to investigate the requirements for a submarine that would be capable of hunting an enemy in the marginal sea ice zone. The new submarine had to be able to move around ice ridges in shallow waters, ascend rapidly into the ice while maintaining forward motion, and withstand impacts against the underside of the ice at three to five knots. Following these scientific studies, a prototype under-ice submarine should be built. No doubt, the under-ice submarine would have an ellipsoidal shape and use materials to give a low acoustical, magnetic, and electrical signature.

"Warfare in the ice," Lyon stressed, "is like jungle warfare. The sea-ice canopy becomes the jungle in which the submarine must live, work, and fight—not just transit." The U.S. Navy had to acknowledge the existence of this ice jungle, the possibility of war in it, and then "settle down to do the technical work necessary to learn the nature of this warfare."[16]

Lyon's arguments failed to stir the interest of the naval hierarchy. Their answer to fulfilling the navy's mission in the Arctic was the SSN-21/*Seawolf*, a huge submarine that was even less capable of operating in the marginal sea ice areas than the 688I-class.[17]

In the fall of 1993, the Undersea Warfare Systems Division of the American Preparedness Association presented Lyon with its prestigious David Bushnell Award for Technical Achievement in Undersea Warfare in a ceremony held at the Naval Undersea Warfare Center in Newport. If the audience expected the usual polite, perfunctory remarks, they were in for a surprise. Lyon began by expressing his appreciation for the "final recognition" of the work done by the small group of scientists, engineers, and technicians that comprised the Arctic Submarine Laboratory. Adopting the working philosophy of "one foot in the submarine, one foot in the sea, and hands in the laboratory," ASL personnel had developed an under-ice capability for U.S. submarines. "Yet," he continued, "we wonder if you realize that the award, in essence, is a Hollywood Oscar for brilliant staging—staging a defense facade, not a warfare capability."

Lyon went on to explain that the laboratory had never solved the problem of how to locate, approach, and attack a submarine in the marginal sea ice areas. "A few years ago," he noted, "we began to turn part of our lab around to again work on the problem of combat in the ice." This effort, however, had now been stopped in the name of downsizing and the consequent realignment of naval laboratories. "In the same week that I received notice of this award," Lyon concluded, "I also received instructions to plan the closure of the Arctic Submarine Laboratory."[18]

Lyon's lengthy career as a naval scientist ended on a deeply disappointing note. The end of the Cold War, he believed, did not signal the end to the dangers faced by the United States in the Arctic. With an unprecedented worldwide proliferation of submarines, with over 250 submarines in the inventories of Third World nations and most of them capable of being modified for under-ice operations, there would surely come a time when the U.S. and Canada would desperately need an arctic defense.

Lyon continued to lobby for the development of a submarine that would be capable of fighting effectively in the marginal sea ice zone. The navy, however, remained unconvinced. Rebuffed time and time again, his final efforts were devoted to attempts to persuade a skeptical navy hierarchy to at least preserve the facilities of the Arctic Submarine Laboratory, and its vast collection of data, reports, and tests, so that they could be available if ever needed.

Although the laboratory on Point Loma has closed, and the fate of its facilities and archives remains uncertain, Lyon's record of accomplishment will endure. Certainly, many individuals contributed to the development of the under-ice submarine, but Lyon was *primus inter pares*—the first among equals. As the citation of his Bushnell award recognized, "Rarely has a single individual so dominated a technology. . . ." His persistent determination, more than any other factor, was the key element in making possible Bishop John Wilkins's seventeenth century dream of an arctic submarine.

*Waldo Lyon suffered a massive heart attack and died on May 5, 1998. The navy agreed to honor his last wish to have his ashes scattered at the North Pole during an arctic submarine deployment.*

# Notes

## Prologue

1. On Cornelius Drebbel, see Alex Roland, *Underwater Warfare in the Age of Sail*, pp. 18–31.
2. On the remarkable Wilkins, see Barbara J. Shapiro, *John Wilkins, 1614–1672: An Intellectual Biography*.
3. Simon Lake, *The Submarine in War and Peace*.
4. Jules Verne, cable to Lake, Aug. 21, 1898, quoted in Simon Lake, "The Development of the Under Ice Submarine," in Sir Hubert Wilkins, *Under the North Pole*, pp. 230–31.
5. Ibid., pp. 203–204.
6. "The Great Stone of Sardis" is reprinted in Richard Gid Powers (ed.), *The Science Fiction of Frank R. Stockton*, pp. 207–426.
7. Lake, *The Submarine in War and Peace*, pp. 259–64.
8. On Anschütz-Kämpfe's background, see the interview with him that appeared in *Süddentsche Sonntagspost*, No. 44, Nov. 3, 1929, p. 19. He details his plans in "Das europäische Eismeer und ein neuer Expeditionsplan nach dem Nordpole," *Mittheilungen der Kaiserlich Königlichen Geographischen Gesellschaft* 44 (1901): 53–73; they are summarized in *Illustrirte Zeitung* 3059 (Feb. 13, 1902), and *Geographic Journal* 17 (Apr., 1901): 435–36. See also Alfred McLaren, "The Arctic Submarine: Its Evolution and Scientific and Commercial Potential," in Sylvie Devers (ed.), *Pôle Nord 1983* (Paris: Éditions du Centre National de la Recherche Scientifique, 1987), pp. 329–41.
9. A. E. Fanning, *Steady As She Goes: A History of the Compass Department of the Admiralty*, pp. 176–80.
10. *New York Times*, Dec. 2, 1958. The best biography of Wilkins is John Grierson, *Sir Hubert Wilkins: Enigma of Exploration*.
11. Vilhjalmur Stefansson, "The History of an Idea," in Wilkins, *Under the North Pole*, pp. 3–51.
12. Wilkins, letter to Senator F. C. Walcott, May 4, 1930, Correspondence of the Chief of Naval Operations, Records of the Secretary of the Navy, Box 4031, Record Group 38, National Archives, Washington, D.C.

13. Charles F. Adams, letter to Lake & Danenhower, Inc., May 27, 1930, in ibid.
14. Sloan Danenhower, "The Arctic Submarine 'Nautilus,'" in Wilkins, *Under the North Pole*, pp. 233–73.
15. Wilkins, "Report—'Nautilus' Submarine Expedition," 1947, in the Papers of Sir Hubert Wilkins, Byrd Polar Research Center, University Archives, Ohio State University, Columbus, Ohio.
16. "Wilkins-Ellsworth Trans-Arctic Submarine Expedition, Summary of Receipts and Expenditures," n.d., Wilkins Papers. Major expenses included $181,351 for modifications to O-12; $68,000 for salaries; and $13,000 for supplies.
17. Sverdrup's obituary appeared in the *New York Times*, Aug. 22, 1957.
18. Sverdrup, letter to Wilkins, Oct. 9, 1930, Wilkins Papers.
19. Danenhower, "The Arctic Submarine 'Nautilus.'"
20. Wilkins, "Report."
21. *New York Times*, Mar. 4 and Apr. 11, 1931.
22. Wilkins, "The Plans of the Expedition," in *Under the North Pole*, pp. 55–199.
23. For the best summary of the expedition, see [Richard J. Boyle], *Sourcebook on Submarine Arctic Operations*. (San Diego: U.S. Navy Electronics Laboratory, 1966). Wilkins, "Report," contains a narrative account of the expedition.
24. See James Calvert, *Surface at the Pole*, p. 141.
25. Wilkins, "Report."
26. H. V. Sverdrup and F. M. Soule, "Scientific Results of the 'Nautilus' Expedition, 1931," in Massachusetts Institute of Technology and Woods Hole Oceanographic Institute *Papers in Physical Oceanographic Institute* vol. II, no. 1 (Mar., 1933) and no. 3 (June, 1933).
27. Wilkins, "Report."
28. Naval Examining Board, "The Wilkins Submarine (Nautilus) Expedition to the Arctic, August 1931—Evaluation Report," n.d. [1947], the Papers of Waldo K. Lyon, in the possession of Dr. Lyon, San Diego, Calif. The evaluation was attached to Wilkins, "Report." Although there is a copy of the "Report" in the Wilkins Papers, the evaluation is not included. Lyon received a copy of the document from the submarine desk in the office of the chief of naval operations in 1947. He believes that the Examining Board likely had been initiated by Rear Adm. A. R. McCann, an under-ice enthusiast who had been a liaison officer for the transfer of O-12 to Wilkins in 1930.
29. McLaren, "The Arctic Submarine;" Michael L. Hadley, *U-Boats against Canada: German Submarines in Canadian Waters*, pp. 170–75.

## Chapter 1. The Challenge

1. Background information on Lyon is from interviews with the author, San Diego, Calif., Dec. 15–17, 1995.
2. Tom Tugend, "A Bruin Alumnus Under Ice," *UCLA Alumni Magazine* (Jan., 1959): 10–12.
3. Lyon's doctoral work is summarized in Lyon and E. L. Kinsey, "Infra-Red Absorption Spectra of the Water Molecule in Crystals," *The Physical Review* 61 (1942): 482–89.
4. "The Reminiscences of Dr. Waldo K. Lyon," Oral History Department, U.S. Naval Institute. This 297-page transcript of interviews conducted by Etta Belle Kitchen, Jan.–Mar., 1971, is hereinafter cited as Lyon oral history.
5. U.S. Naval Ocean Systems Center, *Fifty Years of Research and Development on Point Loma, 1940–1990* (San Diego: Naval Ocean Systems Center, 1990).
6. Waldo K. Lyon, interview with author, San Diego, Calif., Dec. 15, 1995.
7. *Fifty Years of Research and Development on Point Loma*.
8. Arthur H. Roshon, "CTFM Sonar," 1994. Written shortly before his death, Roshon's memoir covers his work on QLA and its later versions. I am indebted to Mrs. Roshon for allowing me access to this informative document. For Operation Barney, see Charles A. Lockwood and Hans Christian Adamson, *Hellcats of the Sea*.
9. Lyon, letter to the director, Radio and Sound Laboratory, Mar. 1, 1945, Lyon Papers.
10. Lyon autobiographical notes, undated but apparently written in 1952, Lyon Papers; hereinafter cited as Lyon notes.
11. On Canadian efforts to deal with German submarines in cold waters, see Marc Milner, *The U-Boat Hunters: The Royal Canadian Navy and the Offensive against Germany's Submarines*, and Hadley, *U-Boats against Canada: German Submarines in Canadian Waters*.
12. Lyon notes.
13. C. R. Moe, letter to Lyon, Sept. 26, 1945, Lyon Papers.
14. Lyon, letter to Moe, Oct. 1, 1945, Lyon Papers.
15. Lyon notes; Lyon oral history.
16. Waldo K. Lyon, John P. Tully, and William M. Cameron, "Recognition of Submarine Targets: A Joint Report by United States Navy Electronics Laboratory and Pacific Oceanographic Research Group," U.S. Navy Electronics Laboratory, 1946; copy in the Lyon Papers.
17. Lyon, letter to author, July 17, 1996.
18. Lyon, letter to author, Aug. 1, 1996.

19. *Fifty Years of Research and Development on Point Loma;* Lyon notes.
20. Lyon notes; *San Diego Daily Journal*, Mar. 21, 1946.
21. Ibid.; Lyon Station Log, Mar. 8, 9, 11, 12, and 13, 1946.
22. Lyon notes.
23. William L. Laurence is quoted in Jonathan M. Weisgall, *Operation Crossroads: The Atomic Tests at Bikini Atoll*, p. 2; Lyon notes.
24. Undated clipping in Lyon Papers.
25. Lyon notes.
26. Lyon oral history.
27. On Highjump, see Lisle A. Rose, *Assault on Eternity: Richard E. Byrd and the Exploration of Antarctica, 1946–47.*
28. Lyon notes.
29. Ibid.; Rose, *Assault on Eternity*, p. 102.
30. Rose, *Assault on Eternity*, p. 107.
31. Lyon notes.
32. Lyon, "The Polar Submarine and Navigation of the Arctic Ocean," Research Report 88, Nov. 18, 1948, U.S. Navy Electronics Laboratory; copy in the Lyon Papers.
33. Lyon oral history.
34. Lyon, Tully, and Cameron, "Recognition of Submarine Targets"; Lyon, letters to author, July 17 and Aug. 1, 1996.
35. Rawson Bennett, director of NEL, letter to Chief, Bureau of Ordnance, July 24, 1947, Lyon Papers.
36. Lyon, letter to Chairman, Facilities Board, NEL, July 13, 1947, Lyon Papers.
37. J. B. Berkley, BuShips, letter to Chairman, Joint Committee on Oceanography, Naval Service, Ottawa, May 21, 1947, Lyon Papers.
38. George S. Field, letter to BuShips, June 7, 1947, Lyon Papers.
39. R. D. McWethy, "Staff Study: Arctic Submarine Operations," Naval War College, Oct., 1951, copy in the Lyon Papers; Lyon, letter to Technical Director, NEL, Dec. 11, 1951, Lyon Papers.
40. Clippings from the *Nome Nugget*, Aug., 1946, with Lyon annotation, Lyon Papers.
41. Bennett, letter to ComSubPac, May 1, 1947, Lyon Papers.
42. Lyon, "Polar Submarine."
43. Lyon oral history.
44. Lyon, letter to Bennett, Aug. 12, 1947, Lyon Papers.
45. In reviewing material in his files in 1996, Lyon wrote comments on a number of documents. This annotation sometimes consisted of a brief sentence; at other times, it went on for several pages. These comments are hereinafter cited as Lyon annotation.
46. On the *Boarfish* expedition, see Lyon, letter to ComSubPac, Aug. 14, 1947;

John H. Turner, "Report of Alaskan Cruise," Aug. 16, 1947; and Lyon, "Polar Submarine"; all in the Lyon Papers.
47. Lyon oral history.
48. Turner, "Report of Alaskan Cruise."
49. Lyon, letter to ComSubPac, Aug. 14, 1947, Lyon Papers.
50. Lyon, letter to Tully, Oct. 1, 1947, Lyon Papers; Lyon notes.
51. Lyon notes; Lyon annotation.
52. Lyon, letter to McCann, Mar. 8, 1948, Lyon Papers.
53. McCann, letter to Lyon, Mar. 31, 1948, Lyon Papers.
54. CNO, letter to BuShips, Mar. 23, 1948; Station Log, Aug. 9, 1948; both in Lyon Papers.
55. O. S. Colclough, letter to Commanding Officer, USS *Carp*, Aug. 9, 1948, Lyon Papers.
56. J. M. Palmer to CNO, "Report on Under-Ice Operations in Arctic Ice Field of Chukchi Sea," Sept. 20, 1948; Lyon, "Polar Submarine"; both in Lyon Papers.
57. Station Log, Sept. 24, 1948, Lyon Papers.
58. Lyon, "Polar Submarine and Ocean Navigation of the Arctic," remarks presented to ComSubPac on Oct. 8, 1948, Lyon Papers.
59. Bennett, letter to ComSubPac, Dec. 3, 1948, Lyon Papers.

## Chapter 2. A New Ocean

1. Lyon, letter to McCann, Jan. 11, 1949, Lyon Papers.
2. Bennett, letter to CinCPacFlt, Feb. 23, 1949, Lyon Papers.
3. Station Log, Feb. 9 and Mar. 15, 1949.
4. John P. Tully, "Operational Report: Project Aleutians," Sept. 30, 1949, Lyon Papers.
5. U.S. Navy Electronics Laboratory, "Interim Summary Report: Oceanographic Cruise to the Bering and Chukchi Seas, Summer, 1949," Oct. 19, 1949, Lyon Papers.
6. Scientific Journal, July 18, 1949.
7. Scientific Journal, Aug. 2, 1949.
8. Scientific Journal, Aug. 3–7, 1949.
9. Scientific Journal, Aug. 8–10, 1949.
10. The walrus skull is in the Lyon Papers.
11. Scientific Journal, Aug. 15, 1949.
12. U.S. Navy Electronics Laboratory, "Interim Summary Report."
13. Scientific Journal, Aug. 23, 1949.
14. Scientific Journal, Aug. 29, 1949.
15. Tully, "Operational Report."

16. Scientific Journal, Sept. 11, 1949.
17. Station Log, Nov. 15, 1949; Chief of Naval Research to Chief of Naval Operations, "Research Program for Arctic Oceanography," Dec. 23, 1949, Lyon Papers.
18. Colclough, letter to CinCPacFlt, Mar. 26, 1949, Lyon Papers.
19. CNO, letter to Chief of Naval Research, Aug. 15, 1949.
20. Bennett, letter to CNO, June 1, 1949, Lyon Papers.
21. R. P. Briscoe, CNO, letter to Chief, BuShips, Nov. 10, 1949, Lyon Papers.
22. A. D. Struble, CNO, letter to Chief of Naval Research, Mar. 3, 1950, Lyon Papers.
23. Bennett, letter to CNO, Mar. 30, 1950, Lyon Papers.
24. CNO, letter to NEL, Apr. 27, 1950, citing ComSubPac, letter to CNO, Apr. 19, 1950, Lyon Papers.
25. Lyon, letter to author, Sept. 12, 1996.
26. Station Log, May 2, 6, and 7, 1950; Bennett, letter to Chief of Naval Research, May 19, 1950; ONR, letter to CNO, June 2, 1950; CNO, letter to ONR, June 17, 1950; all in Lyon Papers.
27. CNO, "Operation Requirement No. SW-01402," June 5, 1950; Lyon memorandum, "Arctic Panel and R.D.B. Meetings," Oct. 23, 1950; both in Lyon Papers.
28. Scientific Journal, Aug. 10, 1950.
29. Scientific Journal, Aug. 11, 1950.
30. Scientific Journal, Aug. 12, 1950.
31. Scientific Journal, Aug. 20, 1950.
32. Scientific Journal, Aug. 22, 1950. For a more sympathetic portrait of Schwartz, see Jonas Williams, "The Arctic's Warming Up," *Collier's* 131 (Apr. 11, 1953): 60–64.
33. Robert D. McWethy, interview with author, Annapolis, Md., Sept. 14, 1996.
34. Scientific Journal, Aug. 24, 1950.
35. Scientific Journal, Aug. 30, 1950.
36. U.S. Navy Electronics Laboratory, "Polar Expeditions and Studies," n.d., Lyon Papers.
37. "Discussion Prepared for USS *Burton Island* (AGB-1) Patrol Report," Scientific Journal, Sept., 1950.
38. Scientific Journal, Sept. 3, 1950.
39. Scientific Journal, Sept. 4, 1950.
40. Chief of Naval Research, "Minutes of ONR's Arctic Oceanography Conference, October 17–18, 1950," Nov. 7, 1950, Lyon Papers.
41. Lyon memorandum, "Arctic Panel and R.D.B. Meetings," Oct. 7, 1950, Lyon Papers.

42. Lyon memorandum, "Trip Report, Alaskan Science Conference and Arctic Expedition Planning," Dec. 1, 1950; NEL, letter to BuShips, Nov. 8, 1950; both in Lyon Papers.
43. CNO, letter to NEL, Jan. 18, 1951; BuShips, letter to CNO, Aug. 4, 1950; ONR, letter to CNO, Aug. 14, 1950; CinCPacFlt, letter to CNO, Nov. 3, 1950; all in Lyon Papers.
44. Lyon, note on CNO to Director, NEL, Jan. 18, 1951, Lyon Papers; Station Log, Feb. 13, 1951.
45. Director, NEL, letter to CNO, Feb. 21, 1951, Lyon Papers.
46. CNO, letter to CinCPacFlt and NEL, Feb. 21, 1951, Lyon Papers.
47. Director, NEL, letter to Commander, Alaskan Sea Frontier, July 2, 1951; Lyon notes, spring, 1951; both in Lyon Papers.
48. Scientific Journal, July 15 and 16, 1951.
49. Scientific Journal, July 27, 1951.
50. Scientific Journal, July 28, 1951.
51. Scientific Journal, July 29, 30, and 31, 1951.
52. Lyon, "Operations Plan for Field Station at Cape Prince of Wales," July 25, 1951; G. L. Bloom, "Facilities and Measurement Program at Wales, Alaska," U.S. Navy Electronics Laboratory Report 445, May 20, 1954; both in Lyon Papers.
53. Scientific Journal, Aug. 2, 1951.
54. Scientific Journal, Aug. 17 and 21, 1951.
55. Scientific Journal, Aug. 23, 24, 25, and 26, 1951.
56. Scientific Journal, Aug. 29, 1951.
57. Scientific Journal, Sept. 4, 1951.
58. Scientific Journal, Sept. 9, 1951.
59. Scientific Journal, Sept. 6, 1951.
60. Scientific Journal, Sept. 12, 19, and 22, 1951.
61. Cliff Barnes, "Preliminary Report: Oceanographic and Hydrographic Data Obtained on Beaufort Sea Expedition, 1951," Nov. 15, 1951, Lyon Papers.
62. Lyon, "Program Notes—Extension of Expedition thru 1952," Sept., 1951, Lyon Papers.
63. Lyon, "Trip Report," Nov. 20, 1951, Lyon Papers.
64. Lyon, letter to Technical Director, NEL, Dec. 11, 1951, Lyon Papers.

## Chapter 3. Icebreakers and Submarines

1. Station Log, Jan. 14, 1952.
2. ComSubPac, letter to CinCPacFlt, Feb. 26, 1952, Lyon Papers.

3. CinCPacFlt, letter to ComSubPac, Mar. 5, 1952.
4. Scientific Journal, Feb. 13, 1952.
5. Lyon, "Experiments in the Use of Explosives in Sea Ice," *Polar Record* 10 (Sept., 1960): 237–47.
6. Scientific Journal, Feb. 13, 1952.
7. Scientific Journal, Feb. 14, 1952.
8. Scientific Journal, Feb. 17, 1952; Lyon, "Experiments in the Use of Explosives."
9. Scientific Journal, Feb. 23 and 24, 1952.
10. Station Log, Mar. 19, 1952.
11. Lyon notes, spring, 1952.
12. Preparations for the expedition are detailed in the Station Log for May, 1952.
13. On the development of SOFAR, see U.S. Naval Oceans Systems Center, *Fifty Years of Research and Development on Point Loma, 1940–1990*, pp. 27–28.
14. CNO, letter to Director of Research and Development, U.S. Air Force, May 2, 1952, Lyon Papers.
15. J. E. Gibson, letter to Lyon, June 12, 1952, Lyon Papers.
16. NEL, letter to ComSubPac, June, 1952, Lyon Papers.
17. Scientific Journal, Aug. 6, 1952.
18. Scientific Journal, Aug. 11, 1952.
19. Scientific Journal, Aug. 12, 1952.
20. Scientific Journal, Aug. 14, 1952.
21. E. H. Maher to CNO, "Preliminary Report of the Beaufort Sea Expedition, 1952," Oct. 4, 1952, Lyon Papers.
22. Scientific Journal, Aug. 23, 1952; Lyon oral history.
23. USS *Redfish*, "Report of REDFISH Participation in Beaufort Sea Expedition," Oct. 7, 1952, Lyon Papers.
24. Scientific Journal, Sept. 5, 1952.
25. Maher, "Preliminary Report."
26. Scientific Journal, Sept. 6, 1952.
27. Scientific Journal, Sept. 7, 1952.
28. USS *Redfish*, "Report."
29. Scientific Journal, Sept. 16, 1952.
30. Scientific Journal, Sept. 22, 1952.
31. Lyon, "Summary Report of Chief Scientist, Joint Canadian–U.S. Beaufort Sea Expedition 1952," n.d. (Oct., 1952), Lyon Papers.
32. Lyon, letter to C. B. Momsen, Oct. 7, 1952, Lyon Papers.
33. Momsen, letter to Lyon, Oct. 14, 1952, Lyon Papers.
34. Scientific Journal, Oct. 9, 1952.
35. Scientific Journal, Oct. 10, 1952.
36. Dundas P. Tucker, letter to CNO, Nov. 14, 1952, Lyon Papers.

37. Lyon, letter to Director, Research Division, NEL, Dec. 30, 1952, reporting on his visit to Washington, Lyon Papers.
38. Lyon, "An Approach to Submarine Warfare in Sea Ice," NEL Report 353, Dec. 11, 1952, Lyon Papers.
39. Lyon, letter to Director, Research Division, NEL, Dec. 30, 1952; Station Log, Jan. 5, 7, 8, 9, 12, and 13, 1953.
40. Tucker, letter to CNO, Jan. 13, 1953, Lyon Papers.
41. Lyon, letter to Momsen, Jan. 16, 1953, Lyon Papers.
42. Scientific Journal, Feb. 16, 1953.
43. Scientific Journal, Feb. 17, 1953.
44. Scientific Journal, Feb. 19 and 20, 1953.
45. ComSubPac, letter to CinCPacFlt, Mar. 11, 1953, Lyon Papers.
46. Tucker, letter to CinCPacFlt, Mar. 20, 1953, Lyon Papers.
47. Lyon, letter to Momsen, Apr. 2, 1953; Momsen, letter to Lyon, Apr. 6, 1953; Lyon Papers.
48. Scientific Journal, May 18–19, 1953.
49. Lyon, letter to Momsen, May 22, 1953, Lyon Papers.
50. CNO, letter to CinCPacFlt, July 16, 1953, Lyon Papers.
51. Lyon oral history.
52. Scientific Journal, Aug. 20 and 25, 1953.
53. Scientific Journal, Aug. 25, 1953.
54. Scientific Journal, Aug. 27, 1953; Lyon oral history; Lyon, "Summary Report of Chief Scientist, Joint Canadian–U.S. Beaufort Sea Expedition 1953," Sept., 1953, Lyon Papers.
55. Scientific Journal, Oct. 22, 1953.
56. H. E. Bernstein, letter to CNO, Nov. 17, 1953, Lyon Papers.
57. BuShips, letter to CNO, Nov. 17, 1953; Lyon annotation; Lyon Papers.
58. ONR, letter to CNO, Jan. 7, 1954; Lyon annotation; Lyon Papers.
59. Bernstein, letter to CNO, Nov. 17, 1953, Lyon Papers. Emphasis is in original.
60. CinCPacFlt, letter to CNO, Jan. 7, 1954, Lyon Papers.
61. CNO, letter to NEL, Feb. 9, 1954, Lyon Papers.
62. Lyon, letter to Bennett, May, 1954, Lyon Papers.
63. Station Log, June 29 and July 2, 1954; Lyon, draft of letter to Bennett, n.d.; Lyon, letter to Field, July 2, 1954; all in Lyon Papers.
64. Scientific Journal, Aug. 1, 1954.
65. Scientific Journal, Aug. 4 and 5, 1954.
66. Scientific Journal, Aug. 8, 1954.
67. Scientific Journal, Aug. 9, 1954.
68. Scientific Journal, Aug. 11, 1954.
69. Scientific Journal, Aug. 12, 1954.
70. Scientific Journal, Aug. 13, 1954.

71. Scientific Journal, Aug. 14, 1954.
72. Alex Armstrong, *A Personal Narrative of the Discovery of the North-West Passage*, p. 453.
73. Scientific Journal, Aug. 14, 1954.
74. Scientific Journal, Aug. 14, 1954; Lyon annotation.
75. Scientific Journal, Aug. 15, 1954.
76. Scientific Journal, Aug. 17, 1954.
77. T. A. Irvine, *The Ice Was All Between*, p. 125.
78. Scientific Journal, Aug. 27, 1954.
79. Scientific Journal, Aug. 31, 1954.
80. Scientific Journal, Sept. 1, 1954.
81. Scientific Journal, Sept. 5, 1954.
82. Lyon, letter to T. G. Jacobs, Sept. 10, 1954, Lyon Papers.
83. Scientific Journal, Sept. 11 and 14, 1954.
84. Scientific Journal, Sept. 21, 1954.
85. Scientific Journal, Sept. 28, 29, and 30, 1954.
86. Lyon oral history.
87. William L. Maloney to CNO, "Preliminary Report of Joint Canadian–United States Beaufort Sea Expedition of 1954," Sept., 1954, Lyon Papers.
88. Jacobs, letter to CNO, first endorsement on Maloney to CNO, "Preliminary Report"; Station Log, Oct. 6, 1954. The photo spread appeared in *Life* 37 (Nov. 29, 1954): 102–107.
89. Scientific Journal, "Notes," Oct., 1954.

## Chapter 4. A Whole New World

1. Lyon, "NEL Submarine Research Facility," NEL Report 336, Oct. 16, 1952, Lyon Papers; Lyon, letter to author, Nov. 16, 1996.
2. See Harvey M. Sapolsky, *Science and the Navy: The History of the Office of Naval Research*, and John C. Reed and Andreas G. Ronhovde, *Arctic Laboratory: A History (1947–1966) of the Naval Arctic Research Laboratory at Point Barrow, Alaska*.
3. Bernstein, letter to ONR, enclosing "Research Project in Sea Ice Physics," Mar. 24, 1954, Lyon Papers.
4. ONR, letter to NEL, Nov. 23, 1954, Lyon Papers.
5. Scientific Journal, Dec. 7–10, 1954.
6. NEL to ONR, "Request for Support of Research Project in Sea Ice Physics," Jan. 21, 1955, Lyon Papers.
7. BuShips, letter to ONR, Feb. 15, 1955, Lyon Papers.
8. ONR, letter to NEL, Apr. 25, 1955, Lyon Papers.
9. Scientific Journal, Dec. 7–10, 1955.

10. Station Log, May 18 and 19, 1955; Lyon annotation.
11. For the background of the DEW Line, see Thomas W. Ray, "A History of the DEW Line, 1946–1964," U.S. Air Force Air Defense Command Historical Study No. 31, June, 1965; copy courtesy of the Air Force History and Museum Programs, Bolling Air Force Base, D.C. Richard Morenus, *DEW Line*, is an anecdotal account of the project.
12. Station Log, June 2, 1955; Lyon annotation.
13. Scientific Journal, July 2, 6, 8, and 10, 1955.
14. Lyon's participation in the DEW Line supply operation is detailed in his Scientific Journal for July, Aug., and Sept., 1955. See also *New York Times*, June 24 and Aug. 16, 1955.
15. *New York Times*, Sept. 11, 1955.
16. *New York Times*, Sept. 22, 1955.
17. Lyon, letter to J. C. Towner, Sept. 15, 1955, Lyon Papers.
18. Lyon, "Trip Report, Arctic Program," Nov., 1955, Lyon Papers.
19. Ibid.; Citation for Distinguished Civilian Service, July 13, 1955, Lyon Papers.
20. Lyon, letter to author, Nov. 16, 1996.
21. NEL, letter to BuShips, July 6, 1956, Lyon Papers.
22. Ibid.
23. Lyon, letter to Butler King Couper, July 25, 1956, Lyon Papers.
24. Couper, letter to Lyon, enclosing note from J. W. Bull, Aug. 3, 1956, Lyon Papers.
25. BuShips, letter to NEL, Aug. 30, 1956, Lyon Papers.
26. Station Log, Sept. 7, 1956.
27. Lyon, letter to Franz N. D. Kurie, Sept. 10, 1956, Lyon Papers.
28. Lyon oral history.
29. Station Log, Oct. 5, 1956; Scientific Journal, Oct. 30, 1956.
30. Scientific Journal, Oct. 30, 1956; Lyon oral history.
31. Lyon, letter to Bennett, Nov. 30, 1956, Lyon Papers.
32. Station Log, Mar. 1, 1957; Scientific Journal, Mar. 5, 1957; Lyon oral history.
33. Lyon, letter to J. M. Phelps, Mar. 11, 1957, Lyon Papers.
34. Ibid.
35. McWethy, letter to Lyon, Apr. 4, 1957, Lyon Papers.
36. BuShips, letter to NEL, Apr. 8, 1957, Lyon Papers.
37. ComSubPac, letter to BuShips, June 7, 1956, Lyon Papers.
38. BuShips, letter to ComSubPac, Apr. 10, 1957, Lyon Papers.
39. BuShips, letter to NEL, Apr. 26, 1957, enclosing Blair, "Visit to Naval Electronics Laboratory," Apr. 22, 1957, Lyon Papers.
40. Station Log, Apr. 26, 1957; Blair memo, n.d. [late Apr., 1957], Lyon Papers.

41. CNO, letter to BuShips, May 3, 1957, Lyon Papers.
42. BuShips, letter to CNO, May 10, 1957, Lyon Papers.
43. BuShips, letter to NEL, May 10, 1957, Lyon Papers.
44. McWethy, letter to Lyon, May 13, 1957, Lyon Papers.
45. Lyon, letter to Blair, June 18, 1957, Lyon Papers.
46. Ibid.
47. CNO, letter to CinCLantFlt, July 1, 1957, Lyon Papers.
48. William R. Anderson (with Clay Blair, Jr.), *Nautilus 90 North*, p. 48; Anderson, letter to author, Feb. 19, 1997.
49. Station Log, June 30, 1957; Lyon, letter to Anderson, July 2, 1957, Lyon Papers.
50. Anderson, letter to ComSubLant, July 5, 1957, Lyon Papers.
51. Blair, "Arctic Submarine Material," U.S. Naval Institute *Proceedings* 85 (Sept., 1959): 39–45.
52. Anderson, *Nautilus 90 North*, pp. 63–64; Anderson, "Results of Reconnaissance Flight, 7 August 1957," Lyon Papers. In *Nautilus 90 North*, Anderson included Lyon on the flight. Also, he stated: "There was far less open water than I had been led to believe." His contemporary report of the reconnaissance mission, however, indicates that his memory was faulty on both counts.
53. Anderson, *Nautilus 90 North*, p. 69.
54. Quoted in ibid., pp. 69–70.
55. Ibid., p. 71; Scientific Journal, Aug. 19, 20, and 21, 1957.
56. Scientific Journal, Aug. 28, 1957.
57. Scientific Journal, Sept. 1, 1957.
58. Lyon oral history
59. Anderson, *Nautilus 90 North*, pp. 83–84; Frank L. Wadsworth, who served on *Nautilus* as a junior officer in 1957, letter to author, Sept. 11, 1997.
60. Scientific Journal, Sept. 2, 1957.
61. Anderson, *Nautilus 90 North*, pp. 85–89; Wadsworth, letter to author, Sept. 11, 1957.
62. Scientific Journal, Sept. 2, 1957.
63. Scientific Journal, Sept. 3 and 4, 1957; Anderson, letter to author, Apr., 1997.
64. Anderson, *Nautilus 90 North*, p. 94.
65. Ibid., p. 95; Lyon, "Submarine Exploration of the North Pole Region: History, Problems, Positioning and Piloting," in Sylvie Devers (ed.), *Pôle Nord 1983* (Paris: Éditions du Centre de la Recherche Scientifique, 1987), pp. 313–28.
66. Wadsworth, letter to author, Sept. 11, 1997.
67. Scientific Journal, Sept. 5, 1957; Anderson, *Nautilus 90 North*, p. 95.
68. Scientific Journal, Sept. 5, 1957; Lyon oral history.

69. Ibid.; Anderson, *Nautilus 90 North*, pp. 96–97.
70. Scientific Journal, Sept. 5, 1957.
71. Scientific Journal, Sept. 6, 1957.
72. Scientific Journal, Sept. 7, 1957.
73. Scientific Journal, Sept. 8, 1957.
74. Ibid.; Anderson, *Nautilus 90 North*, p. 98.
75. Scientific Journal, Sept. 10, 1957.
76. Scientific Journal, Sept. 11–19, 1957.
77. Lyon oral history; Lyon, "Senior Scientist's Report to C. O. NAUTILUS," Sept., 1957, Lyon Papers.

## Chapter 5. Operation Sunshine

1. *Washington Post and Times Herald*, Oct. 30, 1957; *New York Times*, Oct. 30, 1957; Robert A. Divine, *The Sputnik Challenge*, p. vii; Lyon notes.
2. NEL, letter to CNO, Oct. 11, 1957, Lyon Papers.
3. Scientific Journal, Nov. 7, 1957.
4. Anderson, letter to CNO, Dec. 4, 1957, Lyon Papers.
5. ComSubRon 10, letter to CinCLantFlt, Dec. 17, 1957, Lyon Papers.
6. Anderson, letter to Lyon, Dec. 18, 1957, Lyon Papers.
7. McWethy, letter to Lyon, Dec. 30, 1957, enclosing "Proposed Program for Arctic Submarine Operations," Dec. 30, 1957, Lyon Papers.
8. Calvert, letter to Lyon, Jan. 2, 1958; Anderson, letter to Lyon, Jan. 6, 1958; Lyon, letter to McWethy, Anderson, Calvert, Couper, and Blair, Jan. 13, 1958, enclosing "1958 Submarine Operations, Arctic Ocean"; all in Lyon Papers.
9. Anderson, *Nautilus 90 North*, p. 104; William R. Anderson, interview with author, Great Falls, Va., Feb. 2, 1997.
10. Ibid., pp. 106–108.
11. Scientific Journal, Feb. 13 and June 18, 1958.
12. Scientific Journal, Feb. 14, 1958.
13. Scientific Journal, Feb. 17, 1958.
14. Scientific Journal, Feb. 24, 1958; Lyon memorandum, "Travel Orders for Operation Sunshine," Dec. 15, 1958, Lyon Papers.
15. Scientific Journal, Apr. 8, 1958; BuShips, letter to CNO, Mar. 11, 1958, Lyon Papers. See also Lyon, "Submarine Exploration of the North Pole Region," in Devers (ed.), *Pôle Nord 1983*, pp. 313–28; and Shepherd M. Jenks, "Navigating Under the North Pole Icecap," U.S. Naval Institute *Proceedings* (Dec., 1958): 62–67.
16. Scientific Journal, Apr. 8, 1958.
17. Scientific Journal, Apr. 10, 1958.

18. Scientific Journal, Apr. 25 and 28, 1958.
19. Anderson, *Nautilus 90 North*, pp. 124–28.
20. Scientific Journal, May 13, 1958; Lyon, "Senior Scientist's Report," Aug. 25, 1958, Lyon Papers.
21. Lyon oral history.
22. Scientific Journal, May 13 and 14, 1958.
23. Scientific Journal, May 22, 1958.
24. Ibid.; Anderson, *Nautilus 90 North*, p. 129–30.
25. Lyon, letter to Phelps, May 26, 1958, Lyon Papers.
26. Scientific Journal, June 2, 1958.
27. Scientific Journal, June 4, 1958; Anderson, *Nautilus 90 North*, pp. 136–37.
28. Scientific Journal, June 5, 1958.
29. Scientific Journal, June 7, 1958; Anderson, *Nautilus 90 North*, pp. 133–35.
30. Scientific Journal, June 8, 1958; Anderson, *Nautilus 90 North*, p. 144.
31. Anderson, *Nautilus 90 North*, p. 145.
32. Scientific Journal, June 15, 1958.
33. Anderson, *Nautilus 90 North*, pp. 156–57.
34. Scientific Journal, June 15, 1958.
35. Scientific Journal, June 16, 1958.
36. Scientific Journal, June 17, 1958.
37. Anderson to CNO, "Report of Operation SUNSHINE I," June 28, 1958, copy in Lyon Papers; Anderson to CNO, "Commanding Officer's Personal Diary for Operation SUNSHINE I," June 28, 1958, Record of the Naval Aides to the President, 1953–1961, Box 24, Eisenhower Library.
38. Anderson, "Diary."
39. Anderson, *Nautilus 90 North*, pp. 166–68; Anderson, "Diary."
40. Scientific Journal, June 18, 1958.
41. Anderson, "Diary."
42. Anderson, *Nautilus 90 North*, p. 172.
43. Lyon, "Senior Scientist's Report," June 19, 1958, Lyon Papers.
44. Anderson, letter to all hands, June 21, 1958," in Anderson, "Report of Operation SUNSHINE I."
45. Scientific Journal, June 30, 1958.
46. Scientific Journal, July 1, 1958.
47. Calvert, letter to Lyon, July 8, 1958, Lyon Papers.
48. Lyon annotation.
49. Lyon, letter to Phelps, July 3, 1958, Lyon Papers.
50. Scientific Journal, July 19, 1958.
51. Scientific Journal, July 20, 1958.
52. Scientific Journal, July 22, 1958.
53. Anderson to CNO, "Final Report of NAUTILUS Transpolar Voyage," Aug. 25, 1958, copy in Lyon Papers.

54. Scientific Journal, July 24, 1958.
55. Scientific Journal, July 28, 1958.
56. Ibid.
57. Anderson, "Final Report"; Scientific Journal, July 30, 01958.
58. Anderson, "Final Report."
59. Ibid.; Scientific Journal, Aug. 2, 1958.
60. Scientific Journal, Aug. 2, 1958; Anderson, *Nautilus 90 North*, pp. 213–14.
61. Scientific Journal, Aug. 3, 1958; Anderson, *Nautilus 90 North*, pp. 222–23.
62. Scientific Journal, Aug. 4, 1958.
63. Scientific Journal, Aug. 5, 1958.
64. Ibid.
65. Scientific Journal, Aug. 7, 1958.
66. *New York Times*, Aug. 9, 1958.
67. Ibid.
68. Scientific Journal, Aug. 12, 1958.
69. *New York Times*, Aug. 26, 1958.
70. *New York Times*, Aug. 28, 1958; Scientific Journal, Aug. 27, 1958.
71. Lyon, "Senior Scientist's Report," Aug. 25, 1958, Lyon Papers; *New York Times*, Aug. 13, 1958.

## Chapter 6. *Skate*

1. *New York Times*, Sept. 24, 1958; Albert G. Mumma, letter to Lyon, Sept. 25, 1958, Lyon Papers.
2. Lyon, letter to Mumma, Sept. 29, 1958, Lyon Papers.
3. Lyon annotation.
4. The account of *Skate*'s patrol is based on Calvert to CNO, "Report of Patrol Number One," Sept. 22, 1958, copy in the Lyon Papers; hereinafter cited as *Skate* 1958 Patrol Report.
5. John H. Nicholson, interview with author, La Jolla, Calif., Nov. 18, 1996; Arthur D. Molloy, "Daily Log, U.S.S. SKATE, July 29 to August 24, 1958," Lyon Papers.
6. Calvert, *Surface at the Pole*, p. 86. In naval documents, the indentification of the drifting station is spelled "Alfa." In U.S. Air Force documents and IGY publications, "Alpha" is used. As the station was operated by the U.S. Air Force, I have adopted "Alpha."
7. On the background of Alpha, see William M. Leary and Leonard A. LeSchack, *Project* COLDFEET: *Secret Mission to a Soviet Ice Station*, pp. 29–42.
8. Molloy, "Daily Log."
9. On the scientific mission of *Skate*, see E. C. LaFond, "Arctic Oceanogra-

phy by Submarines," U.S. Naval Institute *Proceedings* 86 (Sept., 1960): 90–96.
10. Lyon annotation on *Skate* 1958 Patrol Report.
11. Scientific Journal, Oct. 8, 1958; Lyon, letter to Phelps, Oct. 21, 1958, Lyon Papers.
12. Scientific Journal, Oct. 10 and 15, 1958; NEL, letter to ComSubRon 10, Dec. 22, 1958, Lyon Papers.
13. CNO, letter to CinCLantFlt, Dec. 30, 1958, Lyon Papers.
14. Lyon oral history. For the calculations, see Lyon's Scientific Journal for 1959, and Lyon, "Senior Scientist's Report: Winter Patrol, USS SKATE," Apr., 1959, Lyon Papers.
15. Scientific Journal, Feb. 11, 1959; ComSubRon 10, "Operation Order 4–59," Feb. 18, 1959, Lyon Papers.
16. Calvert, letter to Lyon, Sept. 27, 1996, Lyon Papers.
17. Scientific Journal, Mar. 2–14, 1959.
18. Calvert to CNO, "Report of SKATE March 1959 Arctic Patrol," Apr. 7, 1959, copy in the Lyon Papers. The trailing wire antenna, A. L. Kelln has noted, advanced SSBN communications "by tens of years." Kelln, letter to author, Mar., 1996.
19. Ibid.
20. William H. Layman, "*Skate* Breakthrough at the North Pole," U.S. Naval Institute *Proceedings* 85 (Sept., 1959): 32–37.
21. Calvert denies Lyon's suspicions. "I never had anything but the warmest feelings toward Waldo as a member of our team," he recalls. "The *Nautilus* experience was never a problem for me. I felt we were in on a different aspect of the arctic effort." Calvert, letter to author, Feb. 6, 1998.
22. Kelln, letter to author, Mar., 1997.
23. Scientific Journal, Mar. 15, 1959.
24. Lyon oral history; Calvert, letter to Lyon, Sept. 27, 1996, Lyon Papers.
25. Scientific Journal, Mar. 16, 1959; *Skate* 1959 Patrol Report.
26. Scientific Journal, Mar. 17, 1959; Calvert, *Surface at the Pole*, p. 179.
27. Calvert, *Surface at the Pole*, pp. 180–83.
28. Ibid., pp. 135–44, 158–59.
29. Scientific Journal, Mar. 17, 1959; Calvert, *Surface at the Pole*, pp. 184–87; Layman, "*Skate* Breakthrough at the North Pole." Sixteen years later, on May 4, 1975, Lyon was on board USS *Bluefish* (SSN 675) when it surfaced at the North Pole. In accordance with the last wishes of Lady Wilkins, he committed her ashes to the sea. See Lyon, "Submarine Exploration of the North Pole Region," in Devers (ed.), *Pôle Nord 1983*, p. 323.
30. *Skate* 1959 Patrol Report.
31. Ibid.
32. Ibid.

33. Ibid.
34. Scientific Journal, Mar. 20, 1959.
35. Calvert, *Surface at the Pole*, pp. 188–93; *Skate* 1959 Patrol Report.
36. Scientific Journal, Mar. 21, 1959.
37. Scientific Journal, Mar. 22, 1959.
38. Calvert, *Surface at the Pole*, pp. 193–96, 200–208.
39. Lyon oral history; Calvert, *Surface at the Pole*, pp. 196–99.
40. Scientific Journal, Mar. 24, 1959; Lyon, "Senior Scientist's Report: Winter Patrol, USS SKATE."
41. Scientific Journal, Mar. 25, 1959.
42. Scientific Journal, Mar. 28, 1959.
43. Scientific Journal, Apr. 7, 1959; Calvert, *Surface at the Pole*, p. 210.
44. Lyon, "Senior Scientist's Report: Winter Patrol, USS SKATE."

## Chapter 7. *Sargo*

1. Lyon annotation.
2. CNO Ser. 0115P31, Mar. 19, 1959; CNO Ser. 0387P31, Aug. 7, 1959; both in Lyon Papers.
3. Roshon, "CTFM Sonar."
4. Scientific Journal, Aug. 15, 1959; Nicholson, interview with author, La Jolla, Calif., Nov. 18, 1996.
5. Scientific Journal, Aug. 15, 1959.
6. Lyon, letter to Phelps, Sept. 9, 1959, Lyon Papers.
7. Lyon, "Scientific Plan," Oct. 28, 1959, Lyon Papers.
8. Scientific Journal, Jan. 3, 5, 6, and 8, 1960; Roshon, "CTFM Sonar."
9. Scientific Journal, Jan. 11, 12, and 13, 1960.
10. Roshon, "CTFM Sonar"; Nicholson, Journal, Jan. 18, 1960. I am grateful to Admiral Nicholson for making available the journal that he kept during the cruise.
11. Nicholson, Journal, Jan. 18, 1960.
12. Scientific Journal, Jan. 19–23, 1960; Nicholson, Journal, Jan. 21, 1960.
13. Nicholson, Journal, Jan. 22, 1960.
14. Nicholson, Journal, Jan. 25, 1960.
15. Ibid.
16. Nicholson to CNO, "Report of January-February Arctic Cruise, USS SARGO," Mar. 3, 1960, copy in Lyon Papers; hereinafter cited as *Sargo* Patrol Report.
17. Ibid.
18. Ibid.
19. Ibid.

20. Scientific Journal, Jan. 28, 1960.
21. *Sargo* Patrol Report.
22. Ibid.
23. David A. Phoenix, letter to author, Mar. 6, 1997.
24. *Sargo* Patrol Report; Scientific Journal, Jan. 29, 1960.
25. Nicholson, Journal, Jan. 29, 1960.
26. Scientific Journal, Feb. 1, 1960; *Sargo* Patrol Report.
27. Scientific Journal, Feb. 1, 1960; *Sargo* Patrol Report.
28. *Sargo* Patrol Report.
29. Ibid.; Scientific Journal, Feb. 2, 1960; Nicholson, Journal, Feb. 3, 1960; Roshon, "CTFM Sonar."
30. *Sargo* Patrol Report.
31. Roshon, "CTFM Sonar."
32. *Sargo* Patrol Report.
33. Lyon, message to NEL, Feb. 5, 1960, Lyon Papers. *Sargo* was unable to send this message until Feb. 9, 1960.
34. Lyon notes; Lyon, "Senior Scientist's Report, SARGO Winter Pacific Arctic Patrol, 1960," Mar. 3, 1960, Lyon Papers.
35. *Sargo* Patrol Report.
36. Nicholson, Journal, Feb. 7, 1960.
37. Roshon, "CTFM Sonar"; *Sargo* Patrol Report.
38. *Sargo* Patrol Report; Scientific Journal, Mar. 5, 1960.
39. *Sargo* Patrol Report.
40. Scientific Journal, Feb. 10, 1960.
41. Scientific Journal, Feb. 12, 1960.
42. Ibid.; Nicholson, Journal, Feb. 12, 1960; Roshon, "CTFM Sonar."
43. Scientific Journal, Feb. 14, 1960.
44. *Sargo* Patrol Report.
45. Ibid.
46. On T-3, see Leary and LeSchack, *Project* COLDFEET, pp. 22–27, 30–31, 68.
47. *Sargo* Patrol Report.
48. Ibid.; Nicholson, Journal, Feb. 19, 1960; Nicholson, "Sargo," in James Dugan and Richard Vahan (eds.), *Men Under Water*, pp. 53–67.
49. Scientific Journal, Feb. 19, 1960.
50. Roshon, "CTFM Sonar"; Nicholson, interview with author, La Jolla, Calif., Nov. 18, 1996.
51. *Sargo* Patrol Report; Lyon oral history.
52. *Sargo* Patrol Report; Scientific Journal, Feb. 19, 1960.
53. *Sargo* Patrol Report; Scientific Journal, Feb. 20, 1960.
54. *Sargo* Patrol Report.
55. Ibid.

56. Ibid.; Nicholson, Journal, Feb. 23, 1960; Scientific Journal, Feb. 23, 1960. The U.S. Coast and Geodetic Survey's investigation of the Bering Sea and Bering Strait in 1960–1962 resulted in the publication of H. O. Chart 5819A.
57. Scientific Journal, Feb. 24, 1960.
58. Scientific Journal, Feb. 25, 1960.
59. Lawrence R. Daspit, message to *Sargo*, Feb. 25, 1960; H. G. Hopwood, message to *Sargo*, Feb. 25, 1960; both in Lyon Papers.
60. *New York Times*, Mar. 5, 1960.
61. Scientific Journal, Mar. 6, 1960; Lyon annotation.
62. *Sargo* Patrol Report.
63. Lyon, "Senior Scientist Report, SARGO Winter Pacific Arctic Patrol, 1960."
64. *Sargo* Patrol Report.
65. Ibid.
66. Lyon oral history.
67. William K. Yates, letter to author, Mar. 24, 1997.
68. American Society of Naval Engineers, "Citation," Apr., 1960; Lyon annotation.

## Chapter 8. Closing the Circle

1. David S. Boyd, letter to Lyon, Apr. 19, 1959, Lyon Papers.
2. Lyon, letter to Boyd, Apr. 23, 1959; Lyon, letter to Daspit, Apr. 23, 1959; both in Lyon Papers.
3. Calvert, letter to CNO, May 13, 1959; Calvert, letter to Lyon, May 21, 1959; both in Lyon Papers.
4. Scientific Journal, May 30, 1959; ComSubLant, second endorsement on Calvert's letter to CNO, June 15, 1959, Lyon Papers.
5. Lyon, letter to Phelps, June 22, 1959, Lyon Papers.
6. ComSubLant, letter to NEL, June 15, 1959; NEL, letter to ComSubLant, July 15, 1959; both in Lyon Papers.
7. CNO, letter to CinCLantFlt, Aug. 14, 1959, Lyon Papers.
8. Quoted in George P. Steele, SEADRAGON: *Northwest Under the Ice*, p. 62.
9. CNO, letter to NEL, Mar. 15, 1960, Lyon Papers; Scientific Journal, Mar. 28, 1960.
10. Lyon, "Scientific Plan," July 15, 1960, Lyon Papers.
11. BuShips, letter to Portsmouth Naval Shipyard, Apr. 8, 1960, Lyon Papers.
12. Steele, SEADRAGON, p. 68.
13. Ibid., pp. 72–75.

14. Station Log, June 15, 1960; H. E. Ruble, director, U.S. Navy Underwater Sound Laboratory, letter to Lyon, June 30, 1960; and CNO, letter to CinCLantFlt and CinCPacFlt, May 21, 1960; both in Lyon Papers.
15. Station Log, Mar. 18, 1960; Scientific Journal, July 28 and 29, 1960; Lyon annotation.
16. ComSubLant, Operation Order 11–60, July 15, 1960; Lyon, "Senior Scientist's Report, U.S.S. SEADRAGON (SSN 584), Arctic Cruise, August–September 1960," Sept. 17, 1960; both in Lyon Papers.
17. Steele, SEADRAGON, p. 103.
18. Steele to CNO, "Report of August–September 1960 Arctic Cruise (SUBICEX 3–60)," Sept. 14, 1960, copy in Lyon Papers; hereinafter cited as *Seadragon* Patrol Report.
19. Roshon, "CTFM Sonar."
20. *Seadragon* Patrol Report; Steele, SEADRAGON, p. 117.
21. *Seadragon* Patrol Report; Roshon, "CTFM Sonar."
22. Roshon, "CTFM Sonar"; Steele, SEADRAGON, p. 117–18.
23. Steele, SEADRAGON, pp. 118–19.
24. Ibid., pp. 121–28; Scientific Journal, Aug. 11, 1960.
25. Steele, SEADRAGON, p. 137.
26. *Seadragon* Patrol Report; Steele, SEADRAGON, p. 138.
27. Johnathan C. Schere, letter to author, Apr. 1, 1997.
28. Ibid.; Steele, SEADRAGON, pp. 138–39; Roshon, "CTFM Sonar." Steele later defended his actions: "I selected the iceberg detector predicted 400–440 foot draft iceberg as just the right size. I selected the depth to go under it, 700 feet, to cover a very large error even though my trials of sonar against another submarine at known depth, and on the earlier shallow draft berg, indicated that it made a good *ballpark* draft measurement. Despite Schere's comment to the contrary, I really did understand that and . . . I always carefully weighed the danger to the ship. Risk taking is part of seagoing." Steele, letter to author, Jan. 27, 1998.
29. Steele, SEADRAGON, p. 140.
30. Scientific Journal, Aug. 13, 1960; *Seadragon* Patrol Report; Steele, SEADRAGON, p. 142.
31. Steele, letter to author, Jan. 27, 1998.
32. *Seadragon* Patrol Report.
33. Steele, SEADRAGON, pp. 157–60; Scientific Journal, Aug. 16, 1960.
34. Steele, SEADRAGON, p. 159; Scientific Journal, Aug. 16, 1960.
35. *Seadragon* Patrol Report.
36. Steele, SEADRAGON, p. 164.
37. *Seadragon* Patrol Report.
38. Steele, SEADRAGON, p. 174; Roshon, "CTFM Sonar."
39. *Seadragon* Patrol Report.

40. Ibid.; Steele, SEADRAGON, p. 187.
41. Steele, SEADRAGON, p. 201.
42. *Seadragon* Patrol Report.
43. Steele, SEADRAGON, p. 218.
44. Roshon, "CTFM Sonar." This incident is neither mentioned in Steele, SEADRAGON, nor *Seadragon* Patrol Report. Roshon declined to identify the conning officer.
45. *Seadragon* Patrol Report.
46. Steele, SEADRAGON, p. 230; Steele, letter to Vice Adm. Richard W. Mies, Oct. 7, 1997, copy provided to the author by Admiral Steele; *New York Times*, Aug. 26, 1960.
47. Steele, SEADRAGON, p. 235.
48. Ibid., p. 237.
49. Scientific Journal, Aug. 26, 1960; *Seadragon* Patrol Report.
50. Steele, SEADRAGON, p. 239.
51. Ibid., p. 242.
52. Roshon, "CTFM Sonar." See also Anderson, *Nautilus 90 North*, pp. 127–28.
53. Steele, SEADRAGON, p. 243.
54. Ibid., pp. 244–46.
55. Roshon, "CTFM Sonar."
56. Scientific Journal, Sept. 3, 1960.
57. *Seadragon* Patrol Report.
58. Ibid.; Scientific Journal, Sept. 5, 1960.
59. Scientific Journal, Sept. 14, 1960.
60. *Seadragon* Patrol Report; Steele, letter to author, Jan. 27, 1998.
61. *Seadragon* Patrol Report.
62. Lyon, "Senior Scientist's Report."
63. Lyon, comments on O. C. S. Robertson, "The Arctic as a Theatre of War," in *Seadragon* Patrol Report; Lyon oral history.

## Chapter 9. Tactics and Weapons

1. Lyon, "Notes—Arctic Program," Jan. 11, 1961, Lyon Papers.
2. Calvert, letter to Lyon, Dec. 5, 1960, Lyon Papers.
3. ComSubLant, letter to ComSubRon 10, Feb. 25, 1961, Lyon Papers.
4. Calvert, letter to Lyon, Apr. 26, 1961, Lyon Papers.
5. NEL, letter to CNO via BuShips, Jan. 20, 1961, Lyon Papers.
6. ComSubPac, letter to CNO, Mar. 6, 1961, and BuShips, letter to NEL, Mar. 16, 1961, Lyon Papers; Lyon notes.
7. Press clipping, undated, Lyon Papers; Lyon annotation.

8. Calvert, letter to Lyon, Feb. 10, 1961, and Lyon, letter to Calvert, Feb. 27, 1961, Lyon Papers; Station Log, Apr. 14 and 25 and May 15, 1961.
9. ComSubPac, letter to ComSubLant, June 20, 1961, and CNO, letter to NEL, Nov. 16, 1961, Lyon Papers.
10. ComSubLant, letter to ComSubRon 10, Jan. 8, 1962; Joseph L. Skoog, Jr., to CNO, "Report of July–August 1962 Arctic Cruise, SUBICEX 2–62," Aug. 28, 1962; hereinafter cited as *Skate* Patrol Report.
11. Skoog, letter to BuShips, Feb. 17, 1962, and ComSubDiv 102, "SUBICEX 1–62 Report," July 13, 1962, Lyon Papers.
12. Lyon, letter to Director, NEL, Mar. 1, 1962, and D. P. Heritage, Code 2730, NEL, letter to Lyon, Mar. 21, 1962, Lyon Papers. On OMEGA, see J. A. Pierce and A. Shostak, "The Omega Navigation System," *Naval Research Reviews* (Oct., 1966): 22–27.
13. BuShips, letter to CNO, Feb. 7, 1962, Lyon Papers.
14. Station Log, Apr. 1–12, 1962; Lyon annotation; Lyon oral history.
15. Scientific Journal, July 8–12, 1962; Lyon annotation.
16. Wadsworth, letter to author, Aug. 11, 1997.
17. *Skate* Patrol Report; Charles D. Summitt to CNO, "SEADRAGON Report of SUBICEX 2–62," Aug. 28, 1962; hereinafter cited as *Seadragon* Patrol Report.
18. Schere, letter to author, Apr. 1, 1997.
19. *Skate* Patrol Report.
20. Scientific Journal, July 14, 1962.
21. [Richard J. Boyle], *Sourcebook on Submarine Arctic Operations* (San Diego: U.S. Navy Electronics Laboratory, 1966). Although there is no author indicated for this valuable reference work, it was written by Boyle.
22. *Skate* Patrol Report.
23. Scientific Journal, July 17, 1962.
24. Wadsworth, letter to author, Aug. 11, 1997.
25. *Skate* Patrol Report; Boyle, *Sourcebook*.
26. Scientific Journal, July 24, 1962.
27. Scientific Journal, July 25, 1962.
28. Michael R. Beschloss, *The Crisis Years: Kennedy and Khrushchev, 1960–1963*.
29. *Skate* Patrol Report.
30. Skoog informed CNO on Aug. 28, 1962, that *Skate*'s patrol report contained "deletions and departures from fact as appropriate to conceal the fact that departure from unclassified track was made." He enclosed the true narrative account, marked SECRET, covering the period July 24–30. Both the "fake" pages of the patrol report and the correct ones are in the Lyon Papers.
31. *Skate* Patrol Report; *Seadragon* Patrol Report; Wadsworth, letter to author, Aug. 11, 1997.

32. *Skate* Patrol Report.
33. Scientific Journal, Aug. 2, 1962.
34. *Skate* Patrol Report.
35. Scientific Journal, Aug. 3, 1962.
36. *Skate* Patrol Report.
37. Scientific Journal, Aug. 4, 1962.
38. Norman Friedman, *U.S. Submarines Since 1945*, p. 20.
39. *Skate* Patrol Report; *Seadragon* Patrol Report.
40. Scientific Journal, Aug. 8, 1962.
41. Scientific Journal, Aug. 9, 1962.
42. *Skate* Patrol Report.
43. Scientific Journal, Aug. 10, 1962.
44. *New York Times*, Aug. 8, 1962.
45. *Seadragon* Patrol Report; Scientific Journal, Aug. 10, 1962.
46. *Skate* Patrol Report.
47. Ibid.; *Seadragon* Patrol Report.
48. Scientific Journal, Aug. 17, 1962.
49. *New York Times*, Aug. 23, 1962.
50. Scientific Journal, Aug. 28, 1962.
51. Wadsworth, letter to author, Aug. 11, 1997.
52. Ibid.
53. *Skate* Patrol Report.
54. Ibid.
55. Lyon, "Senior Scientist's Report, USS SKATE–USS SEADRAGON Arctic Cruise 1962," Aug. 23, 1962, Lyon Papers.
56. Ibid.
57. Ibid.
58. Ibid.
59. Francis Duncan, *Rickover and the Nuclear Navy: The Discipline of Technology*, p. 74; Nicholson, interview with author, La Jolla, Calif., Nov. 18, 1996; Dean L. Axene, letter to author, Mar. 26, 1997.
60. Nicholson, interview with author, La Jolla, Calif., Nov. 18, 1996; Friedman, *U.S. Submarines Since 1945*, p. 146.
61. Nicholson, interview with the author, La Jolla, Calif., Nov. 18, 1996; Friedman, *U.S. Submarines Since 1945*, p.146.
62. Lyon notes.

## Chapter 10. The Arctic Submarine

1. Lyon, "Notes—Arctic Program, 1963," Sept. 30, 1962; Elton W. Grenfell, letter to Lyon, Oct. 8, 1962; both in Lyon Papers.

2. Lyon, letter to Jon L. Boyes, Oct. 15, 1962; Boyes, letter to Lyon, Oct. 20, 1962; Boyes, "Draft Seven Year Program," Oct. 26, 1962; CNO, "OPNAV Instruction 03470.4—Submarine Arctic Warfare and Scientific Program," May 27, 1963; all in Lyon Papers.
3. *New York Times*, Jan. 28, 1963; Norman Polmar and Jurrien Noot, *Submarines of the Russian and Soviet Navies, 1718–1990*, pp. 170–71.
4. Duncan, *Rickover and the Nuclear Navy*, pp. 77–98; Friedman, *U.S. Submarines Since 1945*, pp. 143–45; Norman Polmar, *Death of the Thresher*.
5. Station Log, July 15, 1963; Lyon annotation.
6. Lyon, "Arctic Tasks—Summary, January 1963–July 1964," July, 1964, Lyon Papers.
7. Ibid.; Lyon oral history; Lyon and James H. Brown, "Submarine Surfacing Through Sea Ice," June, 1964, Lyon Papers.
8. Capt. N. C. Nash, assistant chief of staff, ComSubLant, letter to Lyon, Jan. 29, 1964, Lyon Papers.
9. Lyon, "Arctic Tasks"; J. M. Snyder, letter to Lyon, May 1, 1964, Lyon Papers.
10. Edward A. Burkhalter, Jr., letter to Lyon, Sept. 16, 1965, Lyon Papers.
11. Lyon, letter to James T. Strong, Sept. 15, 1963, Lyon Papers.
12. Strong, letter to Lyon, Aug. 27, 1963, Lyon Papers.
13. Strong, letter to Lyon, enclosing proposed plan for arctic operations, Dec. 3, 1963, Lyon Papers.
14. Strong, letter to Lyon, Aug. 3, 1964, Lyon Papers.
15. P. A. Beshany, letter to Lyon, enclosing letter from Edward Teller, Sept. 26, 1967, Lyon Papers.
16. Wilson, letter to Lyon, enclosing Beshany memorandum, "Submarine Arctic Operations," Oct. 24, 1967, Lyon Papers.
17. Lyon, remarks in Report of Submarine Conference, Jan. 18, 1966, Lyon Papers.
18. Boyle, letter to Roshon, July 27, 1966, Lyon Papers.
19. Ibid.
20. Jackson B. Richard, letter to Lyon, Aug. 23, 1966, Lyon Papers.
21. Lyon, "QUEENFISH Considerations," Aug. 26, 1966; Eugene Wilkinson, letter to Lyon, Sept. 20, 1966; both in Lyon Papers.
22. ComSubLant, letter to ComSubPac, Nov. 16, 1966, Lyon Papers.
23. Scientific Journal, Feb. 2, 1967; Richard to CNO, "Report of February 1967 Marginal Sea Ice Zone Operations in the Davis Strait," Feb. 19, 1967, hereinafter cited as *Queenfish* Patrol Report.
24. *Queenfish* Patrol Report.
25. Ibid.
26. Ibid.
27. Ibid.; Lyon, letter to author, Mar. 25, 1997.

28. Scientific Journal, Feb. 9, 1967.
29. Ibid.
30. *Queenfish* Patrol Report.
31. Ibid.; Scientific Journal, Feb. 10, 1967.
32. *Queenfish* Patrol Report.
33. Ibid.
34. Scientific Journal, Feb. 11, 1967.
35. *Queenfish* Patrol Report.
36. Ibid.
37. Lyon, "Senior Scientist's Report, USS QUEENFISH (SSN 651)," Feb. 16, 1967, Lyon Papers.
38. Ibid.; Lyon, letter to BuShips, Nov. 26, 1962, Lyon Papers.
39. Lyon, "Senior Scientist's Report, USS QUEENFISH (SSN 651)"; Lyon oral history.
40. Richard, letter to ComSubPac, Apr. 28, 1967, Lyon Papers.
41. Richard, letter to Lyon, Apr. 28, 1967, Lyon Papers.
42. ComSubPac, letter to Richard, June 13, 1967, Lyon Papers.
43. Strong, letter to ComSubLant, Oct. 19, 1967, with ComSubLant annotation, Lyon Papers.
44. Strong, letter to Lyon, Nov. 16, 1967, Lyon Papers.
45. ComSubLant, Operation Order, SUBICEX 1–69, Mar. 7, 1969, Lyon Papers.
46. Scientific Journal, Mar. 25, 1969.
47. Scientific Journal, Mar. 27 and 30, 1969.
48. William M. Wolff, Jr., letter to author, May 7, 1997.
49. Ibid.; Scientific Journal, Apr. 1, 1969.
50. Scientific Journal, Apr. 2, 1969. Unlike later submarines, *Whale* did not carry $O_2$ candles that could be burned to produce oxygen as a backup.
51. Scientific Journal, Apr. 3, 1969.
52. Wolff, letter to author, May 7, 1997.
53. Ibid.; Scientific Journal, Apr. 6, 1969.
54. Scientific Journal, Apr. 10, 1969.
55. Scientific Journal, Apr. 14, 1969.
56. Scientific Journal, Apr. 16, 1969.
57. Scientific Journal, Apr. 18, 1969.
58. Scientific Journal, Apr. 20, 1969.
59. Steven A. White, letter to author, Jan. 4, 1998.
60. Ibid.; Molloy's journal of the cruise is deposited in the Lyon Papers.
61. White, letter to author, Jan. 4, 1998.
62. White recalled this encounter with Schade in ibid.
63. Lyon, "Senior Scientist's Report—U.S.S. WHALE (SSN 638)," Apr. 24, 1969, Lyon Papers.

64. Ibid.; White, letter to author, Jan. 4, 1998.
65. Lyon, "Senior Scientist's Report—U.S.S. WHALE (SSN 638)."

## Epilogue

1. Alfred S. McLaren, "Under the Ice in Submarines," U.S. Naval Institute *Proceedings* 107 (July, 1981): 105–109. McLaren later used the data obtained on the voyage in his doctoral thesis, "Analysis of the Under-Ice Topography in the Arctic Basin as Recorded by USS NAUTILUS in 1958 and USS QUEENFISH in 1970," University of Colorado, 1986. His findings are summarized in Alfred S. McLaren, "The Under-Ice Thickness Distribution of the Arctic Basin as Recorded in 1958 and 1970," *Journal of Geophysical Research* 94 (Apr., 1989): 4971–83.
2. Scientific Journal, Nov. 18, 1970.
3. Lyon, "Senior Scientist's Report, USS HAMMERHEAD (SSN 663)," Dec. 4, 1970, Lyon Papers. See also Lyon, "Submarine Exploration of the North Pole Region."
4. Lyon, "Report by Technical Director, SUBICEX 1–73," Apr. 6, 1973, Lyon Papers.
5. Scientific Journal, May 10, 1973.
6. Lyon, "Senior Scientist's Report, SUBICEX 1–75," May, 1975; Lyon, "Senior Scientist's Report, SUBICEX 1–76," Apr. 28, 1976; Lyon, "Senior Scientist's Report, SUBICEX 1–77," May 13, 1977; Lyon, "Report of Technical Director, SUBICEX 1–78," Oct., 1978; Lyon, "Report of Technical Director, SUBICEX 1–79," May 5, 1979; all in Lyon Papers.
7. Roshon, "CTFM Sonar."
8. Schere, letter to author, Apr. 1, 1997.
9. Boyle, letter to author, May 25, 1997; Lyon, "Senior Scientist's Report, SUBICEX 1–77," May 13, 1977.
10. Roshon, "CTFM Sonar." On the managerial changes for R&D, see Boyle, "Thoroughly Modern Management," U.S. Naval Institute *Proceedings* 98 (Oct., 1972): 34–40.
11. Lyon, "Senior Scientist's Report, SUBICEX 1–76," Apr. 28, 1976.
12. Lyon, "Report of Technical Director, SUBICEX 1–79," May 5, 1979. See also Boyle, "The Fleet Connection," U.S. Naval Institute *Proceedings* 108 (Sept., 1982): 57–61.
13. Polmar, "Sailing Under the Ice," U.S. Naval Institute *Proceedings* 110 (June, 1984): 121–23.
14. Frederick H. Hartmann, *Naval Renaissance: The U.S. Navy in the 1980s*; James D. Watkins, "The Maritime Strategy," U.S. Naval Institute *Proceedings*, supplement, 112 (Jan., 1986): 2–17.

15. Kinnard McKee is quoted in Mark Sakitt, *Submarine Warfare in the Arctic: Option or Illusion?* p. 13. See also Willy Østreng, "The Geostrategic Conditions of Deterrence in the Barents Sea," in Lawson W. Brigham (ed.), *The Soviet Maritime Arctic* (Annapolis: Naval Institute Press, 1991), pp. 201–14.
16. Lyon, "Submarine Combat in the Ice," U.S. Naval Institute *Proceedings* 118 (Feb., 1992): 34–40.
17. On the troubled *Seawolf* program, see James L. George, *The U.S. Navy in the 1990s: Alternatives for Action*.
18. Lyon, "Acceptance Speech of Bushnell Award," Lyon Papers.

# Bibliography

SPECIAL COLLECTIONS

The Operations Archives of the United States Navy, Naval Historical Center, Washington, D.C.

The Papers of Waldo K. Lyon, Waldo K. Lyon, San Diego, Calif.

The Papers of Sir Hubert Wilkins, Byrd Polar Research Center, University Archives, Ohio State University, Columbus, Ohio.

Records of the Naval Aides to the President, 1953–1961, Eisenhower Presidential Library, Abilene, Kans.

Records of the Secretary of the Navy, Correspondence of the Chief of Naval Operations, Record Group 38, National Archives, Washington, D.C.

"The Reminiscences of Dr. Waldo K. Lyon," Oral History Department, U.S. Naval Institute.

BOOKS

Anderson, William R. (with Clay Blair, Jr.) *Nautilus 90 North*. Cleveland: World Publishing Company, 1959.

Armstrong, Alex. *A Personal Narrative of the Discovery of the North-West Passage*. London: Hurst and Blackett, 1857.

Beschloss, Michael. *The Crisis Years: Kennedy and Khrushchev, 1960–1963*. New York: HarperCollins, 1991.

[Boyle, Richard J.] *Sourcebook on Submarine Arctic Operations*. San Diego: U.S. Navy Electronics Laboratory, 1966.

Calvert, James. *Surface at the Pole*. New York: McGraw-Hill, 1960.

Divine, Robert A. *The Sputnik Challenge*. New York: Oxford University Press, 1993.

Duncan, Francis. *Rickover and the Nuclear Navy: The Discipline of Technology*. Annapolis: Naval Institute Press, 1990.

Fanning, A. E. *Steady As She Goes: A History of the Compass Department of the Admiralty*. London: Her Majesty's Stationery Office, 1986.

Friedman, Norman. *U.S. Submarines Since 1945*. Annapolis: Naval Institute Press, 1994.

George, James L. *The U.S. Navy in the 1990s: Alternatives for Action*. Annapolis: Naval Institute Press, 1992.

Grierson, John. *Sir Hubert Wilkins: Enigma of Exploration.* London: Robert Hale, 1960.

Hadley, Michael L. *U-Boats against Canada: German Submarines in Canadian Waters.* Montreal: McGill-Queen's University Press, 1985.

Hartmann, Frederick H. *Naval Renaissance: The U.S. Navy in the 1980s.* Annapolis: Naval Institute Press, 1990.

Hewlett, Richard G., and Francis Duncan. *Nuclear Navy, 1946–1962.* Chicago: University of Chicago Press, 1974.

Hunt, William R. *Stef: A Biography of Vilhjalmur Stefansson.* Vancouver: University of British Columbia Press, 1986.

Irvine, T. A. *The Ice Was All Between.* Toronto: Longmans, Green and Company, 1959.

Isenberg, Michael T. *Shield of the Republic.* New York: St. Martin's Press, 1993.

Lake, Simon. *Submarine: The Autobiography of Simon Lake (as told to Herbert Corey).* New York: Appleton-Century Company, 1938.

———. *The Submarine in War and Peace.* Philadelphia: Lippincott, 1918.

Leary, William M., and Leonard A. LeSchack. *Project COLDFEET: Secret Mission to a Soviet Ice Station.* Annapolis: Naval Institute Press, 1996.

Lockwood, Charles A., and Hans Christian Adamson. *Hellcats of the Sea.* New York: Greenberg, 1955.

Love, Robert W., Jr. *History of the U.S. Navy, 1942–1991.* Harrisburg: Stackpole Books, 1992.

Meigs, Montgomery C. *Slide Rules and Submarines: American Scientists and Subsurface Warfare in World War II.* Washington, D.C.: National Defense University Press, 1990.

Milner, Marc. *The U-Boat Hunters: The Royal Canadian Navy and the Offensive against Germany's Submarines.* Annapolis: Naval Institute Press, 1994.

Mirsky, Jeannette. *To the Arctic! The Story of Northern Exploration from Earliest Times to the Present.* New York: Knopf, 1948.

Morenus, Richard. *DEW Line.* New York: Rand McNally and Company, 1957.

Osborn, Sherard, ed. *The Discovery of the North-West Passage by H.M.S. 'Investigator,' Captain R. M'Clure.* London: Longman, Brown, Green, Longmans, & Roberts, 1856.

Polmar, Norman. *Death of the Thresher.* New York: Chilton Books, 1964.

———, and Thomas B. Allen. *Rickover.* New York: Simon and Schuster, 1982.

———, and Jurrien Noot. *Submarines of the Russian and Soviet Navies, 1718–1990.* Annapolis: Naval Institute Press, 1991.

Powers, Richard Gid, ed. *The Science Fiction of Frank R. Stockton.* Boston: Gregg Press, 1976.

Reed, John C., and Andreas G. Ronhovde. *Arctic Laboratory: A History (1947–1966) of the Naval Arctic Research Laboratory at Point Barrow, Alaska.* Washington, D.C.: Arctic Institute of North America, 1971.

Roland, Alex. *Underwater Warfare in the Age of Sail.* Bloomington: University of Indiana Press, 1978.

Rose, Lisle A. *Assault on Eternity: Richard E. Byrd and the Exploration of Antarctica, 1946–47.* Annapolis: Naval Institute Press, 1980.

Sakitt, Mark. *Submarine Warfare in the Arctic: Option or Illusion?* Stanford: Center for International Security and Arms Control, 1988.

Sapolsky, Harvey M. *Science and the Navy: The History of the Office of Naval Research.* Princeton: Princeton University Press, 1991.

Shapiro, Barbara J. *John Wilkins, 1614–1672: An Intellectual Biography.* Berkeley: University of California Press, 1969.

Stefansson, Vilhjalmur. *The Friendly Arctic: The Story of Five Years in Polar Regions.* New York: Macmillan, 1921.

Steele, George P. SEADRAGON: *Northwest Under the Ice.* New York: E. P. Dutton, 1962.

U.S. Department of the Navy. *Naval Arctic Operations Handbook.* 2 parts. Washington, D.C.: Department of the Navy, 1949.

U.S. Naval Ocean Systems Center. *Fifty Years of Research and Development on Point Loma, 1940–1990.* San Diego: Naval Ocean Systems Center, 1990.

Weisgall, Jonathan M. *Operation Crossroads: The Atomic Tests at Bikini Atoll.* Annapolis: Naval Institute Press, 1995.

Wilkins, Hubert. *Under the North Pole.* New York: Brewer, Warren & Putnam, 1931.

ARTICLES

Anschütz-Kämpfe, Hermann, "Das europäische Eismeer und ein neuer Expeditionsplan nach dem Nordpole," *Mittheilungen der Kaiserlich Königlichen Geographischen Gesellschaft* 44 (1901): 53–73.

Atkeson, Edward B. "Fighting Subs Under the Ice," *U.S. Naval Institute Proceedings* 113 (Sept., 1987): 81–87.

Blair, Carvel Hall. "Arctic Submarine Material," *U.S. Naval Institute Proceedings* 85 (Sept., 1959): 39–45.

Boyle, Richard. "Bound in Shallows and Miseries?" *U.S. Naval Institute Proceedings* 122 (Oct., 1996): 52–55.

———. "The Fleet Connection," *U.S. Naval Institute Proceedings* 108 (Sept., 1982): 57–61.

———. "1960: A Vintage Year for Submarines," *U.S. Naval Institute Proceedings* 96 (Oct., 1970): 35–41.

———. "Thoroughly Modern Management," *U.S. Naval Institute Proceedings* 98 (Oct., 1972): 34–40.

Brown, Gordon V. "Arctic ASW," U.S. Naval Institute *Proceedings* 88 (Mar., 1962): 53–57.

Cadwalader, John. "Arctic Drift Stations," U.S. Naval Institute *Proceedings* 89 (Apr., 1963): 67–75.

Jenks, Shepherd M. "Navigating Under the North Pole Icecap," U.S. Naval Institute *Proceedings* 84 (Dec., 1958): 62–67.

———. "Under the Ice to the North Pole," *Reader's Digest* 74 (Feb., 1959): 103–106.

LaFond, E. C. "Arctic Oceanography by Submarines," U.S. Naval Institute *Proceedings* 86 (Sept., 1960): 90–96.

Layman, William H. "*Skate* Breakthrough at the North Pole," U.S. Naval Institute *Proceedings* 85 (Sept., 1959): 32–37.

LeMarchand, T. M. "Under Ice Operations," *Naval War College Review* 38 (May–June, 1985): 19–27.

LeSchack, Leonard A. "ComNavForArctic," U.S. Naval Institute *Proceedings* 113 (Sept., 1987): 74–80.

Lyon, Waldo K. "Experiments in the Use of Explosives in Sea Ice," *Polar Record* 10 (Sept., 1960): 237–47.

———. "Infra-Red Absorption Spectra of the Water Molecule in Crystals," *The Physical Review* 61 (1942): 482–89.

———. "Ocean and Sea-Ice Research in the Arctic Ocean Via Submarine," *Transactions of the New York Academy of Sciences* 23 (June, 1961): 662–74.

———. "The Submarine and the Arctic Ocean," *The Polar Record* 75 (1963): 699–705.

———. "Submarine Combat in the Ice," U.S. Naval Institute *Proceedings* 118 (Feb., 1992): 34–40.

———. "Submarine Exploration of the North Pole Region: History, Problems, Positioning and Piloting," in Sylvie Devers, ed., *Pôle Nord 1983* (Paris: Éditions du Centre National de la Recherche Scientifique, 1987), pp. 313–28.

McLaren, Alfred S. "The Arctic Submarine: Its Evolution and Scientific and Commercial Potential," in Sylvie Devers, ed., *Pôle Nord 1983* (Paris: Éditions du Centre National de la Recherche Scientifique, 1987), pp. 329–41.

———. "Under the Ice in Submarines," U.S. Naval Institute *Proceedings* 107 (July, 1981): 105–109.

———. "The Under-Ice Thickness Distribution of the Arctic Basin as Recorded in 1958 and 1970," *Journal of Geophysical Research* 94 (Apr., 1989): 4971–83.

McLennan, R. A. "Avoiding the Undersea Surprise," U.S. Naval Institute *Proceedings* 120 (May, 1994): 122–26.

McWethy, Robert D. "Significance of the *Nautilus* Polar Cruise," U.S. Naval Institute *Proceedings* 84 (May, 1958): 32–35.

Merrill, John. "Floating Wire Antennas: Communicating with a Submerged Submarine," *The Submarine Review* (Apr., 1957): 56–63.

Nicholson, John H. "Sargo," in James Dugan and Richard Vahan, eds., *Men Under Water* (Philadelphia: Chilton Books, 1965), pp. 53–67.

Østreng, Willy. "The Geostrategic Conditions of Deterrence in the Barents Sea," in Lawson W. Brigham, ed., *The Soviet Maritime Arctic* (Annapolis: Naval Institute Press, 1991), pp. 201–14.

Pierce, J. A., and A. Shostak. "The Omega Navigation System," *Naval Research Reviews* (Oct., 1966): 22–27.

Plummer, John E. "The Mariner and the Arctic," U.S. Naval Institute *Proceedings* 84 (Oct., 1958): 49–55.

Polmar, Norman. "Arctic Operations," U.S. Naval Institute *Proceedings* 113 (Sept., 1987): 133–34.

———. "Sailing Under the Ice," U.S. Naval Institute *Proceedings* 110 (June, 1984): 121–23.

Strong, James T. "The Opening of the Arctic Ocean," U.S. Naval Institute *Proceedings* 87 (Nov., 1961): 58–65.

Sverdrup, H. V., and F. M. Soule. "Scientific Results of the 'Nautilus' Expedition, 1931," *Papers in Physical Oceanography and Meteorology*, vol. II, no. 1 (Mar., 1933) and no. 3 (June, 1933).

Tugend, Tom. "A Bruin Alumnus Under Ice" *UCLA Alumni Magazine* (Jan., 1959): 10–12.

Watkins, James D. "The Maritime Strategy," U.S. Naval Institute *Proceedings*, supplement, 112 (Jan., 1986): 2–17.

Williams, Jonas. "The Arctic's Warming Up," *Collier's* 131 (Apr. 11, 1953): 60–64.

# Index

Adams, Charles F., xxii
Allis Chambers Company, 17
American Society of Naval Engineers, 182–83
Amundsen, Roald, xxiii
AN/BQS-6B. *See* BQS-6B
AN/BQS-8. *See* BQS-8
Anderson, William R.: decorated by Eisenhower, 131; failed attempt to penetrate Bering Strait, 120–23; makes air reconnaissance of polar route, 116–17, 118, 119–20; mentioned, 133; and 1957 *Nautilus* patrol, 98–99, 100, 101, 102–105; penetrates Bering Strait, 127–29; plans 1958 transpolar cruise, 110–11, 112, 113, 114, 115; reaches North Pole, 130; returns to Pearl Harbor, 124–25; welcomed in New York City, 132
Anschütz-Kämpfe, Hermann, xx–xxi
Applied Physics Laboratory (University of Washington), 250
*Archerfish* (SSN 678), 251
Arctic Research Conference of 1950, 35–37; 1951, 49, 52–53; 1952, 66–67
Arctic Research Laboratory (ONR), 85, 119, 159
Arctic Submarine Laboratory (NEL), 250, 253, 254, 255, 256–57. *See also* Submarine Research Facility
*Argonaut I*, xviii, xix
Armstrong, Alexander, 78–79
*Around the World in 80 Days*, 127
*Atule* (SS 403), 19
Aurand, Peter, 113
Australian Flying Corps, xxi
Axene, Dean L., 223

Bakus, Virginia. *See* Lyon, Virginia (nee Bakus)
Baldwin, Hanson W., 131
ballistic missile submarines (SSBNs): under-ice proposals for, 230–32
Banks, Sir Joseph, 41
Barnes, Clifford A., 40, 49–50, 58, 67
Barrow Sea Valley, 48, 124, 125, 128, 129
Battani, Victor, 7
Battery Whistler, 18, 28, 91, 92, 228
*Baya* (SS 318), 29, 30, 33–34, 74
Bayne, Marmaduke G., 113, 116, 118, 125, 130, 189
Beal, Allan, 228, 149
Bender, Harold, 31–32
Bennett, Rawson: at BuShips, 74–75, 86; chief of naval research, 93–94, 133; director of NEL, 14, 19, 20, 26, 27, 28, 29, 31, 33, 38, 39
Benson, Roy S., 203

Bernstein, H. E., 73–74, 85, 86
Beschloss, Michael R., 215
Beshany, Philip A., 231–32
Betatron, 17, 18
Bienia, John P., 61–62, 71
Bilotta, Joseph P., 138
Blair, Carvel Hall, 96, 97, 98, 99, 100, 112, 115
Blandy, William H., 13
Bloom, Gene L., 40–41, 47, 67, 162
*Bluefish* (SSN 675), 251
*Boarfish* (SS 327), 20, 21–22, 25, 26, 27, 28, 115, 156, 157, 204, 236
Boyd, David S., 125–26, 149, 151–52, 153, 185
Boyes, John L., 226, 229, 230
Boyle, Richard J.: BQN-4 echo sounder, 159, 160, 161, 163, 164, 190, 203; and BQS-8 sonar, 233, 252; hired by Lyon, 208; and 1967 *Queenfish* patrol, 233; and 1970 *Queenfish* patrol, 249; on *Skate*'s flooding casualty, 212–13; and SUBICEX 1-62, 209; and SUBICEX 1-69, 240, 242
BQR-2 sonar, 173, 174, 175, 177, 178, 182
BQS-6B sonar, 235
BQS-7 polyna detector, 190, 203
BQS-8 integrated sonar suit: and 1967 *Queenfish* patrol, 232–34, 235, 236, 237; problems with Hazeltine model of, 251–53; and SUBICEX 1-69, 241, 244, 246, 247; and SUBICEX 2-62, 209, 212, 222
Brewer, Glenn M., 199–200
Brewer, Max C., 119–20
British Imperial Antarctic Expedition (1920–21), xxi
Brockett, William A., 227
Brooklyn Navy Yard, xxiv

Brooks, Daniel P., 158
Buck, Beaumont M. 175–76
Bull, William I., 92
Burdge, Ronald E., 234
Burke, Arleigh A., 94, 113, 115, 124, 130
Burkhalter, Edward A., Jr., 190, 201, 229–30
*Burton Island* (AGB 1), 40, 41, 42–43, 44, 46, 48–49, 52, 56, 57, 58, 60, 61, 63, 64, 66, 67, 69, 71, 73, 75, 76, 79–80, 81, 82, 89, 217, 218, 220
BuShips. *See* U.S. Navy Bureau of Ships (BuShips)
Byrd, Richard E., 14

Cairns, Ernest, 119
Caldwell, Henry H., 202
Calvert, James F.: ComSubDiv 102, 206–207, 209; desire for under-ice patrols, 112, 125, 126, 185, 188–89; Lyon's relations with, 126, 146–47; mentioned, 220, 226; and 1958 *Skate* patrol, 134–43; and 1959 *Skate* patrol, 145–54; recommends Boyle, 208
Cameron, William M., 9, 11, 14, 17, 29–30, 58, 66, 67, 79
Canada–United States cooperation, 9–10, 11, 12, 14, 17, 18–19, 24, 28, 29–30, 34–35, 37, 39, 44–45, 49, 58, 59, 60, 66, 67, 70, 72, 77, 79, 80, 82, 185, 188, 191, 253
Canadian Naval Service, 18
*Cancolim II*, 46, 49, 60
Cape Prince of Wales Field Station, 30–33, 34, 35, 45, 46, 47–48, 49, 52–53, 57, 59, 60, 64, 67, 71, 86–87, 91, 92
*Carp* (SS 338), 25–26, 27, 28, 65, 237
Carsola, Alfred J., 48, 128

Carter, Powell F., 249
*Cedarwood*, 30, 31, 32, 33, 34–35
Central Intelligence Agency, 232
Central Sound Laboratory, 8–9, 10, 11, 17–18, 24–25
Charette, Alfred A., 122
Charles R. Lyon Truck & Repairing Company, 3
Chief of Naval Operations (CNO), Office of the, 6, 14, 25, 28, 36, 37–38, 39, 40, 45, 46, 53, 56, 65, 66, 67, 70–71, 73, 74, 94–95, 97, 98, 109–10, 143, 156, 189, 208, 210, 229, 230, 231, 240, 241, 254
Colclough, Oswald, S., 25, 26, 27, 37, 39, 55
Combs, Thomas S., 115, 118, 125
Couper, Butler King, 73, 88, 91–92, 112
Cramer, Shannon D., 209
Crary, Albert P., 59
Crowley, Robert F., 169
Cruzan, Richard H., 16
Curtiss, Laurence M., 167

Danenhower, Sloan, xxii, xxvi
Daspit, Lawrence R., 113, 115, 116, 118, 125, 126, 130, 159, 181, 182, 188
David Bushnell Award, 256–57
Defence Research Board (Canada), 44, 45, 66, 72
DeLong, George Washington, 153
Delsasso, Leo P., 6, 7
Denebrink, Francis C. 90
Dietrich, Edward O., 164
Distant Early Warning (DEW) Line project, 70–71, 87–90
Divine, Robert A., 109
Doherty, Barbara, 10, 11
Doherty, Lowell, 10, 11
Doherty, Ralph B., 10–11, 12

Drebble, Cornelius, xvii
Drifting Station Alpha, 137–39
Drifting Station Bravo. *See* T-3/Bravo
Drifting Station Charlie, 175

Early, Paul J., 103, 120
Echoscope system. *See* QLA sonar
EDO Corporation, 135–36, 159, 190, 191, 206, 209, 212, 251–52
EDO echo sounder, 135–36, 139, 146, 147, 150
*Ehkoli*, 9
Eielson, Carl Ben, xxi
Eisenhower, Dwight D.: approves DEW Line project, 88; and Operation Sunshine, 113, 118, 131, 134; and *Seadragon* patrol, 202
Electric Boat Company, 232, 235
Electronic Position Indicator (EPI), 46–47, 49, 50–51, 52, 58, 59, 60, 61, 63, 66, 79, 80, 81, 82, 83
Ellis, Joseph W., 5
Ellsworth, Lincoln, xxii
*Entemedor* (SS 340), 209
EPCE (R) 857, 20, 30, 33–34

Felt, Harry D., 95
Ferrall, William E. 160
Field, George S., 18–19, 35, 66
Fisheries Board (Canada), 44
*Flying Fish* (SSN 673), 251
*Fram*, 49
*Fulton* (AS 11), 12, 107

Gates, Thomas S., Jr., 143
Geophysical Institute (Bergen), xxiii
*George Clymer* (APA 27), 30, 31
Germany: under-ice operations during World War II by, xxvii, 9, 45, 53

Germershausen, William J., Jr., 159
Gertrude Alpha sonar transducer, 210, 222
Gibson, Jack E., 59–60
*Grampus* (SS 523), 96
"Great Stone of Sardis" (Stockton), xix
Grenfell, Elton W., 96, 131, 226
*Gurnard* (SSN 662), 251
gyroscopic compass: developed by Anschütz-Kämpfe, xx–xxi

*Halfbeak* (SS 352), 110
*Hammerhead* (SSN 663), 249–50
Hammon, Robert E., 199–200
Harvey, John W., 227
*Hawkbill* (SSN 666), 250
Hayworth, Rita, 12–13
Hazeltime Company, 251–52
Hearst Enterprises, xxii
*Henrico* (APA 45), 12
Holland, John, xviii
Hopwood, Herbert G., 181
Howick, Edward, 47

iceberg detecting sonar: developed by NEL, 156–58, 159; on *Sargo* (1960), 162, 164, 165, 166, 167, 168–71, 172–73, 174, 175, 177, 178–79, 181, 182; on *Seadragon* (1960), 190, 191, 192, 193, 194, 195, 198–99, 201–202, 203
icebreaker expedition of 1951, 44–45, 46, 48–52
ice destructor mines, 175, 180, 200, 204
Icenhower, J. B., 16
Imperial Geographical Society, xx
International Geophysical Year, 90, 98, 138, 175
*Investigator*, 78, 79

Jackson, Henry M., 94
Jacobs, Tyrvell G., 81–82
James, Ralph K., 224
*Jeanette*, xxii, 153, 169
Jenks, Shepherd M., 120, 125, 126
Joint Committee on Oceanography (Canada), 18–19
*Journal of a Voyage for the Discovery of a North-West Passage from the Atlantic to the Pacific* (Parry), 190–91

Kampmeier, Anna Mary. *See* Lyon, Anna Mary (nee Kampmeier)
Kelln, Albert L., 134, 145, 146, 148, 152
Kelly, Leslie D., Jr., 100, 101
Kelly, Neil, 234–35
Kelsey, Frances O., 219
Kelso, Frank, II, 251
Kennedy, John F., 219, 221
Khruschchev, Nikita, 214–15
Kielhorn, William V., 73
Kinnear, Wally, 208
Kinsey, E. Lee, 5–6
Knudsen, Vern O., 5
Korth, Fred, 219
Kurie, Franz N. D., 91, 93, 116

*Labrador*, 72, 80, 81, 82, 191
*Lafayette* (SSBN 616), 230, 240
LaFond, Eugene C., 20, 21, 30, 48, 128, 141
Lake, Simon: participation in the Wilkins expedition, xxii, xxiv, xxvii; plans for an under-ice submarine, xviii–xx
Lake & Danenhower, Inc., xxii
Lake Torpedo Boat Company, xix, xxii
Lalor, William G., Jr., 100

Lamont Geological Laboratory, 138
Larsen, J. B., 162–63
Laurence, William L., 13
Law, Frank G., 95
Layman, William H., 145–46, 149, 152
*Leninsky Komsomol*, 227
Lewis, George H., 218
Liddiard, Glen, 158
*Life* magazine, 83
Lill, Gordon, 86
London Naval Treaty, xxii
LST-1126, 47
LST-1138, 47
Lyon, Anna Mary (nee Kampmeier), 3, 4, 5, 219, 242
Lyon, Charles Russell, 3, 4, 219
Lyon, Gretchen Mary, 3, 4
Lyon, Lorraine May, 11, 13–14, 58–59, 219
Lyon, Russell, 11, 13–14, 58–59, 219
Lyon, Virginia (nee Bakus), 5, 10–11, 13–14, 24, 58–59, 219
Lyon, Waldo K.: awards of, 90–91, 143, 182–83, 217, 219–20, 256–57; background of, 3–6; and *Boarfish* experiments, 19–23; and Cape Prince of Wales field station, 30–33, 35, 45, 46, 47–48, 49, 52–53, 57, 60; and *Carp* experiments, 24–26; and concern over navy's future in Arctic, 255–57; death of, 257; DEW Line adviser, 87–90; early interest in cold water acoustics, 9–10, 11, 14; final arctic patrol of, 254; and funding crisis of 1956, 91–93; on *Hammerhead*, 249–50; and ice break-through tests, 228–29; and *Nautilus* patrol of 1957, 93–95, 97–108; and *Nautilus* patrol of 1958 (Operation Sunshine), 109–32; and 1952 *Burton Island* supply mission, 57–58; and 1952 *Redfish* patrol, 58–64; and 1953 *Redfish* patrol, 70–72; and 1954 *Northwind* expedition, 74–83; and 1967 *Queenfish* patrol, 233–37; and oceanographic expedition of 1949, 29–30, 33–35; and oceanographic expedition of 1950, 38–43; and oceanographic expedition of 1951, 46–49; and organizations problems, 253–54; organizes Central Sound Laboratory, 8–9, 10, 11, 13–14, 18–19, 29–30, 35, 39, 44, 58, 66, 72, 82; participates in Crossroads, 11–13; participates in Highjump, 14–17; plans for arctic pool, 53–54, 83–84, 85–87, 91, 96–97, 206, 208, 228–29; and plans for under-ice war games, 206–208, 209–11; plans to use *Triton* for arctic research, 227, 228, 229; protests funding cuts (1973), 250–51; publishes Research Report 88, 27–28, 29; rejects academic career, 23–24; and relations with Calvert, 146–47; and relations with Nicholson, 182; and 637-class submarine, 225, 226, 233–40, 247–48; and *Sargo* patrol of 1960, 156–83; and *Seadragon* patrol of 1960, 185, 188–89, 190–204; seeks submarine for under-ice experiments, 37–38, 45–46, 52–53, 55–56, 65, 67–69, 73–74; and *Skate* patrol of 1958, 133–34; and *Skate* patrol of 1959, 143–55; and SUBICEX 1–69, 240,

Lyon, Waldo K. (*continued*)
241–45, 247–48; and SUBICEX 1–73, 250; and SUBICEX 2–62, 211–21, 222–23; writes plan for arctic research, 36–37; World War II work of, 7–9

Maher, Eugene H., 56–57, 58, 61, 63, 69
Maloney, William L., 74, 75, 77, 78, 79, 80, 81, 82, 83, 88
Manual Arts High School, 3
Mare Island Navy Yard, 8, 25, 118, 158, 168
Maritime Strategy (1980s), 254–55
Mark 16 torpedo, 136, 141
Mark 18 torpedo, 71–72
Mark 37 torpedo, 218, 219, 220, 222, 241, 245, 247
Mark 48 torpedo, 235, 242, 251
*Mathematical Magick* (Wilkins), xvii
Mathis Shipyard, xxiv
Matson, William, 211
*Maud*, xxiii, 20, 35
McCann, Allan R., 19, 20, 21–22, 23, 24, 25, 27, 28, 29, 46, 55, 70
McClure, Robert, 78, 80
McKee, Kinnard, 255
McLaren, Alfred S., 232–33, 249
McWethy, Robert D.: on *Burton Island*, 41–42, 43; commands Ocean Systems, Atlantic, 241; mentioned, 220; promotes 1958 *Nautilus* patrol, 111–12; proposes 1957 *Nautilus* patrol, 93, 94–95, 96, 98, 99; and *Skate*, 137
Medlin, A. Wayne, 47
*Merrick* (AKA 97), 14–15, 16
Metropolitan High School for Adults, 5
Meyer, Harold, 173
Milikan, S. E., 196

Military Sea Transportation Service, 90
Milligan, C. H., 7
Molloy, Arthur D., 135, 138–39, 191, 196, 246
Momsen, Charles B. 55, 65, 66, 68, 69, 70, 95
Morrell, Richard E., 74, 75, 77, 80, 82
Morse, Leighton L., 12, 31
*Mount Olympus* (AGC 8), 16, 89
Mumma, Albert G., 90, 115, 130, 133

Nansen, Fridtjof, xviii
National Academy of Sciences, 6–7
*Nautilus*/O-12, xxii, xxiii, xxiv–xxvii, 19; diagram of, xxv
*Nautilus* (SSN 571): mentioned, xvii, 206, 207, 211, 237; and 1957 patrol, 93–108, 109, 112; and transpolar cruise, 109–32, 133, 134, 135, 143, 144, 146, 147, 156, 157, 201, 249. *See also* Operation Sunshine
*Nautilus* (Verne), xvii–xviii
Naval Ocean Systems Center, 253
Naval Undersea Warfare Center, 256
*Navy Arctic Operations Handbook*, 28
*Navy Goes Forth*, 192
NEL. *See* U.S. Navy Electronic Laboratory (NEL)
NEL Report 88, 27–28, 29, 37
NEL Report 353, 67–68
*Nereus* (AS 17), 21
Newport News Drydock & Shipbuilding Company, 232
Nicholson, John H.: background of, 158; conducts *Sargo* patrol, 160–81; mentioned, 203; at Op-311, 207, 223–25, 226; praises

scientists and crew, 181–82; prepares for 1960 *Sargo* patrol, 158–60; on *Skate* (1958), 134, 135, 137–38, 146, 158

NK-Variable Frequency echo sounder, 143, 145, 146, 147, 149, 150, 153, 154, 159, 161, 164, 165, 168, 171, 172, 176, 180, 190, 203

North American N6A autonavigator, 116, 125, 137, 142, 160, 180

*Northwind* (WAGB 282), 15–16, 66, 67, 71, 72, 73, 74, 75, 77–78, 79, 80–81, 82, 83, 88, 89, 202

O-12. See *Nautilus*/O-12
oceanographic research expedition of 1950, 38–43
*Oceans: Their Physics, Chemistry and General Biology* (Sverdrup), 15
*Odax* (SS 484), 96
Office of Naval Research (ONR), 35–36, 37–38, 40, 53, 59, 67, 73, 85–86, 87
OMEGA navigational system, 209–10, 221–22
Operational Requirement SW-01402, 40
Operation Barney, 8
Operation Highjump, 14–17
Operation Sunshine: abortive beginnings of, 118–24; initial plans for, 109–18; success of, 124–32

Pacific Biological Station, 9, 11, 14
Pacific Naval Laboratory, 29
Pacific Oceanographic Group, 29–30, 44
Page, William M. 174
Palmer, James M., 25–26
*Parche* (SS 384), 224
*Pargo* (SSN 650), 241–42, 245–47

Parker, Frederick, 157, 164, 169, 170, 174
Parry, William Edward (Sir Edward), 41, 190, 196
PC-795, 9
Peary, Robert, 244
*Pensacola* (CA 24), 13
Phelps, John M., 92, 94, 116, 119, 188
*Philippine Sea* (CVA 47), 17
Phoenix, David A., 166–67, 169, 245
*Pickerel* (SS 525), 158
*Pintado* (SSN 672), 251
*Plunger*, xviii
*Pogy* (SS 266), 42
Polaris missile: its proposed use in Arctic, 111
"Polar Queen" (B-50), 42
"Polar Submarine and Navigation of the Arctic Ocean" (NEL Research Report 88), 27–28, 29, 37
Polmar, Norman, 254
Portsmouth Navy Yard, 96, 97
Potter, John, 223, 224
Project Aleutians, 34–35
Project Crossroads, 11–13
Project 572, 88. See also Distant Early Warning (DEW) Line project
Project Icicle, 59
Project Lincoln, 70. See also Distant Early Warning (DEW) Line project
*Protector*, xix–xx
Pulitzer, Joseph, xix

QLA sonar, 8, 17, 20, 21, 22, 23, 24, 25–26, 27, 58, 60–61, 70, 203, 117–18, 124, 156, 157
*Queenfish* (SSN 651), 232, 233–40, 249, 252

**INDEX** 299

RAFOS sonar navigation system, 142
Ramage, Lawson P., 19, 224–25
Reagan, Ronald, 254
"Recognition of Submarine Targets" (Lyon, Tully, Cameron), 17
*Redfish* (SS 395), 58, 59, 60–62, 63, 64, 65, 67, 70, 71, 72, 95, 98, 100
Research Defense Board, 53
Revelle, Roger, 7, 11–12, 18, 36–37, 40, 67
Richard, Jackson B., 233, 234, 235–36, 237, 240
Rickover, Hyman G., 93, 94, 115, 131, 132, 158
Riedel, Alfred, xviii
Robertson, O. C. S., 72, 82, 162–63, 185, 191, 192, 194–95
*Rockbridge* (APA 228), 12
Root, E. F., 188
Rose, Lisle A., 16
Roshon, Arthur H.: on *Boarfish*, 21; and BQS-8 sonar, 233, 251–53; develops iceberg-detecting sonar, 156–58, 159–60; praised by Nicholson, 182; on *Sargo* patrol, 162, 169, 170, 177, 178; on *Seadragon* patrol, 191, 192, 198–99, 201–202
Rowray, Rexford N., 47, 57, 99, 103, 107, 118, 119, 120–21, 126
Royal Canadian Air Force, 83, 196
Royal Geographical Society, xxi
Russell, George L., 74
Russell, R. D., 18

*Sailing Directions for Antarctica* (U.S. Navy Hydrographic Office), 15
Sanders, Fred, 29, 35, 44, 49, 58, 66
*Sargo* (SSN 583): collides with ice, 177–78; and end of experimental program, 183; iceberg detecting sonar for, 156–58, 185; mentioned, 203, 204, 211, 229; and 1960 winter patrol, 160–76, 178–81, 189, 202, 209, 251; prepared for 1960 winter patrol, 158–60
Schade, Arnold F., 247
Schatzberg, Walter E., 77, 145
Schere, Jonathan C., 191, 194–95, 212, 251, 252
Schwartz, John R., 40, 41–43, 49, 56
Scripps Institution of Oceanography, xxiii, 20, 36, 40
Seaborg, Glenn, 4
*Seadragon* (SSN 584): and icebergs, 192–96; makes Northwest Passage, 196–98; mentioned, 213, 229, 232, 240, 246; and plans for Northwest Passage by, 170, 189–191; reaches North Pole, 198–200; results of 1960 patrol by, 203–205, 206; and sound transmission study, 200–202; and SUBICEX 1–73, 250; and SUBICEX 2–62, 210, 211–12, 216–17, 218–19, 220–21
Seaton, Lewis A., 199–200
*Seawolf* (SSN 575), 95
*Sedov*, 49
*Sennet* (SS 408), 15–17, 19, 20, 26, 27, 71
Shackleton, Ernest, xxi
*Silversides* (SSN 751), 254
688I-class submarine, 255, 256
688/*Los Angeles*-class submarine, 255
637/*Sturgeon*-class submarine, 223–25, 226, 229, 232, 233, 240
*Skate* (SSN 528): mentioned, 204, 206, 207, 208, 229, 249; and 1958 patrol by, 110, 112, 115, 126, 132, 133–43; and 1959 patrol by, 143–54, 156; and proposed use to make Northwest Passage, 185,

188–89; and SUBICEX 1–62, 209; and SUBICEX 1–69, 241, 242, 244, 245; and SUBICEX 2–62, 210–11, 211–23, 227, 242, 251
Skoog, Joseph L., Jr., 211, 212–18, 220–22
Snyder, J. M., 229
Solberg, Thorvalt A., 37, 38
SOSUS underwater detection network, 241, 245, 254
Sound Firing and Ranging System (SOFAR), 59–60, 61, 62–63
Soviet Union: arctic interests of, 83, 98, 111, 253; and atomic testing in Arctic, 214, 218; and Maritime Strategy (1980s), 254–55; and possible arctic submarine, 66, 73, 74; and submarine reaches North Pole, 227; and under-ice operations during World War II, xxvii
Special Studies Branch (NEL), 39
Sperry Corporation, 191
Sperry Depth Detector, 237
Sperry Mark 19 gyrocompass, 100, 104, 127, 249
Sperry Mark 23 gyrocompass, 104, 137, 142
Sperry Ship's Inertial Navigation System (SINS), 191, 204, 214, 244, 249
*Sputnik*, 109, 113, 114, 131, 134
SQS-4 sonar, 192, 193
SSN-21/*Seawolf*, 256
Stanford University, 6
*Staten Island* (AGB-5), 161, 162–63
Steele, George P.: conducts sound transmission study, 200–203; investigates icebergs, 192–96; makes Northwest Passage, 196–98; mentioned, 220, 246;

prepares for *Seadragon* patrol (1960), 189, 190–91; reaches North Pole, 198–200; on results of *Seadragon* patrol, 203–204
Stefansson, Vilhjalmur, xxi, xxii, 190
Stefansson Collection of Arctic Literature, 190
Stelter, Frederick C., III, 171–72, 176
*Stickleback* (SS 415), 11
Stockton, Frank R., xix
Storkersen, Storker T., 49
Strong, James T., 230–31, 240–41
*Sturgeon* (SSN 637), 232, 252
Submarine Cold Weather and Arctic Material Program (SCAMP), 99–100
Submarine Ice Exercise 1–62 (SUBICEX 1–62), 109, 211; SUBICEX 1–69, 240–48; SUBICEX 1–73, 250; SUBICEX 1–75, 251; SUBICEX 2–62, 211–22
Submarine Research Facility (NEL), 28, 171, 206, 208, 228–29
Submarine Signal Company, 7
Subsafe program, 227, 228, 230
Summer Study Group (MIT), 88
Summitt, Charles D., 211, 216, 217, 218
Supex Laboratories, 3
Sverdrup, Harald U., xxiii, xxvi, 15, 20–21, 105, 168, 228
Sweetwater Calibration Station, 10
*Swordfish* (SSN 579), 211

T-3/Bravo, 59, 60, 61, 62, 175–76, 185, 189, 191, 194, 200, 201, 202, 243, 244
*Tambor* (SS 198), 9
*Tautog* (SSN 639), 252
Teller, Edward, 231

Thatcher, E. W., 18
Thompson, Llewellyn E., Jr., 219
*Thresher* (SSN 593), 223, 227, 242
Towner, George C., 89
TRANSIT satellite navigation system, 230
*Trigger* (SS 564), 100, 101, 102, 106, 107
*Triton* (SSN 586), 223, 226, 228, 229
Tucker, Dundas P., 39, 46, 66, 67, 68, 69
Tucker, James G., 167
Tully, John P., 9–10, 11, 12, 14, 17, 23, 24, 29–30, 34–35, 44, 49
Turner, John H., 21–22
*Tusk* (SS 426), 209
*Twenty Thousand Leagues Under the Sea* (Verne), xvii–xviii, xxiv

U-262, xxvii
University of California at Berkeley, 4, 7
University of California at Los Angeles, 4, 5, 6, 23–24, 39
University of California Division of War Research (UCDWR), 7, 8, 11
University of Washington, 52
University of Washington Oceanographic Laboratory, 40
UQC-1 underwater communications system, 21, 216, 218, 220
UQN-1 fathometer, 190
UQS-1 sonar, 136, 139, 145, 150, 153, 157, 159, 160
U.S. Air Force, 36, 59, 70, 88, 175
U.S. Coast and Geodetic Survey, 46
U.S. Coast Guard, 36, 74–75
U.S. Naval Academy, 42
U.S. Navy Bureau of Engineering, 6
U.S. Navy Bureau of Ships (BuShips), 11, 18, 36, 53, 67, 73, 85, 87, 91–92, 93, 94, 95, 96, 97, 143, 144–45, 156, 157, 207–208, 251–53
U.S. Navy Electronics Laboratory (NEL), 11, 12, 14, 24–25, 27, 35, 36, 37, 38–39, 40, 45, 46, 49–52, 53, 56, 59, 66, 67, 68, 69, 73, 83–84, 115–16, 143, 170–71, 181, 189, 207, 210, 228, 234, 237, 251
U.S. Navy Examining Board, xxvii
U.S. Navy General Board, 29
U.S. Navy Gun Factory, 17–18
U.S. Navy Hydrographic Office, 15, 53, 86
U.S. Navy Material Command, 250, 253, 254
U.S. Navy Radio and Sound Laboratory, 6, 7, 8, 10, 11
U.S. Navy Underwater Sound Laboratory, 176, 191
U.S. Shipping Board, xxii
U.S. Weather Bureau, 64

Verne, Jules, xvii–xviii, xix, xxiv
*Volador* (SS 490), 70
Volse, L. A., 57, 58
Von Thenen, William A., 170

Wadsworth, Frank L., 104, 210–11, 213–14, 221
Wagner, Richard, 132
Wakefield, John, 223–24
Walker, Archie C., 47, 126, 127
Walker, Frank, 113, 115, 118, 125
Walker, Thomas L., 174
Warder, Frank B., 115, 125, 126, 145, 149, 188
Warner, A. H., 5
"Waters of the North Siberian Shelf" (Sverdrup), 20
Watkins, James D., 254–55
Watson, Robert, 157, 158

*Whale* (SSN 638), 241, 242–45, 247, 252
White, Stephen A., 242, 246–47
Wilkins, Charles W., 97, 100–101
Wilkins, John, xvii, xxi, xxviii, 257
Wilkins, Lady, xxiv, 274*n* 29
Wilkins, Sir Hubert: ashes of, scattered at North Pole by *Skate*, 148–49; background of, xx; and conflict with Lake, xxii–xxiv; mentioned, 49, 99, 228; and polar expedition, xxiv–xxvii, 105, 106; promotes a polar submarine expedition, xxi–xxii
Wilkinson, Eugene P., 93, 94, 95, 98, 134, 158, 233
Wilkins Transpolar Expedition of 1930: conduct of, xxiv–xxvi; mentioned, 19, 105; plans for, xxi–xxiv; results of, xxvii

Wilson, James B., 231
Wittmann, Walter I.: adviser to DEW Line project, 89; on *Pargo*, 246; on *Sargo*, 160, 174; on *Seadragon*, 191, 192, 195; on *Skate* (1958), 134; on *Skate* (1959), 146, 150; and SUBICEX 2–62, 211
Wolff, William M., Jr., 242–45
Woods Hole Oceanographic Institution, xxii, 36
Wuerker, Alexander W., 77, 78, 79, 81
*Wyoming*, xxiv

*Yancey* (AKA 93), 14, 15, 16
Yates, William K., 160, 164, 179, 182

Zhil'tzov, Lev, 227